D1171943

Engineering Animals

Engineering Animals

HOW LIFE WORKS

Mark Denny
Alan McFadzean

THE BELKNAP PRESS OF
HARVARD UNIVERSITY PRESS
Cambridge, Massachusetts
and London, England
2011

Printed in the United States of America

Library of Congress Cataloging-in-Publication Data

Denny, Mark, 1953–
Engineering animals : how life works / Mark Denny and Alan McFadzean.
p. cm.
Includes bibliographical references and index.
ISBN 978-0-674-04854-6 (alk. paper)
1. Physiology. 2. Animals—Adaptation.
3. Animal ecophysiology. I. McFadzean, Alan, 1958– II. Title.
QP31.2.D46 2011
591.7—dc22 2010051355

Mark dedicates this book to a mammal,
Jane Denny, with love

Alan dedicates this book to his pack:
Anita, Kirsty, and Gordon

Contents

Engineering Animals

Prologue

You are a Great Ape. As it happens, both authors are also animals belonging to this family; we share our two homes with other hominids, plus mammals from the canine and feline families. Unintentionally we also share our homes with a wide variety of other animals, mostly from the classes Insecta and Arachnida. We might have begun by saying "You are a biped" and then proceeded, in a mathematically progressive way, to state how we shared our homes with quadrupeds, hexapods, and octapods. However, referring to our readers as "apes" rather than as "bipeds" makes a point. You were probably a little taken aback—metaphorically slapped across the face—which would not have happened had we called you "bipeds."

Why is this so? Both statements are true, yet we hominids of the species *Homo sapiens* (no doubt all our readers are from this species—we would be very interested to hear from readers who are not) sometimes think that we are superior to the rest of the animal kingdom and so resent being reminded of our lineage. In some ways, humans *are* superior, in other ways we are not. Viewed with scientific objectivity—and from a certain pragmatic perspective more often found in engineering than in science—we can estimate how humans stack up, in biological capability, against other animals. We have brains that are uniquely capable (as far as we know) of abstract thought and language; this ability sets us apart. We have pretty good eyesight, by the standards of the rest of the animal world, though not the best. Some of our other senses are dull or lacking: dogs would hold up their noses at our ability to smell (meaning our ability to *sense* smell); owls wouldn't hear of our acoustic abilities. Our skeletons are standard issue, though perhaps not as well adapted to bipedal locomotion as many quadruped skeletons are to four-legged locomotion. Octopuses—if they were capable of human arrogance—might regard us as inferior, because we can't even *sense* polarized light, let alone emit it. Bats might regard themselves as occupying the top of the evolutionary tree, because their sonar is better than the most hi-tech of ours.

This book is about the wonderfully varied and astonishing capabilities of animals. We look at animal design from the standpoint of engineers. Skeletons are marvels of engineering as well as of evolution. Stated differently, evolution has provided animals with support structures that are—that have to be—well engineered. Birds are well engineered for flying. Pigeons are flying remote sensors with capabilities we can appreciate from an engineering perspective: they have celestial navigation, wideband acoustical receivers, hi-res optical receivers, and magnetic sensors. There is a small fly that can locate the source of a sound very precisely, even though the fly is much smaller than the wavelength of the sound it hears—a difficult feat. Albatrosses travel vast distances across the southern oceans while expending very little energy—they exploit wind shear using a technique known to glider pilots as *dynamic soaring*. These last two example—two small pieces, we feel bound to say—of the extensive research that we have pursued in writing this book have been turned into a couple of pedagogical papers written to show university science students how the animal world employs good (and novel) engineering principles to achieve certain goals. In this book you will get the results of our investigations without the math.

What qualifies us to write a book about animal engineering? Well, we are animals—have been for decades—and we both have many years experience as research engineers. Both of us worked for multinational aerospace companies designing radar and sonar algorithms for military and other remote-sensing applications. Both were trained as physicists, and this combination (science and engineering) plus a long shared history of mathematical and computer modeling of natural phenomena provides us with a pragmatic perspective on how things work. Applying our engineering experience to the world of animals has shown us, again and again, how well adapted and well made animals are for the roles they play—they are what engineers would call "mature technology." Looking at animals in this way—analyzing them as engineering structures—in no way reduces our sense of wonder at their diversity, adaptability, and astonishing capabilities. We are confident that reading our book will refine your appreciation of Great Apes like us and of other, more distant, relatives.

PART ONE

Structure and Movement

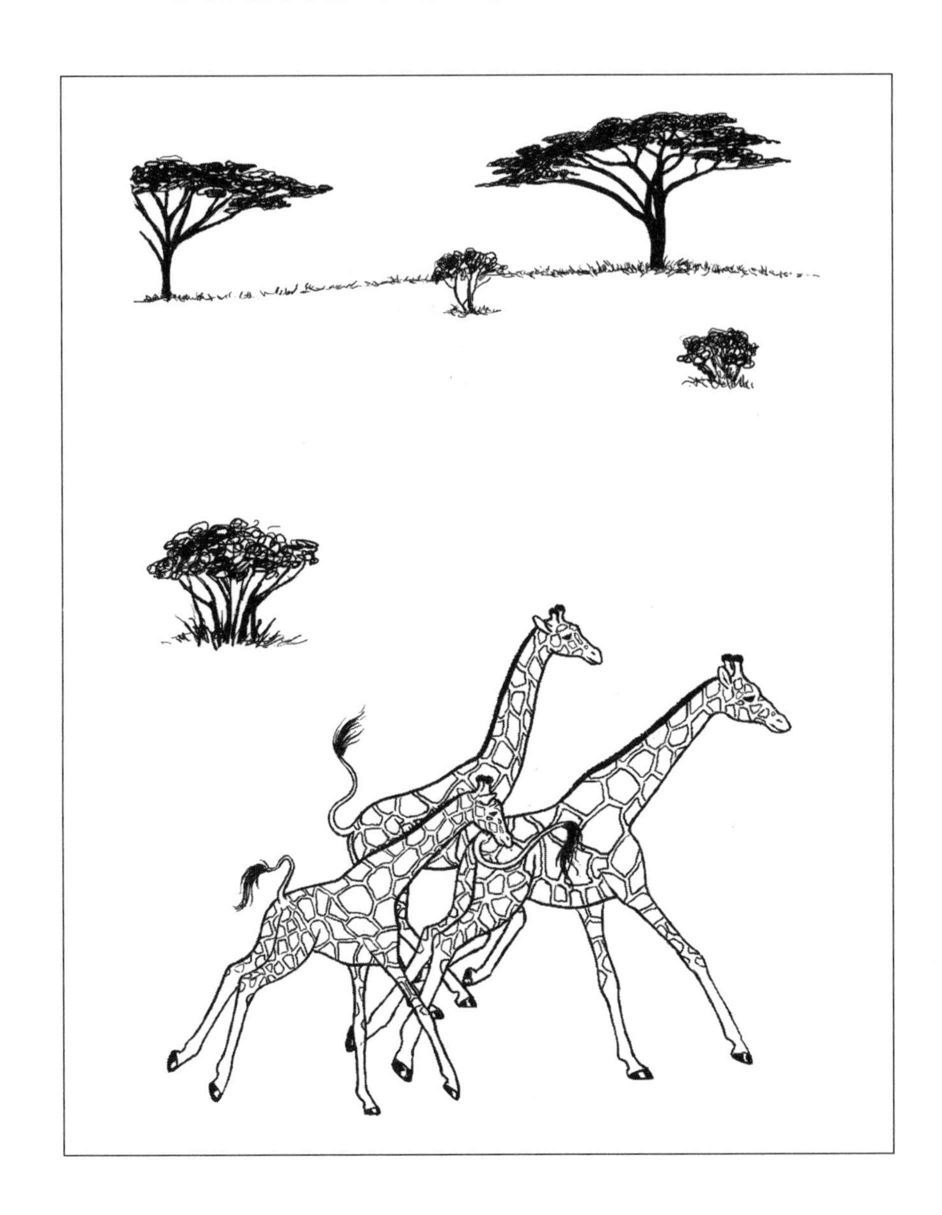

1

Go with the Flow

Energy is a familiar concept in the everyday lives of most people, even if they are unable, when asked, to conjure up the physicists' precise definition. We tire during exercise and feel refreshed after an "energy drink." Our dog runs along the beach for no reason, so it seems, other than the joy of life—he is "full of energy." Physicists tell us that energy is a real thing, like matter, but outside the realm of Einstein's relativity, it is not a material. Instead, it is a property of a material and can take many forms: heat energy, chemical energy, kinetic energy (the energy of motion), and so forth.

Energy flow—the subject of this first chapter—refers to the transmission of energy from one place to another; in our case we will consider the chemical energy that is contained in food flowing into an animal, or energy moving from one ecosystem to another. Energy flows out of an animal when it dies and into another animal or plant that eats it, or into the atmosphere as heat. Energy flow is an important consideration when we seek to understand how nature is organized.

Energy flow influences animals at all levels and in every aspect of existence: cells and ecosystems, behavior and structure, size and shape. This pervasive influence may affect a single species uniquely—perhaps through a peculiar adaptation—or it may affect all species so that they exhibit universal features with surprising consistency. In this chapter we will see how that basic engineering concept—the flow of energy—influences different aspects of animal life.

Solar Power

The sun is a fairly average star, located in a middle class suburb of the Milky Way galaxy. Its third planet (the blue-green one) is warmed by a tiny fraction of the truly prodigious power emitted by the sun: about one part in a billion. This solar power mostly takes the form of electromagnetic radiation, and the amount of this radiation falling on Earth's upper atmosphere is about 330 calories per square meter per second ($330 \text{ cal} \cdot \text{m}^{-2} \text{ s}^{-1}$).[1] Power is energy per unit time. So, every square meter of Earth's upper atmosphere is gently bathed by 330 calories of energy every second. This solar energy is absolutely vital to life —without it, life would not have evolved on Earth. More importantly—and here is the central message of this chapter—energy flow influences the structure of all living things; it determines the way that animals move and behave, the way they evolve, and their physical form. Every aspect of animal (and other) life is dominated by the flow of energy.

How much of the sun's energy reaches Earth's surface? Clouds reflect some of it back into space, while the atmosphere absorbs or scatters some more. Furthermore, for most places on Earth, the sun is not directly overhead, and at any given time, half of Earth is sunless.

EXTREMOPHILES

Immediately we are faced with an exception to the rule—life, with all its spectacular variety, is like that. There are some very odd forms of animal life at the bottom of deep oceans, where sunlight does not penetrate. Much of life in the benthic zone does, in fact, depend indirectly on sunlight: it feeds on detritus raining down from the surface. However, near hydrothermal vents and cold seeps there are animals that exist without light or heat from the sun, and they have done so for countless generations, quite independently of life elsewhere on Earth.

Hydrothermal vents occur at the fault lines that separate tectonic plates in the deep oceans of the world, for example the mid-Atlantic ridge. Hot sulfur-rich gases from deeper within the Earth vent in copious quantities, and the geothermal and chemical energy they bring has been harnessed by certain bacteria that live symbiotically in the bodies of animals that populate these regions. The best known of these strange animals is the giant tube worm *Riftia pachyptila,* which grows to 8 ft (2.4 m) in length. It is white with a red "head"—like a lit cigarette. Even bigger are the 10 ft *Lamellibrachia* tube worms that populate cold seeps and live for 250 years. Cold seeps, again usually located so deep below the ocean surface that light will not reach them, are regions where hydrocarbons seep up from below. At these sites, communities of animals and bacteria (more than 100 species) form ecosystems that thrive without light.

Because the hydrocarbons represent stored solar energy from previous geological epochs, the cold-seep creatures do depend indirectly on sunlight. Hydrothermal vent ecosystems are truly independent of photosynthesis, however. These animals, living without light in regions that are very hot and very cold—within the space of a few meters—and at depths where water pressure is enormous, seem wholly alien. They are not: they share the same DNA code as the rest of life on Earth, and so must share a common origin, way down the evolutionary tree. So, even hydrothermal vent ecosystems may have their origins in life forms that once relied on photosynthesis, although today their descendants do not.

In one year, on average, the surface of our planet receives about 10% of the energy that reaches our upper atmosphere from the sun. How much of this solar energy is harnessed by life? Ecologists reckon on the following rough values. The primary producers—photosynthetic plants (on land) and phytoplankon (on the sea surface)—convert about 1% of the electromagnetic energy that reaches them into useful stored chemical energy to fuel life on Earth. Now we turn to the *trophic pyramid* and the ecologists' 10% rule (see Figure 1).[2]

Photosynthesizing plants make up the basement level of our energy pyramid. Herbivores, the animals that feed on plants and constitute the first-floor trophic level, convert about 10% of the plant energy into herbivore energy. Small carnivores, who feed on herbivores and form the second-floor trophic level, convert about 10% of the herbivore energy into small carnivore energy. Large carnivores on the top floor, who feed on their smaller brethren, convert about 10% of the energy locked up in small carnivores. You will appreciate that the 10% rule is, like weather forecasts, only approximately true. Many thousands of food chains exist in nature, with different herbivores and small and large carnivores. Animals use energy for respiration, circulation, digestion, locomotion, temperature regulation, and nervous function, but an awful lot of energy is dissipated as heat. This wastage, from one trophic level to the next,

FIGURE 1 The basic trophic levels, drawn approximately to scale. Photosynthesizing plants occupy the basement floor, providing about 10% of their energy to first-level occupants (herbivores—here represented by a caterpillar), who in turn provide 10% of their energy to second-level occupants (secondary carnivores—here an insectivorous bird). These carnivores in turn provide 10% of their energy to third-level occupants (primary carnivores—here a cat).

is not always a rigid 90%; it varies from about 80% to 95%. Why so much wastage? Physicists would blame the Second Law of Thermodynamics. Converting energy from one form to another always generates some wasted heat as a by-product. We can see other reasons for inefficiency without invoking the Second Law. Photosynthesizing plants are not sensitive to all wavelengths of light emitted by the sun; herbivores cannot eat all plant material—some is indigestible; carnivores produce energy-rich feces (which other creatures—*detritivores*—may consume). As we will see, even the very act of digestion takes energy—a significant fraction of an animal's metabolic rate, in fact. All these aspects of life represent inefficient uptakes of energy.[3]

Fox Populi

As a first illustration of how energy flow influences animal life, we take the 10% rule at face value and show how this inefficient flow of energy or power from the sun to top-level carnivores limits carnivore population density. We saw that about 10% of the solar energy reaching the upper atmosphere is available on the surface; about 1% of this is converted into usable energy for photosynthesizing plants; about 10% of plant energy is converted to herbivore use; about 10% of herbivore energy is converted for use on the second floor (small carnivores). Finally about 10% of small carnivore energy makes it up to the top floor. So, large carnivores convert about one part per million of the solar energy that bathes our upper atmosphere. How does this limited energy conversion restrict the large carnivore population density? Consider a 1 km square on the surface of Earth: our calculations show that this area will yield about 29,000 kcal of energy to be shared out among the large carnivores—of all types—each day. A 100 kg (220 lb) carnivore requires about 7,000 kcal of energy per day, and so the maximum large carnivore population density is about four animals per square kilometer. In fact, this number is much greater than actual carnivore densities, for several fairly obvious reasons. Climate limits the density of photosynthesizing plants, so most regions of Earth's surface are not completely covered with green leaves. Plant defenses reduce the impact of herbivores. Disease restricts animal numbers; territoriality and other modes of behavior limit population densities. Lion density in an African game reserve is about one per 7 km^2, whereas wolf populations in northern Canada vary between about one per 20 km^2 and one per 500 km^2. It is easy to see why the actual numbers are lower than our upper limit; the main point of this calculation is to show that there is a strict upper limit, and it is imposed by energy flow.[4]

Carnivores that are smaller than lions (but still occupy the top floor of the trophic pyramid) eat less, and so their population densities are higher, at approximately 24 per km^2 for 10 kg animals. We find for example that the population density of Red Foxes (*Vulpes vulpes*) in Poland averages one per km^2, rising to twice that number on farmland and five times as many in suburbs. Again, these numbers are less than our strict upper limit, as expected. Given the rough nature of our calculation, the upper limits we find are very credible.[5]

The idea of relating population density to available energy may be extended to other trophic levels. It is trickier to work out numbers, because there are generally more species at lower trophic levels and the available energy has to be shared out between them, so the actual population density of a given species will be much less than the estimated maximum. Furthermore, the population density of first-floor animals (herbivores) is limited by predation by second-floor species, as well as by energy availability. Energy flow is by no means the whole story, but you can see how it is always a significant factor in determining the behavior and form of animal species. Later in this chapter we show how the idea of energy flow leads to *scaling laws* amongst animals. These seemingly universal, or nearly universal, laws relate different aspects of animal form and behavior and apply across species of many different shapes and sizes. Here we have seen a simple example of how energy flow relates body size to population density.[6]

Foxes illustrate another way that energy flow influences animal life. Locomotion requires energy. For example, a study shows that the Kit Fox (*Vulpes macrotis*) spends about 20% of its energy budget on moving around (they can cover 30 km per day). These data were gathered outside their breeding season, so we can assume that most of that movement was associated with hunting. So, 20% of the foxes' energy is spent obtaining more energy (and a further 20% of any animals' energy budget is spent digesting its food). Clearly, all carnivores have similar dilemmas: they must expend energy to gain more energy. Natural selection has equipped them with optimized strategies to obtain the maximum food for the minimum effort. Equally, natural selection will have optimized their prey's abilities to avoid being caught. The hunters improve their chances of a successful kill if they improve their eyesight, hearing, and their brains (for cooperative hunting, perhaps, or ambushes). They must decide when to give up a chase and when to initiate one, weighing the odds of success against the cost in expended energy. Prey species similarly are driven to evolve superior predator-avoidance strategies: better eyesight and hearing, faster legs, cooperative lookouts who communicate the presence of predators to the rest of the herd, and so on. In most cases the length of food chains—the height of the trophic pyramid—is limited by energy availability to two or three links. We have seen that most of the energy at one trophic level does not make it to a higher level: the energy cost of locomotion (in particular of catching prey) accounts for part of this wastage.[7]

A recent study shows yet another way in which energy flow influences carnivore evolution. Small carnivores (less than about 15 kg) tend to hunt prey that is much smaller than themselves—invertebrates or small vertebrates. Large carnivores, however, tend to hunt prey roughly their own size: domestic cats hunt mice, but lions take wildebeest. The transition weight of predators (the point at which they switch from small to large prey) can be predicted from energy considerations. We can readily see why this might be the case:

- Catching a mouse requires little energy, but the gain is small.
- Catching a wildebeest requires lots of energy, if you're big enough to do so, but the gain is large.
- A large predator uses a lot of energy to chase prey, even if the prey is small.
- A domestic cat can't catch a wildebeest and it isn't worth the effort for a lion to chase a mouse.

The same analysis predicts a maximum size for mammalian land predators of about 1 ton. The largest land predator in the world today, the Polar Bear, weighs about half a ton. The largest extinct terrestrial mammalian predator (the Short-Faced Bear) is thought to have weighed about a ton.[8] So it seems that the size of terrestrial predators and of their prey is determined by energy efficiency.

Hare Today, Gone Tomorrow

So the natural world has partitioned itself, in terms of energy distribution, into trophic layers. This distribution is far from simple and is far from static. Had space permitted, we could show you reams and reams of charts that ecologists and biologists have put together, displaying their discoveries about the complex interactions among the species that make up an ecosystem. Our trophic pyramid itself is a simplification, although, like all the best simplifications that scientists construct, it is a useful and insightful concept. It is a simplification because many real animal species do not fit neatly into a single trophic level. For example, omnivores, such as pigs, piranhas, and people, can change trophic floors with ease—herbivores one day and carnivores the next. Rats, rheas, and raccoons also fit into this extended group of animals.

This complexity of animal behavior is reflected in their interactions. We have seen that first-floor animals (herbivores) munch away at plant life, absorbing perhaps 10% of the plants' stored energy in the process. Second-floor carnivores munch away at the herbivores. It is worthwhile stressing the dynamical nature of this predator-prey interaction. We rarely see stable populations in equilibrium—unchanging year in and year out. Though stability may be possible in a mathematical sense (we can construct math models of interacting populations where stability reigns), it is rarely true in the real world. Nature is red in tooth and claw: the rabbit does not actually want to be eaten by the fox, just to satisfy our desire to neatly compartmentalize species into convenient niches. So our rabbit does its best to run away from the fox. Countless generations of evolution have increased rabbits' abilities to run fast, precisely so they can avoid foxes and other predators. Foxes have evolved in parallel, and so we have a tension—an evolutionary struggle—affecting the success or failure of predator and prey populations. This struggle takes place on two very different time scales: the fluctuations of predator and prey populations takes place over a season or a few years, whereas the evolutionary honing of predator hunting skills and prey evasion capabilities takes hundreds of thousands of years.

It used to be thought that there was one classic case of predator-prey population dynamics that was well understood mathematically. It arose from the detailed statistics gathered by the Hudson Bay Company, in Canada, of Snowshoe Hare (*Lepus americanus*) populations. Pelt-trading records for these animals and of their main predator, the Canadian Lynx (*Lynx canadensis*), were kept for more than a century (see Figure 2). These records display a fluctuation in both predator and prey populations, with peaks occurring every decade or so. The peaks in lynx population lagged behind the hare population peaks, and the troughs lagged behind the hare population troughs. It is easy to understand this behavior (illustrated in Figure 3a). When the hare population is high, the lynxes feed well and their population increases. More lynxes mean more predation, so hare numbers fall. Fewer hares mean a food shortage for the lynxes, so the lynx population falls. Fewer lynxes mean less predation, so the hare population rises, and the cycle begins again. Mathematically, this understanding of the dynamical relationship between hare and lynx populations is expressed in the Lotka-Volterra equations (from which the graphs of Figure 3 were constructed). Static equilibrium of populations is replaced by fluctuating populations that nevertheless exhibit a dynamic equilibrium—for example, the average populations change only slowly.

The reason this hare-lynx model is a classic—studied by every ecology student and reported in most biology textbooks—is twofold. First, the unique and detailed records kept by the Hudson Bay Company are unusually long and complete. Second and more importantly for our purposes here, the connection between prey and predator species seemed unusually pure—uncontaminated by other species. Hence the Lotka-Volterra model of the hare-lynx population dynamics needed to consider only these two species and no others. Most species interact much more widely, with dozens or hundreds of other species that together form a complex web, a seething ecosystem of mutually dependent populations. A small subsystem, such as lynx and hare, is mathematically tractable, whereas a significantly more complicated system involving hundreds of species is not. So, here we have a rare case of nature behaving simply.

FIGURE 2 Canadian Lynx versus Snowshoe Hare. Place your bets. . . .

Oh no, we don't. It is true that the hare and lynx populations are strongly dependent on each other, but the world of each species is much broader than our simple two-species model suggests. Detailed studies now show us that hare populations fluctuated in the characteristic predator-prey manner even in areas where there are no lynx.[9] Well, perhaps this variation is due to fluctuations in the Snowshoe Hare's food supply. After all, the hare can be regarded as a predator of the grasses, bog birch, soapberry, and other plants on which it feeds, and so the plant populations should also fluctuate. In fact, the plant populations do fluctuate, but this is not caused solely by Snowshoe Hare predation. Insects, Red Squirrels, Spruce Grouse, and moose also feed on these plants. Let us see whether we can include these other predators by adding a randomly fluctuating component to the hare food-plant population: a graph

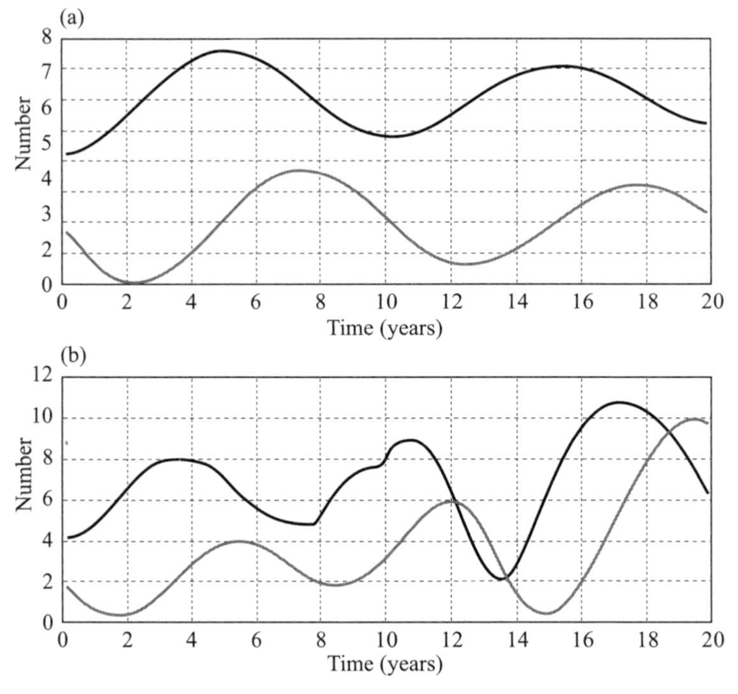

FIGURE 3 Predator-prey population dynamics. Here population density (numbers per km^2) is plotted as a function of time (years). (a) In a simple two-species model, the populations of Snowshoe Hare (black lines) and Canadian Lynx (gray lines) oscillate. The predator lynx oscillations lag the prey hare oscillations in a regular and predictable manner. (b) Adding a third component—here a randomly fluctuating density of grass, on which the hare feeds—leads to much less regular fluctuation of populations.

of the results from this extended model is shown in Figure 3b. The hare and lynx populations now fluctuate more wildly and irregularly.

Yet even this model is an oversimplification. The same study shows that lynxes eat or are eaten by squirrels, Red Foxes, Coyotes, wolves, and Wolverines. The moose that share food plants with the Snowshoe Hares are eaten by the wolves that also interact with lynxes, and so forth. So the simple model is too simple (seemingly straightforward linear food chains are in reality complex and interwoven food webs); Lotka-Volterra equations can apply, but real ecosystems are usually far too complicated for us to be able to extract enough details to run a mathematical model. Consider the simplest extension to the hare-lynx system, including only one other species (say, grass). The question arises: is the Snowshoe Hare population controlled by the lynxes (the "top-down" model) or by the grass (the "bottom-up" model)? It seems that the answer is: both. We are not surprised by this complexity in the fight for food—it is part of the energy flow between interacting species.

Energy Flow and Locomotion

For a lynx to catch a hare—or for any carnivore to catch any animal prey species at a lower trophic level—it needs to move, and to move faster than its prey. Locomotion is a key characteristic of animal life, and here, too, we will see that energy flow exerts a considerable influence. Animal locomotion is a big subject that merits a chapter all to itself (Chapter 3), but it is appropriate here to show how energy flow affects the way in which animals move.

We expect that animal movement (whether in water, on land, or in the air; powered by fin, leg, or wing) has been honed by evolution to be as energy efficient as possible. Indeed so, and to an astonishing degree. Thus a shark's skin may minimize hydrodynamic drag, and a flying albatross gains a little extra aerodynamic lift from ocean waves. We discuss more of this kind of optimization in Chapter 3. Here we show a more general way in which energy flow governs legged motion. Two-legged animals walk, hop, skip, or run. Quadrupeds can walk, trot, bound, pace, canter, pronk, or gallop in one of two different ways. Six-legged animals have even more combinations of leg movements. For some animals, such as horses and dogs, it has been shown that the type of gait an animal chooses when moving at a given speed is the most energy-efficient gait for that speed. Furthermore, the speed at which an animal changes gait is determined by the energy cost. Thus, a horse switches from walk to trot when the speed exceeds a certain value, because trotting at the higher speed is less energetic than walking at that speed. Going up through the gears, a horse finally breaks into a transverse gallop at the speed at which this gait consumes less energy (as measured by oxygen intake) than any other running gait. Human gait at a slow pace is a stiff-legged walk, which we adjust without thinking about it into a bent-leg walk when the gradient becomes steep enough and into a run when our speed is high enough.[10]

In Chapter 3 we will gain insight into how and why legged motion is the way it is. A few facts, gleaned from detailed studies of horse locomotion, will suffice here to show that energy flow plays a significant role. The legs of running animals are strong but are as light as they can be (think of horses or deer). The light weight minimizes energy wasted when swinging an airborne leg into position (airborne legs contribute nothing to propulsive force: only legs pushing against the ground can propel an animal forward). Horses' leg tendons act like springs that return 90–95% of the energy stored in them, which suggests strongly a pogo-stick model of energy flow during locomotion.[11] This model is most easily outlined by considering the pronking (sometimes also called "stotting") gait of quadrupeds. Thomson's Gazelles, Springboks, and deer will switch to a pronking gait in certain situations when being pursued by a predator. This peculiar gait consists of all four legs hitting the ground in unison; the animal bounces along like a four-legged kangaroo or hopping bird. A pogo stick provides a one-legged analogy. On contact with the ground, energy stored in the spring is released as forward and upward kinetic energy of motion, which then is converted back into stored spring energy as the stick spring compresses—the spring compression also acting to soften the landing. Gazelles may pronk when pursued to gain a better view of the predator or of the ground ahead, or they may be taunting the predator. ("See how fit I am—you can't catch me!" Cheetahs pursuing gazelles abandon the chase more often when the gazelle pronks.) We have noticed deer in the forests of Canada and the United States pronk away from us when startled—perhaps this gait reduces the chance of legs snagging in undergrowth. Whatever the reason, the gait is an impressive exhibition of energy conversion.

Cubus ridiculus and Geometrical Rules

We have seen how the flow of energy largely determines the structure of ecosystems via food webs and the trophic pyramid. We have seen a few tantalizing hints about the important role of energy flow in determining animal behavior (hunting strategies, gaits) and structure (leg shape). There are many, many other examples we could have chosen to illustrate our point, but we prefer a broad brush approach with only a few detailed examples rather than a comprehensive survey. Behavior: we might have discussed how warm-blooded animals huddle together in cold weather to limit heat loss, or how they shiver—making use of the inefficiency of muscles to generate heat; migration is largely determined by limited energy resources; flight differences between birds—soaring, gliding, flapping at high and low altitudes are almost entirely governed by energy considerations. Structure: we mentioned in passing the light weight of running legs; we said nothing about the shape of birds' wings; we said nothing about scaling rules. Biologists have written down rules to describe how the structures of animals vary in relation to one another or to environmental conditions. This subject is too important to skirt with only a throwaway line, and so we devote the next few sections to showing how these rules came about and why they are such hot topics (and somewhat controversial, being not entirely cut and dried) among today's research biologists.

Consider an animal new to science, shown in Figure 4. This creature, *Cubus ridiculus,* is here revealed to the world for the first time. *C. ridiculus* was previously unknown to zoology for the simple reason that it resides in the deep, luxuriant undergrowth of our artist's fertile imagination and browses on the

FIGURE 4 Animals new to science—a family of *Cubus ridiculus.* The juveniles are shaped like a $1 \times 1 \times 1$ cube, whereas Mom is $2 \times 2 \times 2$, and Pop is $3 \times 3 \times 3$. These creatures conveniently demonstrate Rubner's Law.

pedagogical thoughts of two science authors. This animal is warm blooded and will serve to illustrate the first of our rules: *Rubner's Law,* dating from 1883. This law is an interesting introduction to the subject of *allometric scaling,* because the derivation is almost certainly wrong, though Rubner's Law itself is probably right in other contexts. Hmm, perhaps an explanation is in order.

Let us say that *C. ridiculus* lives in a cold climate and so, being warm blooded, needs to conserve heat. Ecologists note that many of the young animals die from hypothermia, and the females must stay in the family nest to keep warm, but the males seem inured to the cold. The reason is plain to a biophysicist. Warm-blooded animals generate heat throughout their bodies while losing it only through their skins. Thus, heat generation is proportional to body volume, whereas heat loss is proportional to body surface area. *C. ridiculus* is uniquely capable of conveying to us the variation of body surface area and volume with body size. Note that the youngsters have a surface area of 6 and a volume of 1, so that their area to volume ratio is 6. Mom is a $2 \times 2 \times 2$ cube with a surface area of 24 and volume of 8. Her area to volume ratio is 3. So the ratio of heat lost to heat generated is less for Mom than for the youngsters: she is better able to withstand the cold climate. Pop does even better; a $3 \times 3 \times 3$ cube, he has a surface area of 54 and a volume of 27, giving him an area to volume (or heat loss to heat generation) ratio of only 2.[12]

Here is *Bergman's Rule,* well known to be true for a wide variety of animal species: for a given species, animal bodies increase in size with increasing latitude. Black Bears in the Canadian north are bigger than Black Bears in California, because northern Canada is colder than California. Our family of *C. ridiculus* has shown us why this rule is widely obeyed across the animal kingdom: increased body size among *endotherms* (warm-blooded animals) means more body heat is retained. Bergman's Rule is a simple consequence of geometry, once we recognize that heat loss is a surface phenomenon, whereas heat generation is a volume effect. Now consider *Allen's Rule:* animals are rounder in cold climates than they are in warm climates. That is to say, the cold-climate version has shorter limbs and tends to be more rotund, even if it is the same weight (which, according to Bergman's Rule, it won't be). So, the southern desert jackrabbit is more gangly and scrawny than the chubby Arctic Hare. Allen's Rule is also a simple consequence of geometry, as can easily be seen by considering a relative of *C. ridiculus.* This southerly cousin, *Planus preposterus,* is the same size as his northern relative but of a different shape. The adult male is of dimension $3 \times 1 \times 9$ and so has the same volume as a $3 \times 3 \times 3$ adult male

C. ridiculus. However, his surface area is larger: 78 instead of 54. He will consequently lose more heat. This is bad in a cold climate, but may be a distinct advantage in a hot southerly desert. So, cold-climate animals are rounder than warm-climate animals.[13] Geometry links with heat flow to influence animal shape and size.

Rubner and Kleiber

We have mentioned allometric scaling; now it is time to explain this term and to show why it is an important subject for biologists who are trying to understand why animals are built the way that they are. Consider two humans, say, a small child and a large man. They have different weights and different heights. For a person whose weight is in between that of the child and the tall man, we expect that her height will also be in between. We expect height to increase with weight—not proportionally, but it will increase. This is an example of scaling: height scales with weight. Stepping back and viewing the bigger picture, we find that, across species, the size of many animals scales with their weight. We can plot data for many species and show that size scales with weight in the same way, described in the next paragraph. Not just height—many other measurable anatomical or physiological factors, such as heart beat rate, or lifespan, scale with the body weight of an animal. This fact is surely telling us something about a common principle that underlies the structure of many types of animals. Biologists do not yet fully agree among themselves on just what that principle is, but progress is being made. Here we give a few examples of candidate scaling theories and of the underlying principles that they invoke.

Rubner developed his rule based upon the following observation. The rate at which a resting mammal consumes energy—its *basal metabolic rate*—is proportional to the heat that it loses: this must be true, more or less, because otherwise the animal would cook itself or freeze. We have seen that heat is lost through the surface, and so the basal metabolic rate is proportional to surface area. The weight of an animal is proportional to its volume, and so we have a connection between an animal's metabolic rate and its weight. We know that the surface area of an object—a cube, a sphere, or whatever—is proportional to the object's volume raised to the two-thirds power,[14] and so basal metabolic rate M_b is proportional to weight W raised to the two-thirds power: $M_b \propto W^{2/3}$ (the symbol $\propto$ reads "is proportional to"). This is Rubner's Law, derived from geometrical area-volume considerations of mammal heat generation and loss. It is an example of an allometric law, or scaling rule, because it tells us how two properties of an animal change as the scale changes. So Rubner's Law states that if animal A is 8 times heavier than animal B, then the basal metabolic rate of animal A is expected to be about 4 times that of animal B, because $8^{2/3} = 4$. An underlying principle (heat generation throughout a volume and heat loss through a surface) has led Rubner to a scaling law.

The same kind of relationships can be derived for an animal's strength to weight ratio because, for example, the strength of a leg is proportional to its cross-sectional area, whereas the weight it supports is proportional to the animal's volume, so we expect that leg thickness must increase faster than size. Scale up a mouse to the size of an elephant, and its legs will break, because their strength increases with scale less rapidly than the weight they must support. In fact an elephant's legs are proportionally thicker than a mouse's legs.[15]

The strength to weight relationship, derived from geometry, works well, but in fact the metabolic rate to weight relationship may not be quite right: the same allometric laws may apply to cold-blooded creatures, who do not maintain a constant temperature, and perhaps also to plants and to unicellular creatures. How can a heat loss argument explain the very similar connection between M_b and W that is observed for these species? It cannot. Numerous measurements of the weights and metabolic rates of many different animal species have led to *Kleiber's Law*, which states that $M_b \propto W^{3/4}$. Max Kleiber announced this law (really the result of many observations) in 1932. Some biologists consider that Kleiber's law is a better description of nature than is Rubner's Law, and so they seek to develop theories that will lead to the $3/4$ scaling of Kleiber instead of the $2/3$ scaling of Rubner.[16] These attempts are controversial, because the data are ambiguous: some data favor Kleiber, other data are consistent with both Kleiber and Rubner.

It is worthwhile pointing out the nature of the data that biologists have gathered over the decades—the data that are used to establish the actual scaling relationship between animal basal metabolic rate and animal weight. These data are from hundreds of different animal species of all shapes and sizes, covering twenty orders of magnitude, from tiny bacteria to the largest of animals.[17] A small subset of this data is shown in Figure 5. As you can see, the data are statistical, meaning that few species lie exactly on the mathematical line defined by Kleiber's Law or Rubner's law. Include the many hundreds of data points corresponding to the many hundreds of animal species that have been measured, and we obtain a cloud of data points that cluster quite closely around a straight line. The cluster is sufficiently tight that some people say it follows Kleiber's Law, though other biologists hold out for Rubner's Law or for something in between.[18] A complication arises in that the data measure basal (i.e., resting) metabolic rate rather than animals' metabolic rate while active, which is different and perhaps more relevant (but also more difficult to measure). For this reason the data and their statistical significance are hotly debated in the biological research community. Life is statistical, and statistics are messy —but we have to live with it.

Whether the scaling factor for metabolic rate is $^2/_3$ or $^3/_4$ or something in between, it is clear that metabolic rate increases more slowly than body weight. This fact has consequences for individuals and for entire populations. Small animals must eat more, relative to their body size, than large animals must, to maintain their metabolic rate. A meadow vole must eat six times its own weight

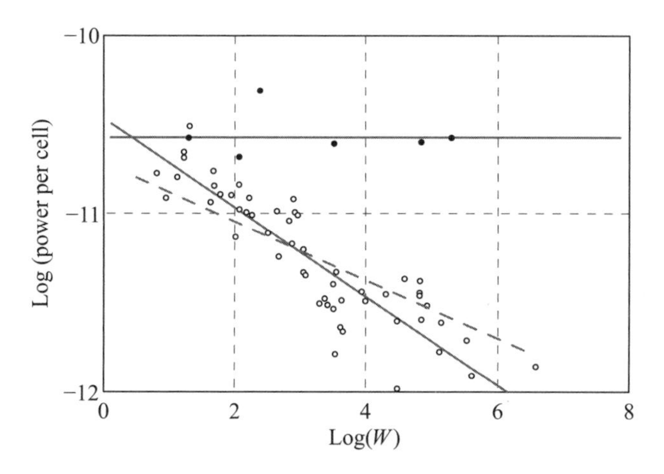

FIGURE 5 Metabolic rate of in vivo mammalian cells versus organism mass W (open circles). Here logarithms are plotted, so that the scaling law appears as a straight line in the data. Cell power is measured in watts and organism mass in grams. Cell power is equivalent to organism metabolic rate per unit weight and so, according to Kleiber's Law, should scale as $W^{-1/4}$, shown as the sloping gray line. Kleiber's law fits the data better than does Rubner's law (dashed line) for this particular set of data, though Rubner's law is not ruled out. The same mammal's cells grown in vitro (filled circles) show no dependence on organism mass (horizontal gray line). Data from West et al. (2002). Copyright © 2002 National Academy of Sciences, U.S.A. Reproduced with permission.

of plant matter each week, whereas a white rhinoceros (with a similar diet) eats only one-third of its weight each week.[19] Because small animals eat more, they defecate more and in general their throughput of energy is greater. One side effect is that small animals accumulate dietary or airborne toxins more than do large animals. On the level of populations, we can anticipate that large animals will have greater biomass per unit area of land than do small animals of similar diet. This prediction concerning population mass density, based on the allometric scaling law for the metabolic rate of individual animals, is borne out by observation.[20]

So far we have examined the manner in which energy is distributed among animal species in an ecosystem, and how the amount of energy that an individual animal requires depends on its size. Now we must examine the manner in which energy is distributed inside each animal, because several recent theories, invoking different principles, claim that it is the geometry of the internal distribution networks that is responsible for the allometric scaling laws.

Life's Distribution Networks

We will look at one theory that suggests that the internal energy distribution networks of animals is organized on *fractal* principles. A second theory states that the network shape is not necessarily fractal, but it is determined by the cost of pumping fluid around the network. Another theory claims that the network shape chosen by nature maximizes access of animal tissue to the pumped fluid. These theories all suppose branching networks, but the details differ. The matter is controversial, to the extent that a prominent biology research journal recently published a special issue to air the different theories and the data that each claims as supporting evidence. Here we show how the fractal theory leads to a prediction of Kleiber's law, but other theories can lead to this law or to slightly different scaling behavior. The matter will not be finally settled until better data become available.[21]

Let us pretend for a moment that you are Edward Russet Warba, a couch potato who loves to eat real potatoes. You grow your own in your garden in Tucson, Arizona. Now, southern Arizona is not renowned for its high annual rainfall, and thus you must frequently water your potato crop. Couch potatoes are not renowned for their energy, and so you wish to set up an irrigation system that will automatically water your spuds. Your first attempt at a water distribution system, shown in Figure 6a, doesn't work very well. The water pressure at the start of the pipeline (a hose), close to the water source, is much greater than the pressure at the end of the pipeline, so the distribution of water is uneven. Also, the hose must be able to withstand high pressures. Your second attempt (Figure 6b) is much more successful. Here you have split up the pipeline into a network in such a way that the distance from water source to potato plant is the same for all plants. The water pressure is consequently the same at each sprayer outlet, and so all plants receive the same amount of water. Also, no particular section of the network is subjected to especially high water pressure. Another bonus of your new system: if you decide to quadruple the density of potato plants, you do not have to haul out your irrigation pipes and install a new system from scratch (four times as long); instead you simply add another layer of network (16 new "H" sections added to the termini of the network shown in Figure 6b).

This example shows the advantages of self-similar (fractal) networks. You may already have guessed where we are going with this idea. Nature has a distribution problem to solve in the body of every animal: how to get resources (food, oxygen) from a central distribution point (stomach, nose) to the individual cells of the body, all of which require the energy contained in these resources. The fractal theory suggests that nature has hit on the idea of fractal networks. That is to say, many different distribution ideas will have been tried, but the sieve of natural selection has weeded out the inefficient systems and left us with branching networks like that of Figure 6b. Blood is carried in you through three circulation networks to distribute oxygen and nutrients to all parts of your body and to remove waste products. (The leaves of plants similarly contain a network for the distribution of nutrients.) Fractal-like networks are the most efficient form of distribution system (except for very small organisms), because gases and nutrients are exchanged across surfaces, and so a large surface area at the point of exchange is a good idea. Your lungs form a network that maximizes surface area (at about 70 m^2), and a fractal-like structure permits this large area in a small volume. "Indeed, a fractal geometry would be Nature's strategy for enhancing the surface to volume ratio."[22] We now illustrate why fractal-like structures are optimum in this sense.

Consider Figure 6c. A fractal network branches in a systematic way, so that it covers a wide area. Mathematically, if we continue the branching down to an infinitesimal length scale, then every point in the square is covered—so our essentially linear network is in fact two dimensional. Similarly, scrunched up paper becomes something that is, in a mathematical sense, between two and

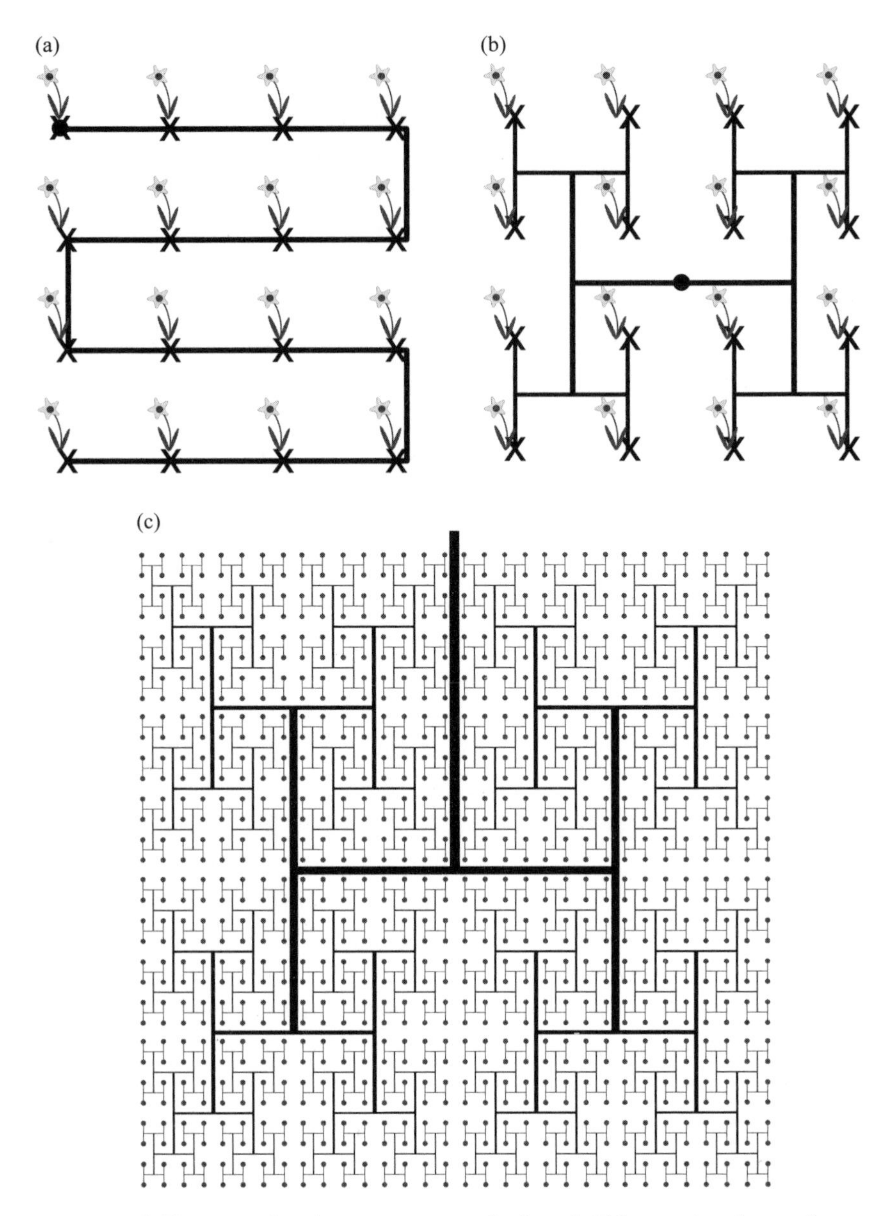

FIGURE 6 Two water distribution systems and a fractal. (a) Linear distribution from source (black circle)—this system is not very good, because the pressure varies for each sprinkler head (X). (b) A distribution network—this one is much more efficient and can be extended to cover a greater area more economically than can the linear system. (c) Five levels of the H fractal; continued to infinitesimal scale, it covers every point in a square.

three dimensions, whereas a flat sheet is clearly two dimensional. Another example of fractal dimensions: "the total area of two-dimensional sheets filling three-dimensional washing machines clearly scales like a volume rather than an area."[23] What we are seeing here is a way of extending lines into two dimensions (Figure 6c) or surfaces into three (washing, lungs). In fact, not all fractals increase dimensionality by one. The H-network of Figure 6c is an example of a tree fractal of dimension two, but some other tree fractals do not entirely fill a plane and so have dimension between one and two. Similarly, some tree structures that extend into three spatial dimensions do not fill the entire volume and so have dimension between two and three. The dimensions of lungs, circulatory systems, and plant vascular systems are close to three. Our derivation of Rubner's Law showed that the geometrical scaling factor for metabolic rate is given by (area dimension) ÷ (volume dimension) and so, because the dimension of volume is one more than that of area, we have a scaling factor of $d/(d + 1)$. In the pre-fractal days of Rubner, area dimension was always $d = 2$, and so a scaling factor of $2/3$ was found; the fractal theory of West et al. predicts the area scaling dimension of lungs, circulation systems, vascular systems, and other biological distribution networks as being closer to three, and so the relevant scaling factor for metabolic rate is $3/4$. Because of this prediction, biologists who believe that distribution networks are fractal-like are rooting for Kleiber's law.

If fractal mathematics doesn't work for you, then Figure 7 conveys another theory of distribution networks. One section of a web of pipes is magnified in the figure. In this particular network, the pipe length is a multiple k of the radius r_1; the pipe splits into two pipes of radius r_2 and length kr_2. Imagine air moving through the network at constant pressure.[24] It is not difficult to show that the two radii are related by $r_1 = 2^{1/3}r_2 = 1.26r_2$. This equation is known as *Murray's Law* and has been derived here assuming constant fluid pressure in the pipe, as shown in Figure 7. (A section of fluid in the pipe of length kr_1 is moved into two sections of length kr_2; these volumes must be equal, or else the air pressure would rise or fall along the pipeline. Requiring equal volumes leads to the stated relationship between adjacent pipeline radii.) In fact, though, Murray's Law can be derived from a much more general principle: the network geometry minimizes the energy cost of transmitting fluid through it. Suppose we label the radius of the nth level of pipeline r_n; we see that r_n depends on pipeline level n as shown in the graph. So radii get smaller as air flows farther down the pipeline network. But notice what happens to the surface area of pipes at the nth level: simple geometry reveals that it increases (also plotted in Figure 7).[25] After, say, 30 branchings ($n = 30$), the pipeline network area is 1,000 times larger than at the start. Such area enlargement is exactly what is needed for exchanging gas or nutrients through a surface. Furthermore, the speed of the blood (or air, if we are considering the plumbing of lungs) through the network has dropped a thousand-fold. Why? Because it moves a distance kr_{30} in the same time as it moved a distance kr_1 at the beginning of the pipeline. So, for example, there is a lot of time to exchange oxygen

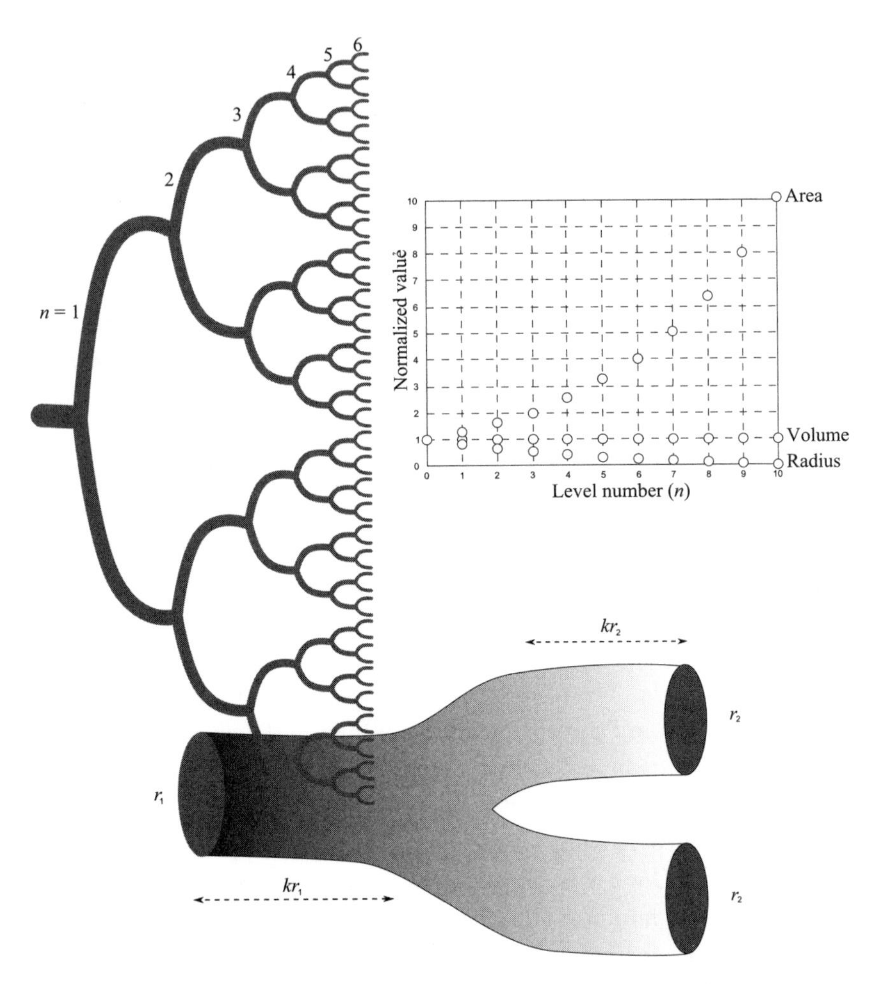

FIGURE 7 A network that exhibits constant volume at each level n will grow in area as n increases, while the pipe radius gets smaller.

and carbon dioxide across the surface of the 30th level of a lung network made up of the array of pipes shown in Figure 7.

Real vascular systems are not confined to spread across a page, essentially in two dimensions as for the pipeline network of Figure 7, and so the gain in area (and reduction in pipeline fluid speed) through the network is more marked than our simple example indicates. The point is made, however: branching network structures display exactly the right geometrical properties for conveying nutrients and gases across surfaces that are confined to a limited volume. The view that real vascular networks are responsible for scaling behavior in animals is given support by the data of Figure 5, which shows the scaling behavior for cells grown in mammals' bodies (in vivo) and in a test tube (in vitro): those grown in vivo scale according to Kleiber's Law (or Rubner's law or something similar), because their nutrients are made available via vascular networks, whereas those cells grown in vitro do not scale, because they were supplied with nutrients in a different manner.

More Allometric Laws

The fractal theory of allometric scaling is not the only theory being debated in the research literature, as we have seen, and it is not universally accepted. Its proponents claim that there is a considerable body of evidence to suggest it might hold water (or blood, or air . . .). Not only does it predict the three-quarters scaling factor for metabolic rate, but it leads to other predictions. For example, it predicts that tree trunk diameters increase as the three-quarters power of tree weight (i.e., as $W^{3/4}$); the mass of brains and the size of aortas in animals also increases as $W^{3/4}$; the lifespan of an animal increases as $W^{1/4}$; its heartbeat rate falls as $W^{-1/4}$. The authors of this theory claim that all the quarter-power allometric scaling laws originate from the geometry of fractal-like energy distribution networks in organisms.[26] They claim that many experimental data support their theory, though opponents say that the data are not nearly so clear cut: the evidence may be consistent with fractals, but it also fits other theories. Whether or not fractal-like networks apply as widely as their advocates claim, they are sure to generate a lot more debate in the technical literature.[27]

We can combine the allometric laws: if lifespan scales as $W^{1/4}$, and heartbeat rate scales as $W^{-1/4}$, then this suggest that the product—the total number of heartbeats during an animal's lifetime, is pretty much a constant (about one or two billion beats—a little higher for humans) independent of animal weight. Multiply metabolic rate by lifetime and divide by body weight, and again we find the result to be independent of weight. Thus the total energy consumed by an animal, per gram of body weight, over the course of its life is constant—it does not depend on the animal's size. Indeed it was precisely this observation made by Rubner (he found a value of 200 kcal per kg—again humans are exceptional, at 800 kcal per kg) that led him to his original law predicting a $^2/_3$ scaling factor.[28]

Homo sapiens—Maybe

Why should we humans be biologically exceptional in the energy that our bodies consume, per kilogram, over a lifetime? Because we are (in relative terms) very brainy animals. Each gram of gray matter consumes about 10 times as much energy as the same amount of muscle or other body tissue—equivalent to the power consumption of a leg muscle running a marathon. Fully 20% of our metabolic rate is due to our energy-hungry brains, even though brains constitute only 2.5% of body weight. Clearly, brains must confer survival benefits (it is easy to see why this might be so—use your brain), and yet the cost is significant. Brain mass is limited by available energy resources; brain functioning must be streamlined to be energy efficient; thus the way that we think is influenced by energy flow.[29]

We do not need to go into the reasons why gray matter consumes so much energy—recall that all highly organized structures consume energy just to maintain themselves. Think of a juggler who balances rotating plates on top of poles; he rushes around twirling each pole to keep the plates balanced. The more poles and plates he maintains in this state, the harder he works. Natural selection has chosen to endow us with large brains, and we have paid the price in other ways: belt-tightening of the energy budget in other parts of our bodies. We do have higher metabolic rates than other mammals, but this is not enough. Other parts of our bodies economize on energy; for example, by reduced growth or locomotion costs, or a higher quality diet (cooking food may ease the cost of digestion, which in most mammals, recall, costs 20% of the energy budget).[30]

Recent research shows that the scaling of mammalian brains with body weight does not follow Kleiber's Law: detailed measurements reveal that brain metabolic rate increases as $W^{0.86}$, with a measurement error on the scaling factor of only 0.03. (Here *W* is brain weight, not whole body weight.) So, brains increase their energy consumption faster than bodies as size increases. Perhaps neurons become more extensive or work harder as brains enlarge. A mathematician would say that your brain has a fractal area of dimension five or six—try bending your brain around that thought without blowing your energy budget.

Feeling the Heat

Finally, a very obvious example of the influence of energy flow on animal structure and behavior. We might have mentioned thermoregulation several times already, but have only skirted this important topic with brief references to shivering, huddling in groups for warmth, and the like. In fact, different animals deal with it in different ways: birds and mammals maintain a nearly constant body temperature (remarkably so, given the environmental and physiological factors that tend to upset temperature); fish, reptiles, and others go with the flow, permitting body temperature to be controlled by the environment.

Endothermic animals maintain their core body temperature within a very narrow range (centered on 37–38°C (98–100°F) for mammals and 40°C for birds). The narrow range enables the brain and certain enzymes to function most efficiently. Temperature is maintained at a constant level via several heat conservation and dissipation mechanisms (e.g., shivering, sweating, and elevating metabolic rate by eating more). Endothermy confers some survival advantages compared to ectothermy: a warm-blooded animal can function over a wider range of ambient temperatures, geographical range is less limited, and longer periods of high energy expenditure are possible. However, warm-bloodedness requires a greater intake of food and water per kilogram of body mass, less of that energy intake is available for growth and reproduction, and small animal body size is restricted because of the Rubner surface heat-loss effect.

In this section we want to introduce you to hibernation and torpor—two extreme adaptations of endotherms that permit them to survive when available energy resources are insufficient to maintain body temperature—but before doing so we should mention in passing some of the structural and behavioral consequences of temperature regulation. Heat can be absorbed or lost via conduction (think of a pig or hippo wallowing in mud), convection (wind cooling), radiation (ectotherms, such as snakes, basking in the sun), or evaporation (sweating). Fur, feathers, and blubber provide insulation, and as we have seen, body shape and size further influence heat loss. Tuna and sharks swim constantly, and so their active swimming muscles continually generate heat; these muscles are located centrally and so maintain core temperature above ambient. A Galápagos marine iguana regulates his heart rate during cold-water dives, shunting blood away from his skin to avoid heat loss—unique adaptations permitting this cold-blooded creature to feed on underwater seaweed. Only the largest males brave the cold waters.

Endothermy permits mammals and birds to populate cold climates where ectotherms cannot survive. In some cases and in very harsh environments endothermy and normal thermoregulation behavior may not be enough. Some large mammals in cold climates indulge in shallow hibernation, during which their body temperature drops about 5°C. Bears (also racoons and chipmunks) do this; they find a well-insulated den and then curl up in a ball, with extremities withdrawn (to reduce heat loss). Such shallow hibernators are capable of waking if disturbed; true hibernators are not. A bear's reduced activity and lowered temperature trim energy expenditure a little. True hibernation reduces energy expenditure drastically; it is a response to cold weather and reduced food supplies—conditions that would require an animal to expend more energy each day than it could acquire. Only three orders of placental mammals are capable of—or need—true hibernation: Insectivora, such as the European hedgehog; Chiroptera (bats); and Rodentia, such as marmots and ground squirrels. These animals are mid-sized: larger animals do not in general need to hibernate, and smaller ones cannot, for geometrical reasons that we have already discussed. Hibernation is a physiological response that enables a warm-blooded animal to sleep through lean times and (hopefully) wake up when good times return. It is a risky strategy, because the individual hibernator is dead to the world and cannot respond to life-threatening situations. Respiration slows to perhaps three breaths per minute, the heartbeat is all but imperceptible, and the animal is not aware of its surroundings. A predator, an extended period of freezing temperatures, or disease may kill the hibernator while it sleeps.

What about very small animals? Surely they are the most susceptible to cold weather and reduced food supplies, for scaling reasons we have already encountered, and so are most in need of hibernation? But they cannot hibernate. Their heat loss is too great. Instead, they undergo a short-term version of hibernation called "torpor." (Torpor usually occurs overnight, and therefore sometimes is known as "noctivation.") The triggers are the same as for hibernation. Thus, a mouse that requires 17 g of food per day for each 100 g of body weight will go into torpor overnight if it is unable to obtain this level of food. The Chilean mouse-opossum (a South American marsupial) responds in the same way to an insufficient food supply; increased frequency of torpor and a reduction of internal organ size both lower maintenance costs.[31]

There is a limit to how small an endotherm can become, even with the strategy of torpor, before Rubner's geometrical heat loss argument makes life very difficult. Hummingbirds (Figure 8) appear to live right on the edge. They are very small and so suffer more heat loss per gram of body mass than do larger endotherms; at the same time they must maintain a very high metabolic rate. So, hummingbirds spend most of their waking hours obtaining high-energy food, such as nectar (two to three times their own weight each day), despite which they are only hours away from starvation. Hummingbirds as a whole could not survive without going into torpor at night.[32] During this state their body temperature plummets (they lack insulating downy feathers), their metabolic rate drops by 95%, and they appear to be dead. They wake up from torpor according to some internal clock and then spend 20 minutes shivering to bring themselves up to normal working temperature. (Here is one advan-

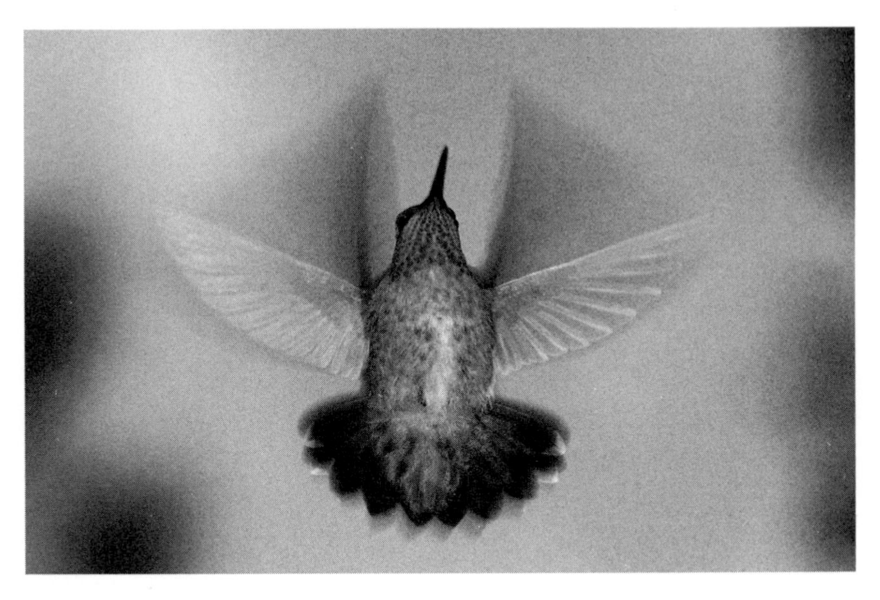

FIGURE 8 Life on the edge: an Anna's Hummingbird (*Calypte anna*). Thanks to Rich Swanner for this image.

tage of being small, and a very good reason why large animals don't become torpid overnight: it would take many hours and a lot of energy reserves to bring themselves up to normal body temperature.)

Hummingbirds may use torpor to conserve energy during migration (yes, some of these little creatures migrate long distances).[33] Other small birds in cool regions of the world may also resort to torpor to see them through the night.

Torpor and hibernation are an extreme response to harsh environments that permits warm-blooded creatures to live in these zones.[34] Being nearly dead for much of their lives is the price they pay—because of energy flow.

2

Structural Engineering

The Bare Bones

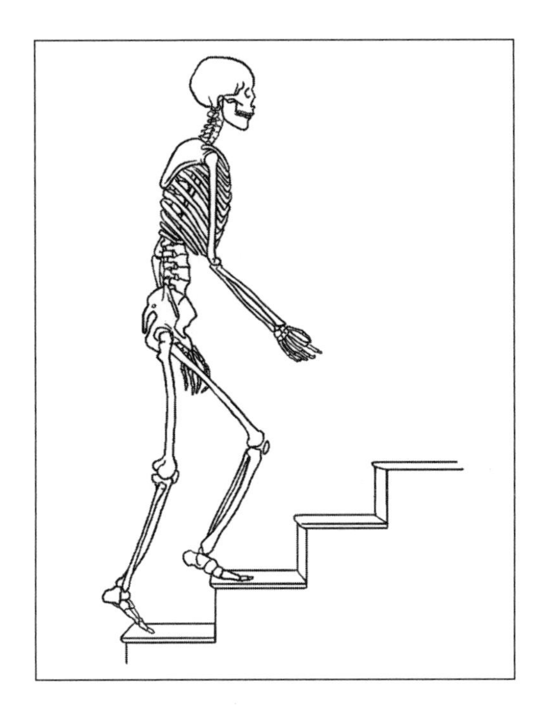

An animal's structure is based on its skeleton, just as a building's is based on its frame. Frames vary enormously and so do skeletons. Many skyscrapers have a central core that supports the weight of the building—a backbone. Other buildings are designed to be supported by the external walls—a carapace, or exoskeleton. Different designs suit different purposes; one of the main impressions that emerge from even a cursory survey of animal anatomy is the structural diversity that exists. A good idea (say, the development of keratin, a tough, fibrous protein) blossoms over eons into a hundred variants (including horns, hair, wool, nails, quills, claws, hooves, scales, feathers, baleen, beaks, and shells), each finely honed by evolution to suit a very specific function.

This wide variability makes it challenging to present an instructive account of animal structural engineering: to be comprehensive would require a multivolume encyclopedia, whereas to present a broad overview of guiding principles risks losing contact with everyday reality. This challenge is exacerbated when we add in the obvious fact that there is much more to animal structural engineering than skeletons. The structure moves, and this requires power—muscles. The muscles must be supplied with energy—a circulatory system. And so on.[1]

The approach we adopt in this chapter sidesteps the challenges, in a sense, because we want to be neither verbose nor vapid, neither terse nor turgid. So, to be both instructive and readable, we present here some basic concepts and engineering principles that all structures (in particular, skeletons and circulatory systems) adhere to, interspersed with very specific examples—some of them extreme. This quantum leap is sometimes stark, and you may have to think a little to see the connection among general idea and specific implementation. For example, fish will take center stage to illustrate the structural diversity that all animal classes exhibit; we will leap from simple beam theory to bird bones; giraffes will tell us about blood circulation, after a brief discourse on fluid flow in pipes. In this chapter more than most you will be relying on the bibliographical references to fill in the gaps, given the wide range of subject matter. Yes, you will find answers to questions here, but mainly you will be given the tools to ask deeper questions.

Fishing for Answers

Biologists suspect that the 1.7 million living species that have been described scientifically are but a small subset of the whole. Half of these are insects—there are more types of beetles (290,000) than there are of plants. The profusion of insects is not surprising. There is a scaling argument that suggests species number N should increase as body weight W decreases: $N \sim W^{-1/3}$, so one group of animals that weighs 1,000 times less than another group should be divided into roughly 10 times as many species (10 times more mouse species than deer species, for example).[2] Of the 55,000 known species of vertebrates, about half (52%) are fish. Why so many? Surely we cannot employ a scaling argument in the sea. There are no niches in a cubic meter of water. Surely the diversity of terrestrial habitats exceeds that of aquatic habitats, and so shouldn't we expect more terrestrial species? In this section we want to emphasize the sheer diversity of animal structures, and we choose fish as our explanatory vehicle precisely because it is less obvious why they are so diverse. In fact, the scaling argument does apply to fish (there are more small fish species than large ones), but evolutionary history and geology also play significant roles.

First, let us turn to the very basic body plans, which go back a long, long time in evolutionary history. Early creatures were likely spherical blobs, but the development of a digestive tract (at its simplest, a tube from mouth to anus) produced axial symmetry. Think of a roughly spherical blob with a hole through it. (This level of symmetry is usually mislabeled as "radial symmetry." To a physicist or engineer, however, a sphere is radially symmetric, whereas a cylinder—or a sphere with a hole through the center—is axially symmetric). Many aquatic animals retain this early axial symmetry (jellyfish, for example). It works well for sedentary creatures for which stimuli and food may float by from any direction (see Figure 9). Such slow or immobile creatures are vulner-

FIGURE 9 A sand dollar (order Clypeasteroida) "test" or exoskeleton displays approximate axial symmetry.

able to predation and so many have developed protective structures, such as spines, shells, or stinging cells. The next evolutionary step reduced symmetry further, from axial to bilateral. Gravity imposes a ventral and dorsal side (bottom and top), but not a left/right asymmetry. So you, all other vertebrates, most other terrestrial animals, fish, and birds are bilaterally symmetric: the left and right halves of your external structure, and much of your insides, are (almost) mirror images. Bilateral symmetry is better than axial symmetry for locomotion—it is difficult to imagine fish swimming or birds flying without it. Bilateral symmetry permits streamlining. It also means that internal organs can be placed in different parts of the body; in particular, sensory organs can be grouped together at the front, where they are needed. Such a clustering of sense organs leads to cephalization—the development of a head and brain.[3] The same engineering principles apply to machines that move. An airplane is bilaterally symmetrical with sensory instruments at the front; this concept is so basic that it goes back a century, to the beginning of manned flight (after a brief phase of lighter-than-air axially symmetric balloons). Animal bilateral symmetry goes back further: half a billion years to the Cambrian period, during which the structural diversity of animal body plans exploded. From bilateral symmetry, it is a small step to segmentation—another very successful evolutionary innovation. Insects are obviously segmented, but so are you (think of your vertebral column). Segmentation is an algorithmically simple extension of body plan design. (It eases the problem of encoding for development of animal structure—the same instructions work for each segment.) Among other benefits, it aids locomotion.[4]

Now for those fish. It is generally reckoned that there are five basic fish body plans, with subdivisions. Predators divide into chasers (e.g., tuna: fast moving and therefore streamlined, often large and usually with a forked tail) and ambushers (e.g., pike: large mouth and with fins concentrated at the back to facilitate rapid acceleration). The surface fish (e.g., Guppies, *Poecilia reticulata*) are small, with upward-pointing mouths. Bottom feeders (e.g., flounder) have flattened bodies and small eyes. Deep-bodied fish (e.g., sunfish) are compressed laterally and usually have big fins and large eyes. Eel-like fish are elongated like a rope or ribbon, with blunt heads that conform to the body cross-section. Why such diversity of size and shape? We might expect bottom feeders to exhibit the same diversity as terrestrial animals, but what about the rest? There are different aquatic habitats: fresh water accounts for one-third of fish species; shallow marine habitats divide into warm and cold; deep marine habitats are pelagic (open sea) or benthic (deep sea beds)—but these five broad divisions are not as distinct or as numerous as terrestrial habitats.

One clue to fish diversity is found in the broad trends of evolutionary development. Early fish were large and fast, with elongated bodies and fins clustered in the posterior half. Later fish were smaller and spinier, with fins more evenly distributed around the body. A phase of rapid (for evolution) development occurred after the breakup of Pangaea, the supercontinent that formed most of Earth's land mass 200,000,000 years ago. The smaller continents created increased coastline length and therefore increased littoral regions

FIGURE 10 Five very different fish designs: tuna (genus *Thunnus*), Porcupinefish (*Diodon nicthemerus*), flounder (e.g., *Hippoglossoides platessoides*), seahorse (genus *Hippocampus*), and freshwater sunfish (genus *Centrarchus*).

and shallow continental shelves. These regions are rich in nutrients and also are bathed in sunlight, and so they are very rich in biomass. The number of fish, and the number of fish species, grew astronomically as a consequence of increased coastlines. Before Pangaea split up, most of Earth's surface was deep sea, and so predators and their prey had to be swift. Afterward, the shallow seas provided coral reefs, seaweed beds, and other nooks and crannies to hide in, so fish became smaller to exploit these new habitats; they and their predators needed maneuverability rather than raw speed, and so fins were redistributed. Spininess is another method of self-defense when speed is taken away.

So the great diversity of fish (see Figure 10) shows that the general form of animal structure depends on habitat and lifestyle but also on the vagaries of evolutionary background and geological change.

The Shell Game

Generally speaking, exoskeletons—external shells—are the chosen method of structural support for smaller animals. In many cases they also act like a suit of armor. Mollusks make their exoskeletons out of calcium carbonate, whereas arthropods (insects, spiders, crabs, and lobsters, among others) choose a different substance: *chitin.*[5] A true exoskeleton is layered, the outer layer being strong in compression and the inner layer strong in tension. This combination is robust. (Again there is an architectural analogy. Large buildings in medieval Europe were often constructed of stone and supported on the inside by a wooden frame. Stone is strong in compression, whereas wood is strong in tension.) Consider an insect exoskeleton as a hard tube; predator jaws attempting to crunch the tube will cause the outer layer to compress and the inner layer to stretch—playing to the strength of the exoskeleton. As well as structural support and protection against predators, exoskeletons serve another purpose. They prevent dehydration, thus permitting many arthropods (for example) to occupy arid ecological niches that would be challenging for other creatures.

There are disadvantages of exoskeletons, however. Heat dissipation would be a problem for many large animals with exoskeletons (recall the scaling arguments of Chapter 1), which is why true exoskeletons are widespread only among smaller creatures. Another major disadvantage is that chitin does not grow, and so the growing occupant of an exoskeleton must shed it periodically and develop a new one.[6]

Bone Engineering

Most, if not all, readers of this book will have an endoskeleton. The main purpose of endoskeletons is to provide mechanical support. Thus, each bone must be strong enough to resist the forces and torques that will act on it, and it must be connected to other bones in such a manner that the proud owner of the skeleton can move freely. The complex design of a skeleton is, consequently, largely a matter of engineering, and so we can reasonably apply our knowledge of mechanical engineering to generate insight into the construction of skeletons. We need not delve into the extensive mathematics of stress analysis to convey this insight. Let us quote just one formula, however, to illustrate how engineers describe mechanical stresses and how such a formula can relate to skeleton biomechanics. It will serve to tell us which aspects of skeletal design we engineers are free to play with and which are fixed by reasons other than stress analysis.

If an architectural stone column or a steel pylon is subjected to a vertical load (a weight), then it may buckle, if the load is big enough. The equation from structural engineering that predicts the maximum load a column can support without buckling is the classic *Euler formula:*

$$F_B = \frac{kEI}{L^2}.$$

Other formulas exist for other types of applied stress, such as bending or twisting. In the Euler formula, F_B is the maximum force that the column can withstand without buckling, L is the column length, and k is a dimensionless constant that depends on how the column is attached at both ends. (For example, if the column is planted firmly in the ground, then it has a different k value than an identical column that is attached to the ground by a hinge.) E is *Young's modulus,* a parameter that describes the intrinsic stiffness of a material: stone is stiffer than pastry, and so stone has a (much) bigger Young's modulus than pastry, which is one reason why the Parthenon columns are made of stone and not pastry. The last remaining parameter in Euler's formula is I, a geometrical parameter that goes by the ungainly (and, to a physicist, confusing) name of "area moment of inertia." It is also known as "second moment of area." This parameter describes how the shape of a column (square cross-section or circular, hollow or solid, and so forth) influences its strength.

Substituting the parameter values for a particular column into Euler's formula tells an engineer whether a given column will support a given weight. If not, then she can increase the column strength by making it shorter, attaching it differently, constructing it from a stronger material, or changing its shape.

Don't sweat the math—we will not be testing you on the arcane names of parameters or exposing you to the many, many formulas that exist in mechanical engineering to predict the strength of a column or beam that is subjected to tensile, compressive, or shear forces, or to torsion. Euler's formula is given here only to show how engineering analysis can be put to use in describing skeletons (clearly, leg bones can be considered as columns supporting a load). Thus, for example, we see that long bones are more prone to buckling than are short bones—a fact that the reader may verify by playing with a strand of dry spaghetti.

Two bones in a skeleton will be subjected to different forces, and these differences will influence their structure. For example, there are different types of connections among bones: ball and socket joints (your shoulder) provide a different k parameter than will hinge joints (elbows or knees). These are functional rather than mechanical constraints: ball and socket joints permit more freedom of movement; hinge joints support greater loads.[7] Bone length also depends on function. Let us assume, therefore, that k and L in Euler's formula are fixed for a given skeleton.

Young's modulus for bone is quite high—bone is very stiff. It is not as high as that of steel, but biological systems have not yet evolved to the extent that steel skeletons occur naturally—or perhaps the weight of steel renders it less useful than bone. Nature has settled on bone as the material with a high Young's modulus that it can most readily generate. So, E is also fixed.

The shape of bones (represented by the I parameter in Euler's formula) is of greatest interest to us here. We have seen that the other parameters are all constrained. The question of interest is this: what shape should a bone be? Assume that the weight of the bone is predetermined—nature has allocated a certain amount of material to each bone. How should this material be shaped to provide greatest strength? This is truly a question of engineering and not function. The function of a bone determines the type of forces that act on it, the magnitude of those forces, and also the bone length and attachment method, but the best shape for that bone is largely an engineering problem.

Consider the bone cross-section. In Figure 11 we put forward a number of candidate cross-sectional shapes for bones. All these shapes have the same cross-sectional area and so these bones will all have the same weight per unit length. Weight is an important consideration; a light bone requires less energy to move than a heavy one. The trade-off is between weight and strength. The bones shown in Figure 11 vary greatly in their resistance to buckling, because the parameter I of the Euler formula differs greatly for the four bones shown. You may think that a solid bone is stronger than a hollow bone of the same radius, and you would be right. Yet the hollow bone of Figure 11b resists buckling or bending four times better than the solid bone of Figure 11a (we know this from the calculated value of I for these two shapes). The bone of Figure 11d is twenty times better. So which is the best design?

We all know that the large bones in mammals are hollow, so clearly this design must be the best choice. Research has shown that the long hollow bones of mammals are optimized for maximum bending strength-to-weight ratio. The optimum cross-section is a circle with an inside radius that is about 60% of the outside radius. The calculation that gives rise to this prediction is based on several structural design factors, not just the area moment of inertia. Hollow bones are filled with marrow, and the marrow weight has to be taken into consideration when calculating bone weight. Bones are attached to muscles; weak bones require more muscles, and so muscle weight may be a factor.[8]

We can show that considering only area moment of inertia and marrow density (i.e., neglecting the more complicated effects of muscles) is sufficient to yield an optimum bone shape. The results are a little different from the more detailed research, but only a little. Thus, considering only I (and allowing for marrow mass), we calculate the curves plotted in Figure 11e, which show buckling force and bending stress as functions of the ratio r of inside to outside diameter. For buckling force, the optimum is seen to occur for $r = 0.7$; for bending stress, it is about 0.63. Our simplified calculation shows that other stresses, such as torsion (twisting), are best resisted by solid bone. Clearly in many cases of practical interest, torsion stresses on bones are either uncommon or are weaker than bending stresses, and so are not taken into account by Mother Nature. She does take into account bending stress, however; the calculated optimum value is close to our average value of r for real bones (and the full, detailed calculations are even closer).

We can see that the shapes in Figure 11 are not optimum (that in Figure 11b is closest: optimum bone thickness is almost twice as thick as shown). Despite being suboptimum, the cross-sectional shapes in Figure 11a,d are both known in nature, because strength-to-weight ratio is not always the deciding factor.

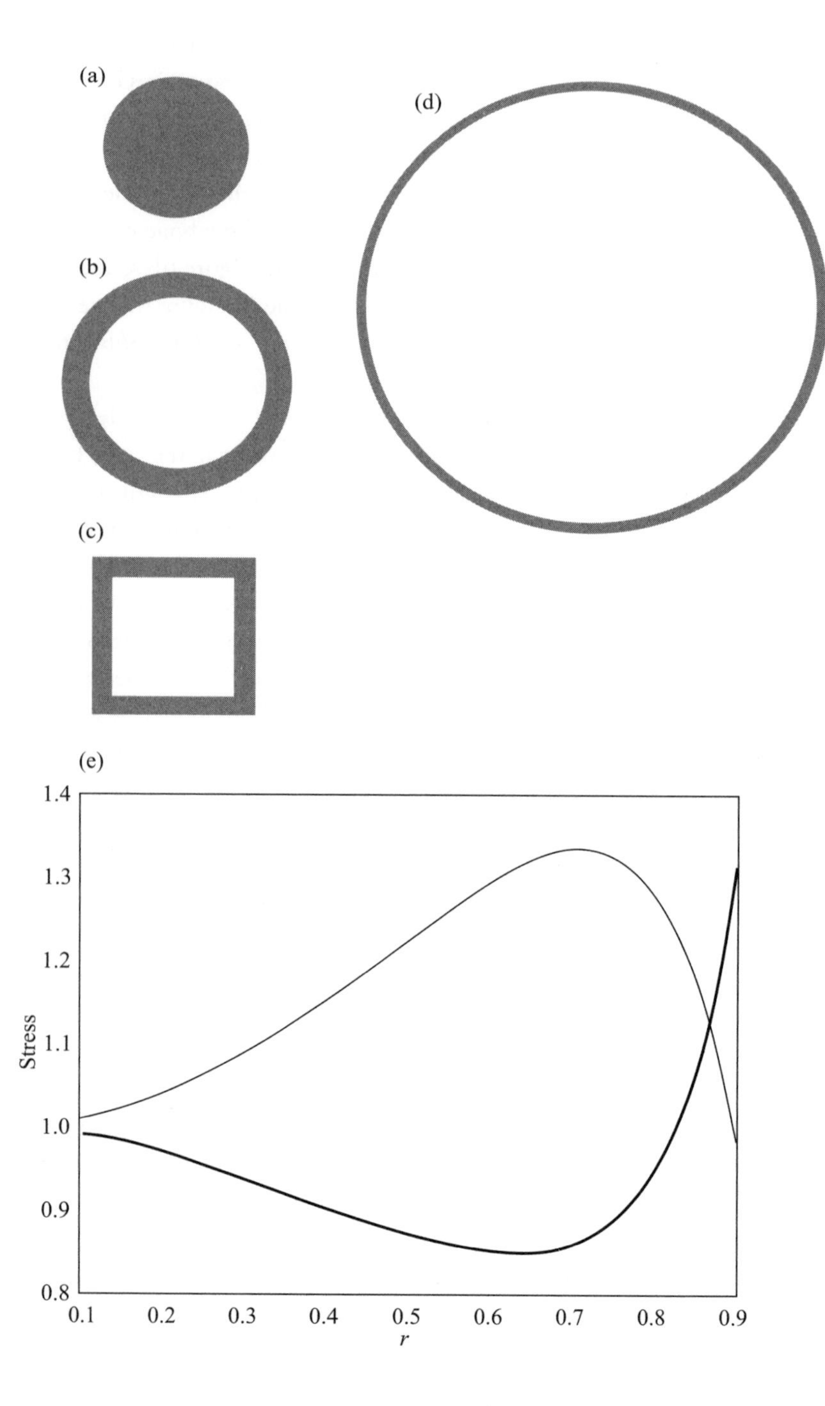

(a)
(b)
(c)
(d)
(e)
Stress
1.4
1.3
1.2
1.1
1.0
0.9
0.8
0.1
0.2
0.3
0.4
0.5
0.6
0.7
0.8
0.9
r

Small bones tend to be solid, because their weight is not great and solid bones are stronger—strength wins out over strength-to-weight. When weight is the overriding factor, thin bones like that in Figure 11d arise. The pteranodon, a giant flying dinosaur of 80 million years ago with a wing span of 9 m (30 feet), required long and stiff bones that were extremely light. Research shows that it did indeed have large diameter and very thin bones.[9] The price paid for such bones is that they are weak in other ways. For example, the bone shown in Figure 11d would be relatively easily crushed by a toothed predator.

What about the bone shown in Figure 11c? Nature appears to have overlooked the possibility of bones with a square cross-section (except, perhaps, in the mythical *C. ridiculus* of Chapter 1). Circular cross-sections are better, because they are equally strong in all directions. A buckling force must occur along the bone's length, but a bending force applies across the bone and can be applied from any angle. The trouble with the square bone is that its resistance to bending is not constant; it depends on the angle of application. The square bone shown in Figure 11c is as strong as that in Figure 11b when the bending stress is vertical or horizontal; at other angles, the bone is weaker (stress concentrates at the corners, which is where the square bone will most likely break). Of course, in nature the direction of bending (and other) forces is variable—an animal in a fight or in a fall can expect to receive blows from all directions. Hence, circles beat squares.

The pteranodon skeletal adaptation is extreme; such extremes also occur in modern flying animals because of the requirement for a lightweight frame. Figure 12 shows examples of the structural adaptations of wings. Consider the albatross and magpie wings shown. These are anatomically very similar; by stretching the magpie wing bones to differing degrees, we could produce a wing that is very like that of the albatross—the magpie readily morphs (in the mind of an anatomist) into an albatross. By so doing, we have converted short, broad wings that enable the woodland magpie to change direction quickly into long, thin wings that carry an albatross over thousands of kilometers of

---

FIGURE 11 (*opposite page*) (a)–(d) Candidate bone cross-sections. All four have the same area and so represent the same weight of bone per unit length. (e) Buckling force (thin line) and bending stress (bold line) for hollow bones of circular cross-section, plotted against r, the ratio of inner to outer bone radius. The best choice of r yields the maximum force required to buckle the bone (corresponding to $r = 0.7$), or the minimum stress induced in the bone (corresponding to $r = 0.63$).

open ocean in gliding and soaring flight. (We will learn more about albatross flight in Chapter 3.) The point is that structural adaptations required for flight are extensive and very specific to the type of flight. In the box, we indicate the kind of structural changes that *you* would have to undergo to become a bird.

The other two wings shown in Figure 12 are clearly very different, because bats and dragonflies approach the problem of flight from radically dissimilar starting points. The bat's wing is not a morphed version of a bird's wing; it is

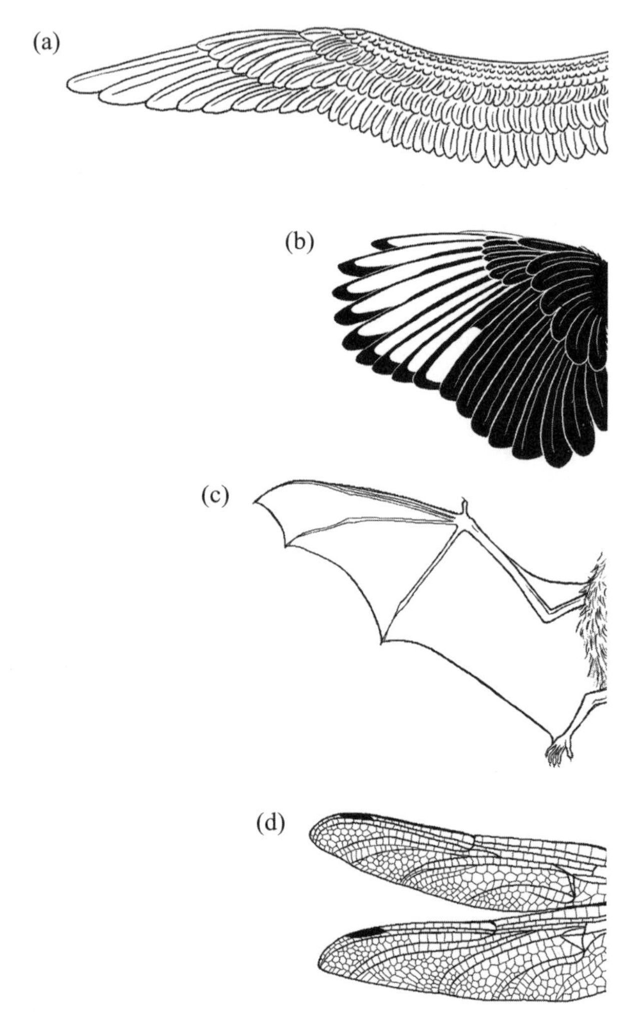

FIGURE 12 Diverse wing shapes: (a) albatross, (b) magpie, (c) bat, and (d) dragonfly.

instead a morphed version of your arm.[10] A hooked "thumb" protrudes from the wing; between this hook and the bat's body we see the "forearm"; the slender bones radiating out from the thumb are "fingers" (the first two finger bones are fused to form the leading edge of the outer wing). The structure of insect wings, as well as their aerodynamic performance, is unlike that of birds or of flying mammals. We will postpone discussion of flight aerodynamics until Chapter 3. Here we mention wings simply as examples of structural diversity arising from different starting conditions (recall the discussion of fish) as well as structural diversity due to different functional requirements.

ON BECOMING A BIRD

Should you wish to become a bird, you will learn that the ability to fly comes at a cost. First, you must get smaller: the scaling arguments of Chapter 1 show that birds need to be small.[11] You will need to be as light as possible, so you must reduce the number of bones, eliminating some, fusing others together, making them all as thin as possible, and even hollowing out the solid bones. Your breast bone needs to grow larger, to anchor those powerful flight muscles. The result: a skeleton that is amazingly light for its strength.[12]

Those flight muscles need a very efficient oxygen supply: heart and lungs are modified accordingly, and even the hollow bones are used to store air in a unique one-way respiratory system, more efficient than that of mammals and four times as large, relative to body size.

Next, you will need feathers for thermal insulation, for display, and of course for flight. Flight feathers are light, strong, and aerodynamic. The complex structure of locked barbs forms vanes that can easily be groomed, so that feathers provide smooth contours around the body. Flight feathers can be adjusted: you can make whatever shape aerodynamics requires of your wings, and your feathers slide over one another to form a smooth, streamlined surface.[13]

Finally, the brain: you will need excellent visual processing, and your sense of balance must be improved. Extra senses will help with navigation. And now, you are a bird.

What Goes Around Comes Around

One feature that we did not have to change when turning you into a bird is the relative volume of blood in your body—about 7%. Is this coincidence?

Biologists doubt it very much: most mammals, from mice to men, have about the same relative volume. Even an octopus has the same fractional volume. There must be strong selection pressure to account for this uniformity. We have seen that there are compelling reasons for circulatory systems to be set up as networks, and at least one of the distribution network theories of Chapter 1 can provide an explanation for the uniformity of blood volume.[14]

Very small animals can distribute nutrients and oxygen via diffusion. All other animals require convection—the nutrients are carried by blood through a network. The circulatory systems that work the best are very broadly similar across widely different classes of animals. Thus, blood is moved through a branching network of tubular vessels, powered by a positive displacement pump (also known as the heart). There are other types of pumps: fluid dynamic pumps, such as propellers, and the wafting, hair-like cilia attached to blood vessel walls, but for high-pressure systems, hearts work best. Blood pressure increases with increasing body size for two reasons. First, the heart must generate pressure to overcome the force of friction between flowing blood and vessel wall (due to blood viscosity)—and vessel wall area increases with body size. Second, pressure is needed to overcome the force of gravity, to raise blood from the heart to the head, and the height difference between heart and head often increases with animal size. A spectacular and extreme case is the giraffe, which we encounter later. Here and in the next section we ignore gravity and look at the work that hearts must do to overcome friction.

The power of a pump is the pressure difference that it generates (i.e., the difference between fluid pressure on the outflow side of the pump and the inflow side) multiplied by the flow rate. Your heart pumps blood around your body with an effective output power of about 1.5 W. This may sound puny, but it adds up. Your heart maintains a flow rate of $5\ \text{l} \cdot \text{min}^{-1}$ (perhaps seven times as much when you are exercising); over a lifetime this is enough to drain a lake 300 m across and 2 m deep. The input power required by your heart accounts for about 10% of your resting metabolic rate. Expressed differently, the flow rate in your circulatory system is such that it takes less than a minute for a red blood cell to complete one circuit around your body. The oxygen requirement of your muscles is largely responsible for your heart rate: more muscular exertion requires more oxygen requires faster blood flow rate requires faster heart rate, and on and on.

Blood pressure is usually measured in units of millimeters of mercury. So, a blood pressure of 100 mm Hg (average human blood pressure while resting) is sufficient to maintain a column of mercury at a height of 100 mm (say,

4 inches). In fact, your doctor will express blood pressure not as an average, but instead as broken down into *systolic* and *diastolic* phases (i.e., referring to a heart that is pumping and one between pulses): 120/80 is normal. The pressure is highest at the pump outlet (the *aorta*—the large artery that carries blood from the left ventricle of your heart) and lowest at the pump inlets (the *venae cavae*—the two large veins that convey deoxygenated blood into the right atrium). Pressure falls as blood moves around the system, dropping due to friction at the vessel walls. Thus veins are thinner than arteries (the fluid pressure in them is lower).

The biophysics of circulatory systems is very complicated for reasons we will let our giraffe explain. This complexity (and the fact that in humans, circulatory system malfunction is a common cause of illness) accounts for the vast amount of research on the subject. We summarize the basics of animal circulatory systems in Figure 13. Primitive animals, such as jellyfish and flatworms, have no circulatory systems—their cells absorb what nutrients they need directly through the cell walls. Other small animals, such as clams, snails, and insects, have an *open circulatory system.* Blood (technically, *hemolymph* for these creatures) is pumped into a cavity, bathing organs directly before draining back into the heart—rather like a water sprinkler. The rest of us have a *closed circulatory system,* in which the blood is confined to blood vessels. Fish possess a two-chambered heart that pumps blood around a single circuit (Figure 13b). Mammals and birds have a four-chambered heart that is a double pump: blood moves around the pulmonary system (heart to lungs and back) and then around the systemic system (heart to body and back again), as shown in Figure 13d. Blood pressure is much higher in the systemic than in the pulmonary system, because the systemic system is longer. Amphibians and most reptiles have a peculiar three-chambered heart, with two atria and a single ventricle (Figure 13c). There is a problem here: oxygenated and deoxygenated blood get mixed together in the ventricle and so may not be pumped along the correct route. This problem is ameliorated by a ridge in the ventricle; the ridge is more pronounced in reptiles and is complete (forming a four-chambered heart) in crocodilians.

Networking

The essence of circulation system engineering can be conveyed to you with a simple physical model. This model is similar in some ways to the pipeline model for lungs discussed in Chapter 1, but there are important differences:

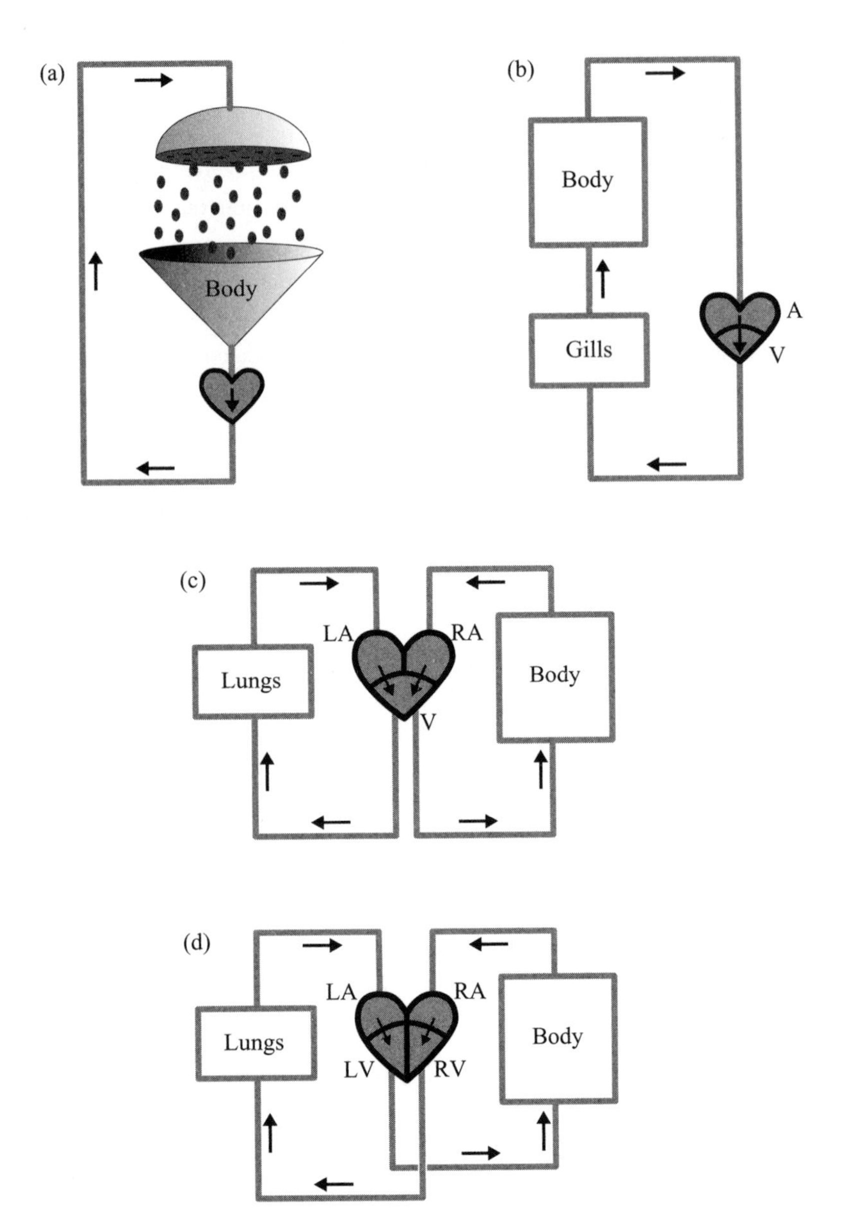

FIGURE 13 Circulation system schematics: (a) open circulation, (b) fish circulation, (c) amphibian and reptilian circulation, and (d) mammal and bird circulation. A, atrium; LA, left atrium; LV, left ventricle; RA, right atrium; RV, right ventricle; V, ventricle.

here the system is closed, and the fluid is incompressible. The simplifications we must make are extreme: without them the analysis becomes much, much more painful. We assume that:

1. A heart pumps smoothly, maintaining a constant pressure between input and output instead of a pulsating pressure.
2. The walls of both arteries and veins are rigid.
3. Blood is a uniform liquid.
4. Blood vessels branch in a binary manner.

Our network is sketched in Figure 14a. An arterial system takes blood from a single aorta, through arteries that branch again and again, and into a capillary network. The blood supplies oxygen and nutrients to the cells of a body—say, a human. The venous system returns blood to the heart via increasingly large veins that culminate in two venae cavae. We adopt typical human values for blood vessel dimensions and blood flow velocity, as shown in Table 1.

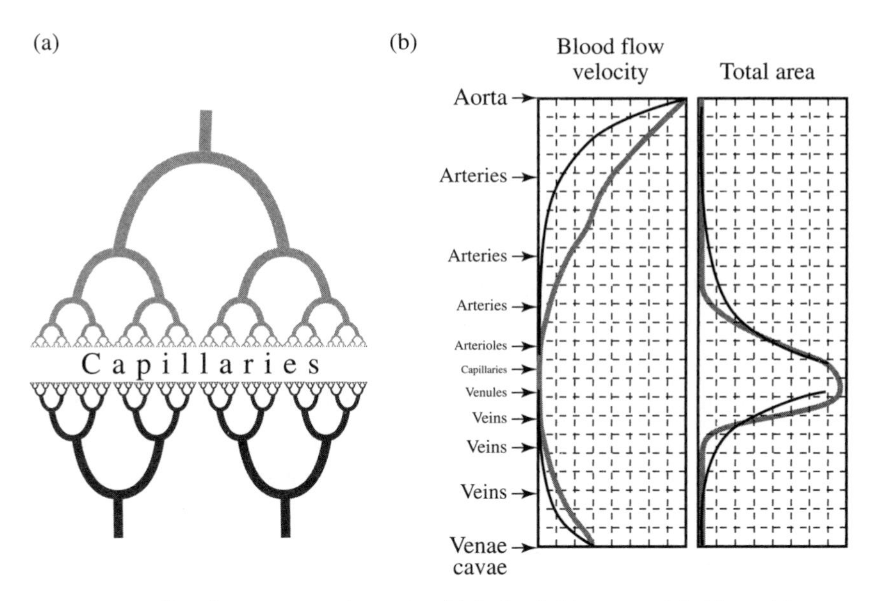

FIGURE 14 Circulatory system dynamics. (a) A simple binary model of branching, from aorta along arteries to capillaries, and then from capillaries through veins to venae cavae. (b) Total cross-sectional area of the circulatory system and blood flow velocity at different stages of the circulation system. Actual values for humans are shown in gray; values predicted by our simple pipeline model are shown as thin black lines.

Table 1 Typical blood flow speed, blood vessel radii, and blood vessel cross-sectional area for humans

	Aorta	Venae cavae	Capillaries
Flow speed ($cm \cdot s^{-1}$)	40	15	0.05
Vessel radius (cm)	1.0	1.1	0.0004
Cross-sectional area (cm^2)	3	4 (×2)	2600

For the systemic circulatory system, Figure 14b shows (gray lines) the total cross-sectional area and the blood flow velocity at any given stage. Black lines show the predictions of our simple model. There is rough agreement between them—the best we can do with such simplifications. Here is how it works. Blood is a liquid and so is pretty much incompressible. This fact has the important consequence that flow velocity varies with vessel size, as shown in Figure 15a. During any given time interval, blood flows faster in a narrow pipe than it does in a wider one. The incompressibility of liquid in a pipeline network is expressed mathematically (the *equation of continuity*) by requiring that the product of vessel cross-sectional area and blood flow velocity be constant.[15] This requirement and the data in Table 1 tell us that, for humans, the simple binary network model in Figure 14a must consist of 32 stages; that is to say, there are 32 branchings from aorta to capillaries, and 32 from venae cavae to capillaries. Let us make the network fractal-like, with the size of blood vessels changing in a regular way from one stage to the next. Say that if the radius of an artery is r, then the radius of the two arteries that it branches into is kr. The number k is taken to be the same for all branchings—this makes the network self-similar and thus fractal-like. (We will soon see that the assumption of fractals is not essential—Murray's Law leads to the same results.) Again using the data in Table 1 and our new knowledge that there are 32 branches, we find that $k = 0.78$. So, at each branching of our binary circulation network, the blood vessels become smaller, but the cross-sectional area of successive stages

FIGURE 15 (*opposite page*) (a) The continuity equation for an incompressible fluid. Fluid flows at the same rate everywhere, which means that $LaV = lAv$, where v and V are fluid velocities. So fluid flows faster in narrow pipes than in wide ones. (b) Blood fluid dynamics. Due to friction at the vessel surface, blood flow velocity is greatest at the center. Individual blood cells spin as a consequence. (c) Spinning blood vessels accumulate at the center, giving rise to axial streaming.

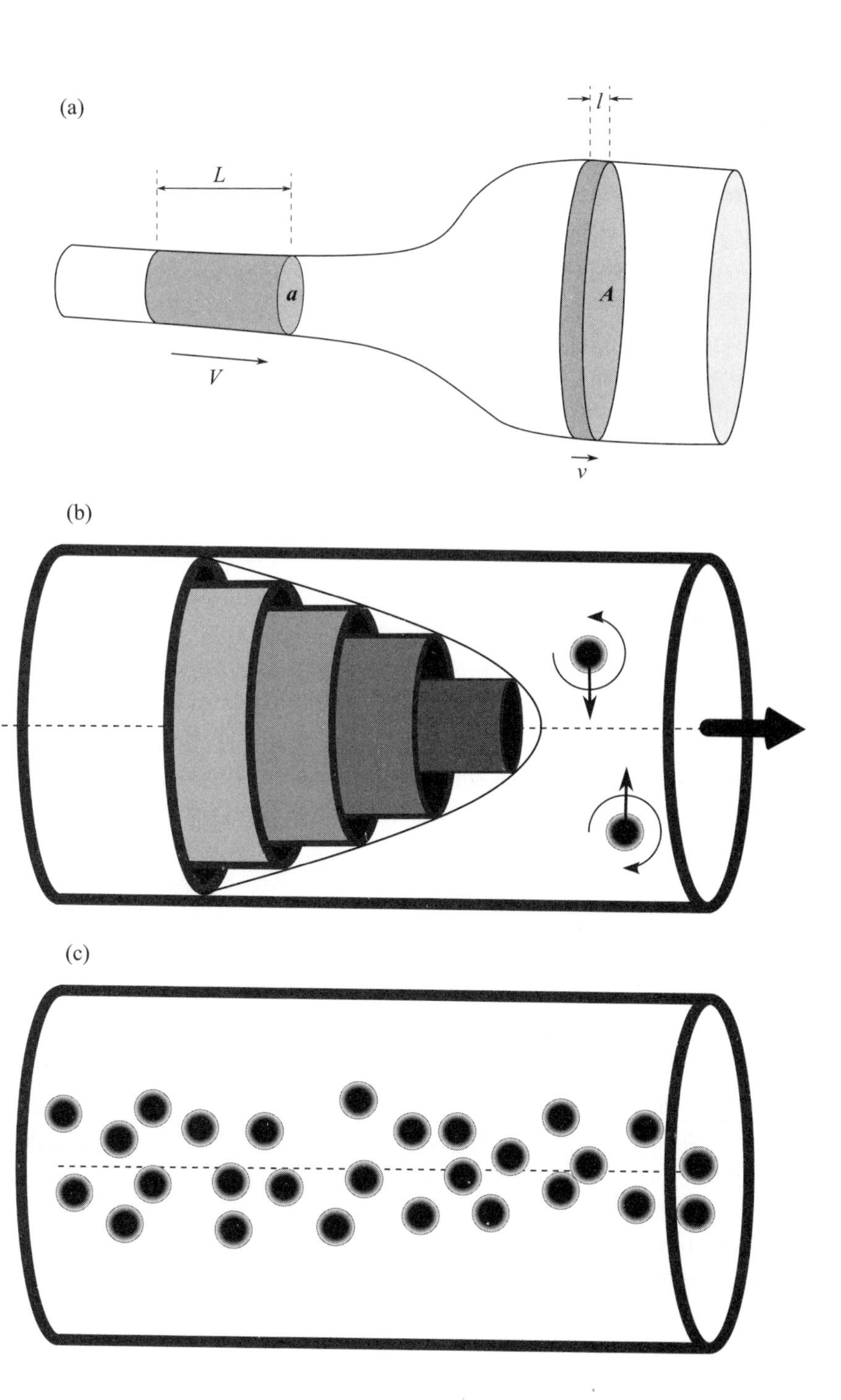
(a)
l
L
a
A
V
v
(b)
(c)

becomes larger (e.g., an artery of radius 1 cm has area 3.14 cm^2; it branches into two arteries of radius 0.78 cm, with a total area of 3.82 cm^2). Area grows with each successive branching as shown in Figure 14b—in broad agreement with real data for humans.

We can make one further prediction with this simple binary network model of the human circulatory system, if we make an assumption about the length of each stage of the pipeline. For a fractal-like network, we would expect the length of a section of pipe to be proportional to the radius, so that the network is self-similar when viewed at different magnifications. Using this assumption and the value of 0.78 obtained for the parameter *k,* we find that the volume of blood in the venous part of the network is about twice the volume of the arterial part. That is to say, our model predicts that two-thirds of your blood is in your veins at any given moment. The real value is about 70% for humans, so the model prediction is pretty close. It turns out that we can change the assumption about pipe length being proportional to radius (instead assuming that the length of each section of pipe is the same, whatever the radius) and a very similar prediction arises, which goes to show, once again, that the fractal assumption is not essential. (As in Chapter 1, we have found that more than one network distribution theory can account for observations.)

Thus our simple model shows that key circulatory system characteristics can be explained as a consequence of branching networks and incompressible blood fluid flow.[16]

Added Complications

The complexity of circulation engineering analysis is due to the difficulty of fluid mechanics as a discipline—particularly organic fluid mechanics. Complications that were glossed over in our simple binary network model of circulation arise when we relax the simplifying assumptions. The first assumption was that the heart pumps smoothly; real world hearts pulse, and so the resulting flow is not as smooth as, say, water flowing in a steady stream through a pipe. Smooth (*laminar*) flow is illustrated in Figure 15b, which shows a fluid velocity profile that varies with distance from the center of the pipe. Friction between pipe and water means that the water velocity is zero at the boundary and increases in a predictable way for water molecules that are closer to the center. At the central axis, water velocity peaks. The curve depicting the water velocity profile in Figure 15b is a parabola; the flow is smooth and steady. Such steady laminar flow is less likely to occur inside an artery, because the fluid is pumped. In addition, the many branchings that occur also churn up the fluid, as do the many valves that exist in arteries and veins to prevent backflow. The result is *turbulent* flow, which appears much more chaotic and requires much greater pressure (and so much greater power) to push fluid through the pipe. Optimization leads to laminar flow in most places in an animal's circulatory system, but because of the nature of the networks some turbulence is unavoidable. Turbulence makes detailed mathematical analysis very difficult.

Our second assumption was that the blood vessels—all the arteries and veins—are rigid. This is not the case, as our giraffe will soon testify. Arteries and (particularly) veins are elastic. They stretch when pressure increases, thus increasing their radii. Blood in veins is under lower pressure than blood in arteries, and thus vein walls can be much thinner, so they stretch more. Arteries expand with pressure in a nonlinear way, which further complicates matters.

Assumption 3 states that blood is a uniform liquid. Not so: blood is far from uniform, and it is not really a true liquid. Human blood contains about 45% cells by volume. Cells influence the viscosity of blood and so influence its resistance to flow. Blood cells tend to clump together when flow velocity is slow, leading to an increase in viscosity. Because of the parabolic velocity profile in a pipe, a spin is imparted to blood cells, as indicated in Figure 15b. This spin in turn induces blood cells to migrate toward the centerline of the pipe, a phenomenon commonly attributed to the *Magnus effect.* The result is *axial streaming,* shown in Figure 15c. For small blood vessels—those with a diameter less than 0.3 mm (about a hundredth of an inch)—something odd happens as a result of axial streaming: blood viscosity decreases. This is odd because we expect, for a homogeneous liquid, that resistance to flow through a pipe will increase rapidly as the pipe radius gets smaller. Blood, however, is not a liquid. There is a liquid component, plasma, which flows "normally" (making plasma a *Newtonian liquid*) with viscosity about twice that of water. The cell content changes blood from a true liquid to a non-Newtonian *pseudoplastic.* Of course, the fact that blood flows more easily in narrow pipes is a good thing for hearts, given that much of an animal's circulatory plumbing consists of narrow vessels.[17]

Our fourth assumption was that blood vessels branch into two parts, not three or ten or a number that changes from stage to stage. Real circulatory networks are much less uniform, which makes detailed predictions more complicated. So, now you see why we opted for a very simple model of circulatory systems.

Giraffe Circulation—A Tall Story

The structural diversity of animals becomes vivid with an extreme example. Giraffes (*Giraffa camelopardalis*) have much in common with humans. Our skeletons morph, we have similar circulatory systems—in short, our structures are related.[18] Anatomists will assert our closeness by pointing out that giraffes have the same number of neck vertebrae (seven) as we do—giraffe necks are longer because the vertebrae are longer. In marked contrast, our more distant relatives, the hummingbirds, have more vertebrae: typically two or three times as many as us mammals. Both of these creatures push the envelope so far as circulatory systems go; theirs is similar to ours but is extreme because of their lifestyles. The frantic hummer and the stately giraffe, representing birds and mammals (the two classes of animal life with the highest blood pressures) both have extreme circulatory systems; we have already mentioned hummers in Chapter 1 and so here will adopt giraffes as our example.

Giraffes browse acacia leaves and gain an advantage over rival herbivores because of their height; adult giraffes are 5–6 m tall (plus a ⅔-m-long tongue). Because of this height—more precisely, because of the great difference in height between heart and brain—giraffe blood pressure has to be significantly higher than ours. The systolic blood pressure in giraffes is no less than 280 mm Hg. Typical mammals should not be constructed with their brain more than about ⅔ m above the heart—so humans are borderline cases. Greater height requires greater pressure, which requires a more powerful heart, so giraffe hearts are huge. Your heart weighs a little more or less than ¼ kg, depending on your gender, and is the size of a fist, whereas a giraffe heart weighs 12 kg and is ⅔ m long.[19]

The pressure that a giraffe heart is required to generate needs specialist adaptations. Systemic blood pressure plus hydrostatic pressure combine to produce a total blood pressure of 500 mm Hg in the lower legs. For a typical mammalian blood vessel, such pressure would force blood through the capillary walls or produce an aneurism. To prevent this from happening, giraffes are equipped with thickened vessels in their legs and a tight sheath of thick skin that resists blood pooling, like a surgical stocking. When a giraffe lowers its head to drink, the change in blood pressure would cause an aneurism or burst blood vessel but for a complex pressure regulation system (the *rete mirabile*).[20] When the same giraffe then raises its head, blood should drain away from the brain, causing it to faint. This does not happen, because the jugular vein (which is 25 mm in diameter in the giraffe) is surrounded by a strong muscular cuff that restricts the flow of blood down this vessel as the head is raised.

It used to be thought that giraffes took advantage of the siphon principle to aid circulation. The idea was that gravity could assist the heart by pulling blood down the neck veins—and so pulling blood up the neck arteries. However, siphons require rigid tubes. The elasticity of real blood vessels means that giraffe jugular veins collapse—are partially flattened—by atmospheric pressure (recall that blood pressure is lower in veins than in arteries), which probably rules out siphoning. The subject is still an area of active research[21] and has, unusually for a scientific debate, generated a poem. Tim Pedley of Cambridge University claims that it will win him no literary awards; however it neatly summarizes giraffe circulatory system challenges:

> The giraffe has a neck of phenomenal length,
> up which blood must be pumped to the brain.
> The consequence is a heart of great strength,
> and collapse of the jugular vein.

3

A Moving Experience

"Cheetahs and beetles run, dolphins and salmon swim, and bees and birds fly with grace and economy surpassing our technology." So wrote the authors of a long (nearly 100 pages) and highly mathematical paper on legged locomotion.[1]

These first two sentences encapsulate the challenges of this chapter. There are many ways that animals, of very different biological architectures, can move around. They may walk, trot, pace, canter, bound, pronk, or gallop away from predators; they may slither, slide, or crawl into burrows; they may flap, soar, or glide in search of a mate; they may swim or jet-propel their way toward dinner. Many of this multitude of means of maneuver can be subdivided: for example, there are two types of gallop and many different types of swimming motion. Flapping flight is generally considered to fall into three different classes. The number of possible gaits that an *n*-legged animal can choose from increases exponentially with *n*. (Even zero-legged animals have developed several different gaits, as we will see.) Despite this extensive variability, we are beginning to understand that there are several universal features that govern the dynamics of animal motion. The details can be very complicated, but we can convey the essentials via a few simple physical models. In this chapter we elucidate these models, using very little mathematics, and indicate how they can lead to universal laws of animal motion.[2]

Locomotion consumes a large fraction of an animal's energy budget however it chooses to move on, over, or under the Earth. Consequently there exists significant natural selection pressure to evolve energy-efficient means of locomotion—indeed, energy efficiency may be the universal feature that determines much of the complexity that we see. A universal principle leading to diversity may seem a contradiction, but actually, a lot of physics and engineering consists of applying simple universal rules to complex and diverse systems. We begin our exposition of locomotion with a simple model of the most familiar form of animal locomotion—human walking—and then show how models of this type can be extended to describe other gaits of bipeds, quadrupeds, and, in general, $2n$-peds. The scaling laws of Chapter 1 return to illustrate the universal character of certain locomotion features, such as the dependence of walking speed on animal mass or the transition speed between canter and gallop. Two aspects of bird flight are considered (again, we prefer approximate and simple to exact and complex)—long-range migration flight and open-ocean soaring. Swimming is akin to flying through water, though there are significant differences as well as similarities to be discussed. Even the dynamics of zero-legged motion (snakes and slugs) has been investigated by the bioengineering community—in this chapter we summarize slithering succinctly.

How to Stand—A User's Guide

Before we can run, we must learn how to walk, and before we can walk, we have to stand. Those readers who are four- or six-legged (or more) will wonder what all the fuss is about. If you have three or more legs, then you are able to stand in a stable manner. Most readers are bipedal, however (as are both your authors, unless they are considered as a single entity), and so standing is a challenge. The tendency to topple must be corrected in some manner. Here we construct a simple mechanical model, illustrating how we stay upright.

Consider Figure 16a. We show an upside-down pendulum with a frictionless pivot at the bottom and a large mass at the top. The pivot is set on wheels we can control, so that the whole structure is able to move left or right at our command. This inverted pendulum anticipates a mechanical system often used to describe biped walking and other gaits, as we will see in the next sec-

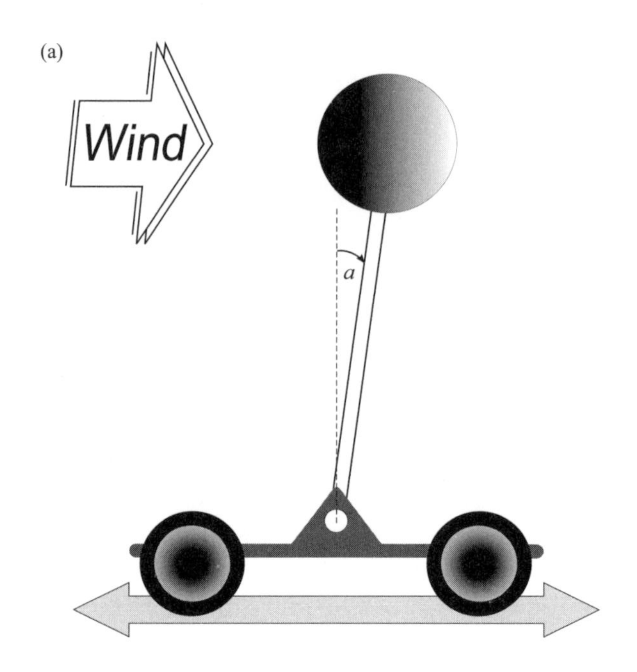
(a)
Wind
a

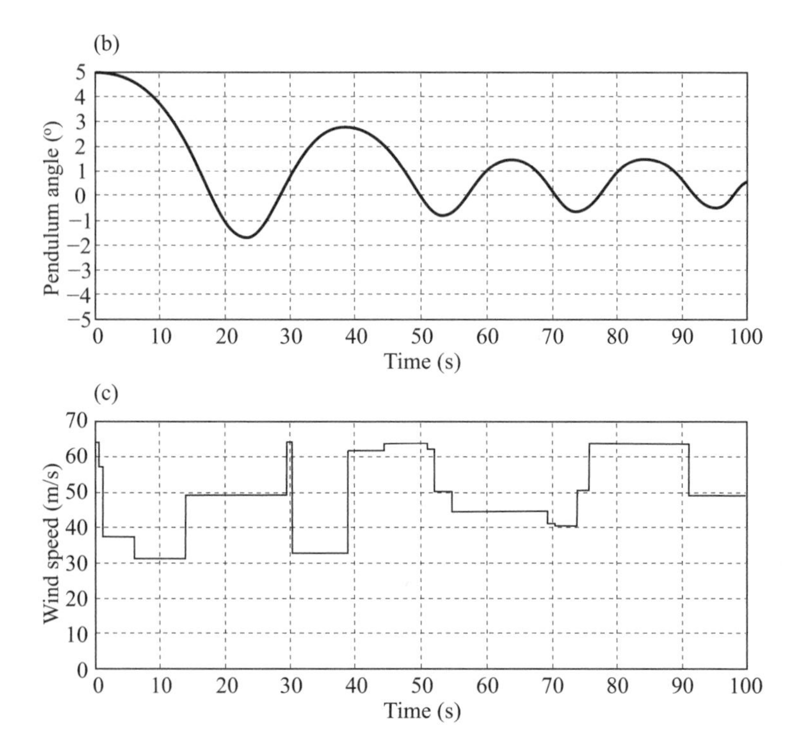
(b)
Pendulum angle (°)
5
4
3
2
1
0
−1
−2
−3
−4
−5
0
10
20
30
40
50
60
70
80
90
100
Time (s)
(c)
Wind speed (m/s)
70
60
50
40
30
20
10
0
0
10
20
30
40
50
60
70
80
90
100
Time (s)

tion. Here the inverted pendulum represents a stiff-legged animal (e.g., yourself when standing); the wheeled base represents the control that your muscles can exert to shift your weight to the left or right. Of course, real animals are much more complex than this simple mechanical model suggests (e.g., they can fall forward or backward as well as left or right), but we have to start somewhere, and simplicity aids comprehension of the essential points.

If the pendulum is perfectly upright (angle a of Figure 16a equals zero), then the system is *unstable,* meaning the pendulum will stay upright unless it is disturbed and then—even if the disturbance is tiny—the pendulum will fall. We don't want this, so we monitor the pendulum angle: if this angle is positive (the pendulum is falling to the right), we hit the accelerator to turn the wheels and move the system to the right. If a is negative, we move the system to the left. This simple algorithm is sufficient to keep the pendulum more or less upright. Even if the pendulum equilibrium is disturbed randomly—say, by a gusting wind—then our feedback control mechanism works to keep things stable: the system is now *dynamically stable.* This, in a nutshell, is how we remain upright, at least when sober.[3]

We say that the pendulum is stable because "the effects of a perturbation are opposed, and therefore small effects remain small." Here opposition is provided by the accelerator in response to the measured pendulum angle (the perturbation). When a system reacts to measured perturbations so as to influence them, then we call it a "feedback" system. In this case the reaction is to reduce the perturbation and so is an example of *negative feedback.*[4]

Note how simple the feedback algorithm is (move to the right/left if the pendulum begins to fall to the right/left). There is an even simpler method of keeping the pendulum upright that requires no algorithm at all, because there is no feedback. If the wheeled base oscillates back and forth rapidly enough, then Newtonian mechanics informs us that the pendulum will remain upright.

FIGURE 16 *(opposite page)* (a) A mechanical model of some standing—in fact, a simplified mechanical model of standing upright. The upside-down pendulum, attached at the pivot to a set of motorized wheels, represents a bipedal animal. The upright position is unstable, and the animal will fall over in even the slightest breeze unless the wheels move to prevent angle a from growing too large. The control of wheel movement is best achieved via a feedback system. (b) Pendulum angle versus time in response to the gusts of wind shown in panel c. (c) The wind speed varies randomly, pushing the pendulum to the left or right, but a simple feedback control algorithm compensates for the gusts, resulting in a pendulum angle that oscillates about the vertical (0°).

It is stable, over a limited range of angles (the range of stability depends on the speed and amplitude of the base oscillations). This *open-loop* system works so long as there is no sudden disturbance that pushes the pendulum beyond the range of stability. The term "open loop" is an unfortunate technical one that makes no sense. All it means is that there is no connection between the perturbation and the wheel movement. The wheel does its own thing without heeding the perturbation. This contrasts with our feedback mechanism—a *closed-loop* system—which is much more adaptable to changing circumstances: perturbation influences response, which influences perturbation. You can see in Figure 16b,c how the pendulum angle recovers equilibrium when the pendulum is disturbed by random gusts of wind. There is still a limited range of dynamic stability—a very strong gust would blow the pendulum down—but that range is much larger than the open-loop stability range. We discuss control and feedback more fully in Chapter 4.

From the foregoing, you might rightly conclude that the stability of our pendulum system (and thus of a standing biped) will be better in systems where the feedback mechanism responds quickly. This is true for standing bipeds and also for moving bipeds and quadrupeds. The inherent instability of legged locomotion is more easily controlled when the animal can react quickly. A tortoise can walk very slowly, because it can keep three legs on the ground at all times: the slow walk is stable. But biped walking and quadruped trotting and galloping, to which we now turn, are inherently unstable.

Walking, Running, and Scaling

You start a walk by intentionally leaning forward—by beginning to fall over. We blithely overlook this simple fact because, once learned, walking is so ingrained. Until you have read the last section, you may have failed to appreciate just what a skillful act you are performing every time you venture outside to collect the mail. Your feedback control mechanism operates autonomously, without informing your conscious mind of the fact. To appreciate your motor skills a little more, imagine a photo of yourself taken mid-stride; now replace your image by a solid life-sized sculpture of yourself in exactly the same captured pose, at the same angle and with the same orientation. The sculpture would almost certainly fall over. This little thought experiment demonstrates the inherent instability of our walking gait. As with standing, biped walking is stable dynamically, not statically, meaning that you don't fall over because you take steps—literally—to ensure that you remain upright.

The basic mechanical model used to describe walking is that of an inverted pendulum, shown in Figure 17a. This model is appropriate for the slow stiff-legged walk of bipeds. Doubled up into two pairs of inverted pendulums, it can be applied to walking quadrupeds. This simple model places the entire mass m of the animal at the top of the pendulum; the pivot is at the other end, where the foot is anchored to the ground. A step corresponds to the inverted pendulum swinging from angle $-\theta_0$ through to angle $+\theta_0$, at which point the other leg takes over, and the same pendulum half-cycle is repeated. A physicist or engineer can easily solve the equation that describes this inverted pendulum movement to obtain both the time t_0 required to take one step[5] and the average walking speed $v = L/t_0$. (L is the distance moved by the body mass during t_0, as shown in Figure 17a.) A dimensionless parameter called the "Froude number" emerges from this analysis: $Fr = v^2/gl$, where g is the acceleration due to gravity and l is leg length.

It transpires that the Froude number pops up in many areas of physics and engineering—it first appeared in the context of fluid flowing past ship hulls—and it is useful for describing how the physics scales with changing lengths. We encountered scaling in Chapter 1, and here it is again. In the context of fluid flow, it shows how the resistance to a ship moving through water changes with ship hull length; here it tells us how walking speed changes with leg length. Calculating the Froude number for humans using leg length $l = 0.9$ m and the average speed that emerges from the inverted-pendulum model yields $Fr = 0.48$. The walking gait of animals of different sizes can be compared when they have the same Froude number. (Because it is dimensionless, the Froude number is not itself dependent on length scale.) Turning this statement around, we find that walking animals all have similar Froude numbers; observations of many different animals of various sizes show that they walk with a maximum speed corresponding to $Fr = 0.3–0.5$. The next speed up from walking is jogging for humans, trotting for dogs, and hopping for crows—these gaits correspond to higher Froude numbers. The transition from trot to gallop for quadrupeds occurs at a still higher Froude number in the range $Fr = 2–3$, whatever the size of the quadruped.[6]

The inverted pendulum model of walking predicts that a human stride length (or that of any legged animal) divided by leg length is proportional to

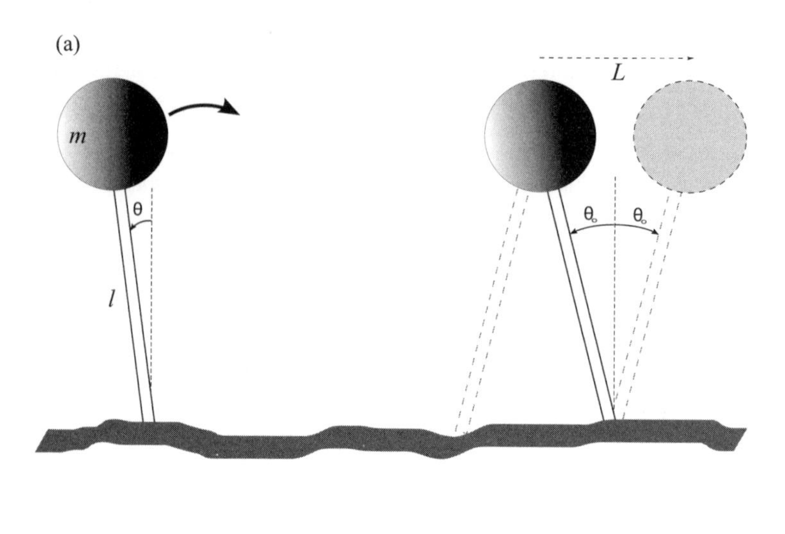

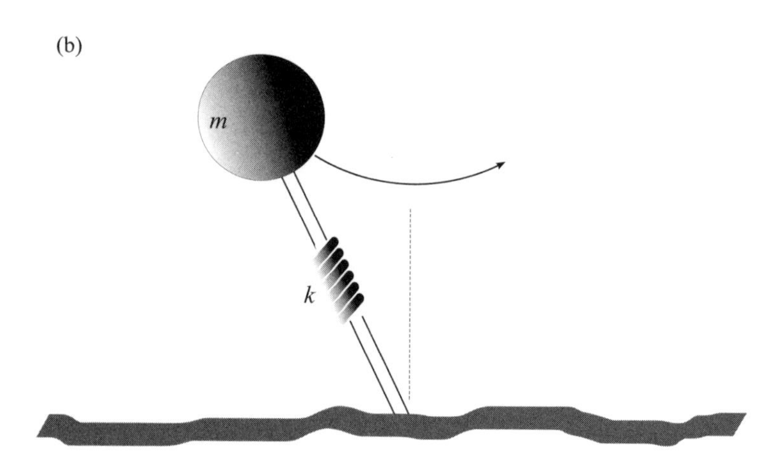

FIGURE 17 Two mechanical models that have been successfully applied to explain legged locomotion. (a) The inverted pendulum works well for bipedal walking (two such pendulums work for quadrupeds). Here, *m* is animal mass and *l* is leg length. During a step of length *L*, the pendulum moves from angle $-\theta_0$ through to $+\theta_0$. (b) The spring pendulum with spring stiffness *k* better explains higher speed gaits, such as running, where the legs bend.

the square root of the Froude number. This prediction has been observed to hold true for many different animals, so the simple inverted pendulum model seems to be on the right track. We can show that it also leads to a prediction for the speed at which the transition is made from walking to trotting or jogging, and from trotting to galloping. We define a walking gait as one in which at least one foot is on the ground at all times. Jogging occurs when the pace is picked up enough to cause a person to be temporarily airborne. For quadrupeds, the equivalent trotting gait occurs when one pair of legs is temporarily airborne; galloping occurs when all four legs are off the ground for a brief instant during each stride. (The basic quadruped gaits are illustrated in Figure 18.) In the inverted pendulum model, bipeds become airborne when the centrifugal force of the rotating pendulum mass exceeds the force of gravity: this occurs for Froude numbers greater than one ($Fr > 1$). The trot-gallop transition is a little more difficult to calculate, because there are more parameters for quadruped locomotion. For example, the two inverted pendulums have two masses corresponding to the front end and the back end of the animal. Also, forelegs and rear legs may be of different lengths. Even so, for a reasonable choice of parameters, we can show that both inverted pendulums are airborne for $Fr > 2.8$. So this simple model tells us that all four-legged animals will change gait from a trot to a gallop when the speed increases enough so that the Froude number exceeds 2.8 or thereabouts. Actual observed values, as mentioned, lie between two and three ($2 < Fr < 3$) and so the inverted pendulum model seems to be a reasonable approximation.[7]

However, there is a better physical model for some gaits of legged locomotion: the spring-pendulum model shown in Figure 17b. The addition of a spring allows for the springy motion of stretched tendons that we see in bent leg behavior when hopping or during pronking and other high-speed quadruped gaits. In Chapter 1 we saw how energy could be stored in the tendons (analogous to the compressed spring) and then released as kinetic energy of motion. (Tendons are efficient energy storage devices: between 90% and 95% of the stored energy is returned as kinetic energy.) Note one other difference between the simple inverted pendulum model and the spring-pendulum model: the first has the animal body at maximum height during the middle of the stride, whereas the second has the minimum height at this point. This difference changes the dynamics, because gravitational energy is stored in the raised body and released when the body is lowered. The spring-pendulum model is more difficult to solve, because it involves two *degrees of freedom*—

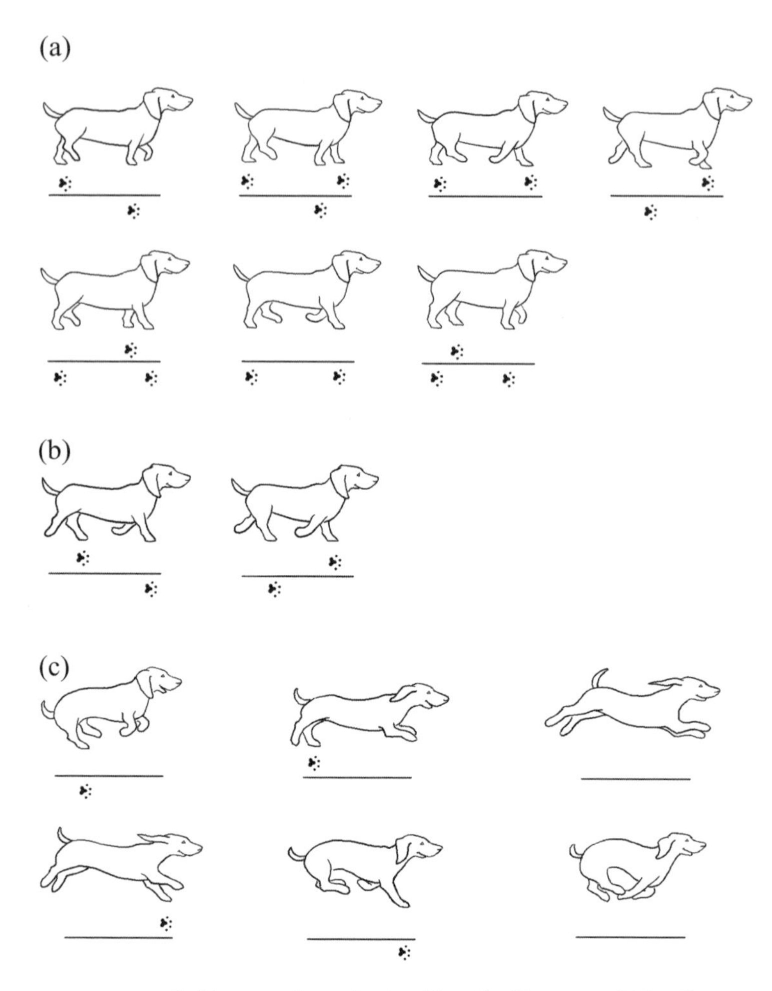

FIGURE 18 Three quadruped gaits: (a) walk, (b) trot, and (c) gallop.

two independent dynamical entities: the pendulum angle and the spring stretching. For the simpler inverted pendulum, there is only one degree of freedom. Add this complexity to the fact that a quadruped needs two such constructions to model a gait (or one spring-pendulum pair and one inverted pendulum pair) and you can see that the analysis becomes difficult. One of the key observations that such an analysis needs to explain is this: energy expenditure per quadruped stride (and per unit mass) seems to reach a minimum value at certain speeds. These speeds correspond to the speeds of the different quadruped gaits (walking, trotting, and galloping) and presumably explain why these gaits are adopted. A horse would expend more energy trotting at 2 $m \cdot s^{-1}$ than it does walking at the same speed; it would expend more energy walking at 4 $m \cdot s^{-1}$ than it does trotting at that speed.[8]

These mechanical models of legged locomotion also predict other scaling behavior that is generally held to be true. For example, the speed at which an animal walks should be proportional to its weight *W* raised to the power of one-sixth ($v \sim W^{1/6}$). Experimental data from many animal species confirm this relationship.

We have examined several key areas of legged locomotion, yet have only scratched the surface of a much-studied phenomenon. We have not discussed the full range of quadruped gaits, for example. Horses naturally have the following distinct, well-defined gaits: walk, trot, canter, and transverse gallop. They can be taught to *pace* (walk with both left legs in phase and both right legs in the opposite phase, like a camel or giraffe), though this is not natural for them—it suits long-legged animals for which trotting or cantering would cause their legs to get tangled up. Dogs switch from a transverse gallop at high speed to a rotary gallop at very high speed. Small mammals have their own gait, the *bound,* when they need a burst of speed. Scaling arguments show us why these small mammals crouch rather than stand erect as large mammals do. Humans can also *skip* (which may be considered to be a bipedal gallop) and *hop*—the preferred locomotion mode for lunar astronauts.[9]

There are two reasons so much effort has been put into understanding how animals move—quite apart from its great variability and intrinsic fascination. First, horse racing is a multimillion dollar industry, and the owners, breeders, and riders naturally want to know how to make horses run faster; horse movement is the most studied of all animals' and has been so since the days when horses were our main means of transportation. Second, the expanding field of robotics is in the throes of developing autonomous legged machines that can rove over terrain inaccessible to wheeled vehicles. So, robotics engineers are interested in how animals move and in how these movements are controlled and coordinated.

Frequent Flyers

The physics of flight is incredibly complicated. It was finally understood about a century ago for the simplest application, which is fixed-winged aircraft. Even so, this simplest case requires very messy equations that are difficult to explain, let alone solve. Because of this difficulty, physicists and aeronautics engineers have, over the decades, tried several different ways to explain the principles of airplane flight to nonspecialists, resulting in confusion and misinformation. These popular-level explanations try to fix on a single physical principle that people can readily appreciate, and you may have heard about Bernoulli's principle, Newton's momentum flux approach, Venturi nozzles, the Magnus effect, the Coanda effect, and so forth. The unfortunate truth is that no single easy-to-understand principle is responsible on its own. The equations that describe fixed-wing flight may reduce to simpler equations that capture, say, Bernoulli's principle, but in so doing, they lose some essential physics. Imagine trying to reconstruct a complicated three-dimensional shape based on projections of its shadow on a screen. From one direction the shadow may look like a pinball machine; from another it may resemble a Duck-Billed Platypus. All projections tell us a partial truth about the object, but none captures the whole image.[10]

Read this short paragraph carefully, because it is as close as we get in this book to a technical explanation of flight aerodynamics. Fixed-wing flight generates pressure differences above and below the wing; the wing shape also generates a downflow of air. These phenomena both contribute to aerodynamic lift. For horizontal flight, this lift force must have the same magnitude as the flier's weight. Wing movement through the air generates vortices, which are shed from the wingtips in complicated patterns; these vortices are largely responsible for aerodynamic drag. The amount of lift and drag that a wing generates depends on the size and shape of the wing (a key parameter is the *aspect ratio*—roughly the ratio of wing length to width), wing cross-section, wing angle presented to the airflow (the *angle of attack*), and many other parameters, such as air density and wing surface roughness.

Now you can relax: you won't need a degree in aeronautical engineering for our discussion of flying animals. Please just bear in mind the complexity of fixed-wing flight and consider that such flight is far, far simpler to describe in detail than is the flapping flight of birds and insects. Why? Some of the reasons will occur to you immediately. For example, the irregular and infinitely adjustable movement of bird wings is a complicating factor. Birds can change the shape and angle of attack of each wing during a wing-beat cycle. Another complication is the increased number of variables needed to describe bird or insect flight: wing-beat frequency, changing body and tail shape, and so forth.

Pity the poor experimental biologist trying to estimate how a bee's wing lift-to-drag ratio (a key indicator of performance) depends on other aerodynamic variables or to understand whether any given bird's flight is being limited by aerodynamic principles or by behavior.

Despite the complexities, a considerable amount of progress has been made over the past few decades in understanding how birds, bats, and bugs fly. This insight is approximate. Here we present some of this understanding—enough, we hope, for you to appreciate the astonishing capabilities of flying animals.

Flapping Flight: Up in the Air

Because bird flight is so much more complicated than airplane flight, the overlap among their technical analyses is only partial at best. Can we apply any of the aeronautical engineers' considerable knowledge of fixed-wing aerodynamics? There is an argument that says we can: the wing flap speed of fast-flying birds is much less than their body speed through the air, and so modeling such birds aerodynamically as fixed-wing aircraft is reasonable, so long as we regard the results as approximate. (If the bird's wings move only a little during the time it takes the bird to travel its own length, then we can approximate the wing as fixed.) For hovering birds this argument fails, and they are modeled better by helicopter aerodynamics. In both cases we will proceed with caution and see the kind of results that are generated by such engineering analyses of bird flights.[11]

The basic physics of bird flight is shown in Figure 19a. Muscles generate wing movement, which provides thrust (the force that projects the bird forward) and aerodynamic lift. These forces are opposed by aerodynamic drag and weight (the force of gravity), respectively. So far, so good. More detailed considerations show that the power generated by a bird should, in theory, look something like the graph in Figure 19b. The curve rises at high speeds, because drag increases with speed, so more power is necessary to overcome it. The curve also rises at low speeds, because a slow-moving bird has difficulty generating lift—think how hard a hovering kestrel has to flap. So, from the overall shape of the power curve, we see there is an optimum power at the minimum of the curve. When flying at the speed v_1 corresponding to this minimum, a bird will consume less power than when flying at other speeds. To minimize power consumption, a bird's speed is thus restricted to certain values. Of course, the detailed shape of the power curve changes from species to species, but it does seem to be the case that many birds fly at certain preferred speeds, corresponding roughly to their power minimum.

There is another optimum speed, shown as v_2 in Figure 19b. The bird might not want to fly at minimum energy: perhaps it is more important to fly as far as possible (as when migrating). This maximum-range speed is analogous to a marathon runner's pace; it is the best speed that a bird should adopt to get as

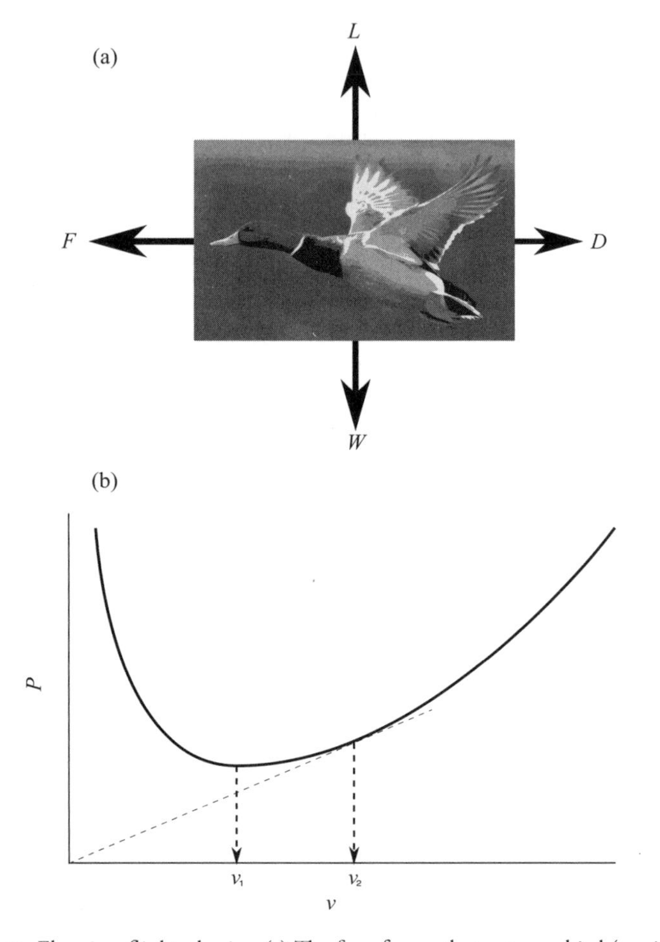

FIGURE 19 Flapping-flight physics. (a) The four forces that act on a bird (or airplane): thrust force F, aerodynamic lift L, aerodynamic drag D, and weight W. (b) The power P that a bird must exert to stay in the air versus air speed v. The speed v_1 corresponds to minimum power per unit speed, whereas v_2 corresponds to minimum power per unit distance traveled.

far as possible on one tank of gas. This speed is not the same as the minimum power speed (at v_1 the bird is consuming minimum power per unit speed; at v_2 it is consuming minimum power per unit distance). Calculations show that v_2 occurs where a line from the origin forms a tangent to the curve, as shown by the dotted line in the figure. Again, there is a body of evidence that supports this analysis: migrating birds do (to a greater or lesser extent) fly at speeds approximating the theoretical v_2 speed appropriate to their species.

Another general result from aerodynamics that is roughly reflected in nature arises by considering the lift force shown in Figure 19a. For a bird in horizontal flight, the lift force must be equal in magnitude (and opposite in direction) to the bird's weight; otherwise the bird would either rise or fall. From aerodynamics we know that lift force is proportional to wing surface area and to the square of velocity: $L \sim Av^2$. Putting these observations together we see that a bird's flight speed scales as $W^{1/6}$.[12] Thus, a 2 kg bird is expected to fly about twice as fast as a 30 g bird. A more detailed scaling calculation leads to the slightly different scaling rule $v \sim W^{0.18}$. Most birds (and airplanes) more or less pay their respects to this scaling rule, but hummingbirds and flying insects do not, when hovering, because hoverers cannot really be analyzed by considering them as little fixed-wing aircraft—they are little helicopters. (They can and must hover because of their small weight, which, as we have seen, implies low flight speed.)

Recall that we characterized terrestrial locomotion via a dimensionless parameter (the Froude number). For flapping flight there is an analogous dimensionless parameter (the *advance ratio*) J, defined as $J = v/(2\phi fl)$, where v is air speed, ϕ is the wing flapping amplitude angle (which measures how deep the wing beat is), f is wing-beat frequency, and l is wing length. The advance ratio is proportional to the ratio of flight speed to wing-tip speed. Birds and insects are dynamically similar when flying with the same advance ratio. So, for example, we expect a slow-flying insect to beat its wings with higher frequency than does a slow-flying bird, which has longer wings.[13]

We can derive another scaling rule from the advance ratio. Because J is dimensionless, and because v scales as $W^{1/6}$ (give or take) and l as $W^{1/3}$, we see that wing-beat frequency scales as $f \sim W^{-1/6}$. So, heavy birds are expected to flap their wings more slowly than do light birds. Once again, this rule is observed, broadly speaking, to hold in nature.[14]

One major complication that arises when attempting a quantitative analysis of flapping flight is that it is so varied. Here we refer not just to the great variety of birds, with their different shapes and sizes, but also to the different modes of flapping flight. Flapping is not usually continuous: many small birds flap intermittently, interrupting the flaps in different ways that lead to three different modes. *Bounding* flight consists of periods of flapping followed by periods with the wings folded against the body. During the flap period, the bird gains lift and thrust; during the folded period, the bird reduces drag, piercing the air like a bullet. Viewed from the side, bounding flight appears sinusoidal: the bird's height varies periodically. Many small, short-winged birds, such as sparrows, favor this mode. *Undulating* flight consists of a period of flapping followed by a period of gliding. It is more widespread than bounding flight and is found among birds with longer wings, such as crows, woodpeckers, and raptors (birds of prey)—but the short-winged European Starling (*Sternus vulgaris*) also has an undulating mode of flight. The third, uncommon mode of intermittent flight is different: it has no rest phase. Instead, the *chattering* flight of birds, such as the Eurasian Magpie (*Pica pica*), consists of continuous flapping at two different frequencies, applied alternately. Detailed theoretical modeling of intermittent flight suggests that there is a savings in power, perhaps 10%, associated with these flight modes.[15]

Another example of the variability in flapping-flight behavior, which makes detailed quantitative analysis difficult, has come to light in experiments where the wings of Zebra Finches (*Taeniopygia guttata*) were clipped a little to artificially reduce their capabilities. It was found that birds with one wing clipped (up to 1 cm taken off the primary feathers) flew as well as normal, unclipped birds. To achieve this feat, the clipped birds adjusted their wing-beat amplitude and frequency: the clipped wing frequency was increased and the unclipped wing amplitude was increased. In contrast, birds with the same degree of clipping, but applied symmetrically to both wings, flew more slowly than normal. Their wing-beat frequency was increased slightly.[16]

We finish this section on flapping flight by noting yet another example of variation, this time concerning groups of flying birds and not just individuals. Many species migrate in flocks, and large birds, such as geese, pelicans, and flamingos adopt a chevron or V formation. Sometimes the V is asymmetrical, like a check mark (√). Are there good engineering reasons such birds adopt these flight formations, or do the reasons lie elsewhere—perhaps to do with pecking order or status within the flock? A lot of ink has been spilled on this single aspect of bird flight. In a nutshell, there does appear to be a theoretical aerodynamic benefit in terms of flight efficiency. This benefit is confirmed by experiments that measured and compared the heart rates of White Pelicans (*Pelecanus onocrotalus*) flying in formation and flying solo. Theory tells us that the airflow around an airfoil (birds' wings are airfoils, albeit complicated ones) generates wingtip vortices that spiral backward as the airfoil moves through the air. These vortices are such that the air is rising just behind and outside of each wingtip. Thus, airplanes or birds that fly in a staggered line, or a chevron or V formation, will receive additional lift. In more detail, the optimum benefit arises for birds that are in the middle of the V, when they are spaced one-quarter of a wingspan apart. The power savings for such birds can be significant (70% in principle). The bird at the apex, and those at the ends, benefit least, which may explain why birds flying in formation on long-haul flights have been observed to swap positions from time to time. One problem with this view of formation flying is that the birds clearly haven't read the theory: they do not appear to fly at optimum separation. Perhaps there are alternative or secondary reasons for adopting a V formation. Most geese species migrate at night and have white rear ends—maybe a V formation or staggered line allows them to see ahead while still being able to follow the bird in front. Many studies on flying formations end with the words "more research is needed here," which says it all.[17]

Albatrosses

We can analyze the flight of birds that glide and soar more confidently and more precisely than we managed for flapping flight, because gliding and soaring is fixed-wing flight, which aerospace engineers understand pretty well. Albatrosses have evolved to glide and soar. Eagles and vultures are also very capable gliders and soarers, but albatrosses are the best. First we explain the why and then the how.

The twenty or so species of albatross all make their living in the southern oceans, mostly in the Roaring Forties and Furious Fifties—those latitudes where the wind blows strongly for most of the year and where there is little land to get in its way. Albatrosses nest on small islands and roam far and wide for food to feed themselves and their offspring. So, these birds must travel long distances: a feeding trip may cover 1,000 km per day, lasting for 10 days. Flapping flight over such distances would consume far too much energy, so albatrosses soar and glide. This mode of flying requires little or no energy; indeed, we will see that it is possible for an albatross to exploit the winds so that it can travel without expending any mechanical energy at all (and only a little physiological energy).

How do they achieve such flight? They are supremely well adapted for soaring flight and have developed a technique of *dynamic soaring* that exploits wind power to the full. Adaptations include large size (the wings of the Wandering Albatross [*Diomedea exulans*] are, at 3.5 m, the longest of any living bird), high aspect ratio wings like a glider (seen clearly in Figure 20), and a unique "shoulder lock" skeletal adaptation that permits them to keep their long wings outstretched with minimum effort.[18]

FIGURE 20 Albatrosses epitomize soaring flight. Their flight trajectories extract energy from the wind and waves. Shown are two Black-Browed Albatrosses or "Mollyhawks" (*Diomedea melanophris*). Image courtesy of Wim Hoek.

Now to the engineering heart of albatross flight: gliding and soaring. When gliding, you are floating on the wind, trading gravitational energy for as much horizontal distance as your wings will allow. The glide angle θ is given in terms of the ratio of drag (D) to lift (L) forces: $\tan\theta = D/L$, so a shallow glide angle (large lift) will permit you to glide a long distance from a given altitude. Albatrosses can attain almost as shallow a glide angle as well-engineered man-made gliders: about 3°. Dynamic soaring is a flying technique that takes advantage of wind shear to extract energy from the wind. Humans in gliders have learned dynamic soaring. In the lee of a mountain, wind speed varies with altitude: it is low in the mountain's shadow and is higher above the mountain. By flying in a spiral—described below—the glider becomes a soarer, gaining height by hitching a lift from the wind.

But there are no mountains in the open oceans, so how can albatrosses manage dynamic soaring? In fact there is wind shear near the ocean surface, where friction reduces the wind speed, resulting in a vertical profile: wind speed increases from zero at the ocean surface to its full value at about 20 m altitude. The nineteenth-century physicist, Lord Rayleigh, first suggested that albatrosses might use the vertical wind speed profile to advantage. By climbing into the wind and diving with the wind behind, calculations show that it is possible to extract energy from the wind and so travel long distances with little effort: this is dynamic soaring. Albatrosses take advantage of other—technically more complicated—aerodynamic effects close to the ocean surface (so close that their wingtips may touch the water). Air displaced by a moving bird bounces off the surface and provides it with extra lift—the *ground effect.* Additionally, ocean waves cause air near the surface to move in a manner that can be exploited by an experienced glider pilot or by an albatross. This effect is known as *wave lift.*[19]

Swimmers

Just as aerodynamics is recruited to help understand flying birds, so hydrodynamics is invoked to help us get to grips (figuratively speaking) with the slippery subject of swimming fish. Both aerodynamics and hydrodynamics are subsets of fluid dynamics and are described by more or less the same equations. To an engineer, swimming is a lot like flying. We need spend only a few words describing the principal differences. The fluids—air and water—are quite different, and so the wings and feathers, muscles and gaits, of flyers and swimmers are accordingly different. Wings become fins; the high density of water (compared to air) means that fins can be much smaller than wings. Another consequence: fish are about the same density as water—the difference is controlled by a gas bladder within many fish—and so most fish, unlike flying birds, do not have to generate lift. Consequently, fin designs differ from wing designs. Swimming is hydrodynamically unstable, just as bird flight is aerodynamically unstable, and for the same reason: instability permits greater maneuverability. Instability requires exquisite control, and fish attain this with many control surfaces.[20] Most fish have pectoral, pelvic, median, dorsal, caudal, and anal fins, used for both thrust and stability. Swimming styles are grouped into two broad categories: burst swimmers and cruisers. (The latter include sharks, who obtain buoyancy not from gas bladders but from hydrodynamic lift generated by pectoral fins. Consequently, sharks must swim continuously if they are not to sink.) The split over which fins do which tasks—thrust or balance—in which fish is still a matter of active research.[21]

Swimming speed in fish scales (pardon the pun) with weight in the same way as flying and running: $v \sim W^{1/6}$. Some biomechanical theories claim this is no coincidence—that the underlying physics dictates such behavior. Fin-flapping frequency scales as for flight: $f \sim W^{-1/6}$. (The usual weasel words apply here: scaling rules apply statistically to fish as a whole; individual species may vary markedly in their behavior.) The transition between swimming gaits or modes of locomotion is governed by the advance ratio, as for flight.[22]

Body shape matters rather more for fish than it does for birds, because hydrodynamic drag, unsurprisingly, is much greater than aerodynamic drag. Shape is adapted for lifestyle so that, for example, very fast swimmers, such as tuna, have a streamlined torpedo shape. The sliminess we feel when handling a fish is due to a special secretion that may lower hydrodynamic drag by reducing vortex formation and thus maintaining laminar flow (i.e., flow with smooth streamlines).

Slitherers and Sliders: Snakes

Snakes have scales that can act collectively like a ratchet, enabling the reptile to pull itself along the ground. When a snake slides forward, the scales lie flat and glide smoothly over the surface, but when part of the snake threatens to slide backward, the scales open out to provide grip. Snakes also take advantage of rocks, bumps, trees, and any other fixed protuberances to push themselves along. The locomotion of snakes is every bit as complicated as that of limbed animals, and they have evolved several distinct gaits (not all of which are used by all snakes). *Rectilinear* motion is the simplest to understand. The snake moves caterpillar-like (though not exactly walking on scales) with its body held straight.[23] It makes progress by contracting and expanding; the different action of belly scales on the surface, described above, ensures that it moves forward. This mode of locomotion is suited to large, heavy snakes, such as boas, that do not care to bend themselves very much or lift themselves off the ground. It is also used by smaller snakes in burrows or confined spaces, or when stalking prey.

Lateral undulation is the most common serpentine movement and the fastest (see Figure 21a). The snake pushes against rocks and the ground with its body twisting sinusoidally—swimming snakes also use this mode. *Concertina* movement is different (Figure 21b). The snake contracts by bending, then straightens to move forward, and then contracts again to repeat the process. Thus progress is made by alternately pushing from the rear (against the ground, a rock, or the sides of a tunnel) and pulling from the front (against the ground, as in rectilinear motion). Concertina mode is often used when space is too confined for lateral undulation. Western Diamondbacks (*Crotalus atrox*) and a number of climbing snakes move in this manner. Many desert snakes, in fact, use an unusual though well-known mode of locomotion—*sidewinding* (Figure 21c). This strange gait is useful on sand and other shifting

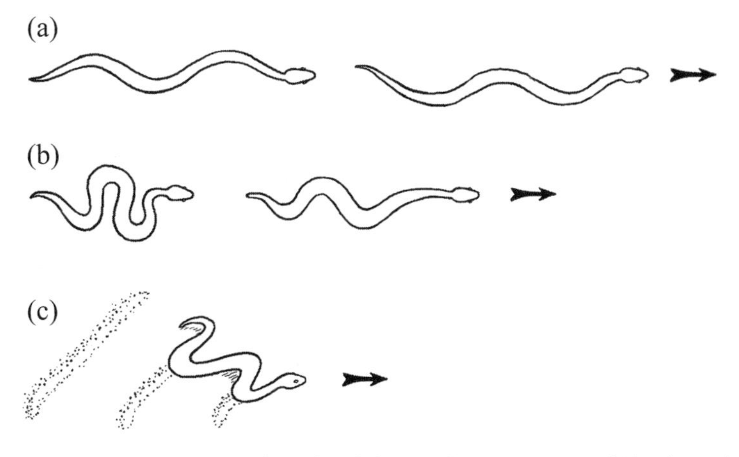

FIGURE 21 Some snake gaits: (a) lateral undulation, (b) concertina, and (c) sidewinding.

ground, where it is difficult to get a grip with scales. A sidewinder moves sinusoidally across the surface, with each of its contact points static—not sliding over the surface. Static friction is greater than moving or kinetic friction, so traction is better. The moving parts of the snake are not in contact with the surface—a serpentine equivalent of stepping.

The different gaits illustrated in Figure 21 require different combinations of muscles applied at different times, as with legged locomotion. In the past it was thought that the energetic cost of snake movement was less than that of walking or running, but now researchers have shown that this is not the case: snake locomotion costs are comparable to those of legged locomotion (and the concertina gait is rather more expensive).[24]

Gastropod Get Up and Go

Garden snails and slugs and their aquatic relatives move in a very different manner from slithering snakes. They get from A to B via *adhesive locomotion*—a unique and ponderous mode of transport that short zero-legged animals have evolved. From a sedate 0.8 $mm \cdot s^{-1}$ they can rev up to a blistering 2.0 $mm \cdot s^{-1}$ when circumstances dictate. Quick they're not, but they're faster than a strawberry (and their other food items), and that is good enough. They merit space in our book because adhesive locomotion is based on some interesting engineering ideas and because it provides an unlikely locomotion solution for creatures with no legs or scales or anything else with which to grasp.

Snails and slugs secrete slimy mucus from their single foot. This mucus provides a thin layer (10–20 microns—say, 1 two-thousandth of an inch) that separates the foot from the substrate over which our slimy friend moves. The foot itself is very flexible. Seen from below, the rim of the foot—via the mucus—hugs the substrate like a suction cup, so a greedy gastropod can climb vertical surfaces to get to food. The foot interior exhibits waves that travel toward the head of the animal—say, a snail—as it moves. The waves travel at about twice the speed of the snail, so that the movement is roughly akin to that of a tracked vehicle—like a tank—as it moves forward (see Figure 22). The top of the tank track moves at twice the speed of the vehicle, while the bottom of the track is stationary on the ground.

You may well wonder how the waves provide locomotion when they are separated from the substrate by slippery mucus. The answer lies in the unique qualities of the mucus. *Rheology* is the study of slow fluid flow, and the flow of

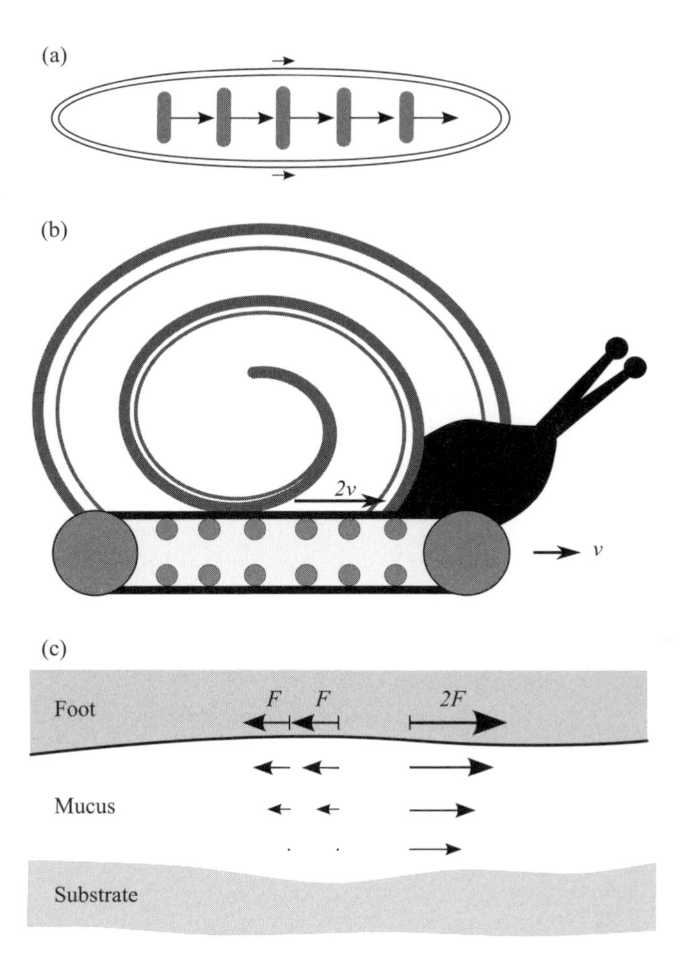

FIGURE 22 Gastropod adhesive locomotion. (a) The foot as seen from below. The rim of the foot holds the animal to the substrate by suction. Waves travel along the center of the foot in the direction of motion at twice the animal's speed. (b) Snail with tank tracks: the sections of track that are in contact with the ground are stationary, whereas those not in contact move forward at twice the snail's speed. (c) The mucus layer seen from the side. Mucus material properties are crucial to adhesive locomotion. Here the foot expands forward and backward with force $2F$ (a net force of zero), and yet, because the forces are distributed differently and because of the mucus deformation properties, there is a net force at the substrate, generating movement.

gastropod mucus is currently a subject of active research by the rheology community. The mucus has a very useful property that is technically known as "shear thinning." Shear refers to the force that acts parallel to the substrate; it tries to tear the mucus apart, because the top and bottom of the mucus move at different speeds. Shear thinning means that the mucus resists shear forces better when the force is low than when it is high—not just proportionally better (mucus moves half as much when the force is halved) but much better—and it happens because mucus viscosity is reduced by deformation. This property means that the mucus is more fluid when stress is increased, which has been shown to account for terrestrial slug and snail locomotion—we assume that it applies for all gastropods (see Figure 22).[25] There is a price to pay for such inventiveness, however: a gastropod expends twenty times more energy in creating the mucus as it does moving. This makes adhesive locomotion much more expensive than legged locomotion.

4

A Mind of Its Own

Animal computing crops up in several chapters. It is intimately bound up with engineering functionality—for example, vision and signal processing, or locomotion and feedback control, as we have seen. Computing merits its own chapter not just because it is important, but also because it exhibits a number of features that may not come up in a discussion of vision processing or feedback control.

Who's in Charge Here?

Let's begin with a simple experiment. Read the following instructions. Then, without any further thought or inner debate, carry them out:

> Put this book down on a flat surface within easy reach. Turn your head away, and close your eyes. Now pick the book up again, open your eyes, open the book, and read on.

Easy, wasn't it? But what did you have to do to achieve this very simple task? Or put another way, what capabilities would we have to give a machine before it could replace you? First, you made a conscious decision to act—in fact, you made several. You decided to put the book down, close your eyes, and pick the book up. These decisions were made in your brain—that's what brains are for. But the decision to act is only the first stage in the control sequence that must be followed to achieve the desired goal. What's more remarkable is that the subsequent stages don't really involve conscious decisions—you make them, somewhere, but you're not really thinking about them. To grab the book with your eyes closed, you had to remember where in three-dimensional space the target book was located as well as knowing the location, orientation, speed, and acceleration of your body, arm, hand, and fingers at all times. You had to send control signals to the mechanical components of your arm, telling them where to move and how fast to move there. You had to constantly monitor and analyze return signals telling you what these components were actually doing at any given instant, allowing you to adjust your control signals so your hand found the book. Once you touched it, you had to recognize that it was indeed the book, then close your fingers around it with just enough force to hold it against the downward pull of gravity as you brought it back in front of your eyes, while simultaneously ensuring your arm's movements compensated for the extra mass with which it was now burdened.

That's a lot of sophisticated control, just to pick something up. And we haven't even considered how much more involved it would have been if you had kept your eyes open and made use of visual cues, which would have to be integrated with the biomechanical signals coming back from your arm. It's enough to make you wonder how you manage to get out of bed in the morning. The point here is that without a control system, a machine does nothing—it is just a sculpture. The same is true of animals. What makes them actually work? What are their control systems? To get to grips with this problem, let's start with a simplistic high-level model of what control systems do. Basically, there are four things going on:

1. Signals are received.
2. Signals are analyzed so that a decision can be made.
3. Encoded instructions are prepared for any organic machinery that will be involved in implementing the decision.
4. Signals containing those instructions are transmitted to the organic machinery, where they will either trigger or inhibit some other kind of activity.

Of course, the nature and complexity of the incoming and outgoing signals depends on what is being attempted. The more information they contain, the more involved the actual decisionmaking process will be.

Now let's replace you with a robot, or more specifically, a robotic arm to handle the tedious task of picking up the book for you. Once we've mastered the mechanical aspects of building it,[1] the simplest way to control the arm would be to tell it to move to a predefined position and then operate its fingers in a predefined way to grab a predefined book. It could then swing back to another predefined position, open the fingers, and release the book. In fact this is what the earliest industrial robots did—picked predefined things up from a predefined location and put them down in another predefined location. "Predefined" is very much the operative word here—useful in the context of a factory but very limited. For example, such a robot cannot adapt to a change in circumstances. If the book isn't in exactly the right position and orientation, the fingers may miss or may fail to get a good grip. If the book is the wrong size, shape, or weight, it could all go horribly wrong. Human engineers have developed an elaborate theory to explain and control the operation of machines. Not surprisingly, it is called "control theory." We touched on it in Chapter 3 and the next section should tell you all you need to know. According to control theory, our predefined robotic arm is an example of an open-loop control system in which there is no connection between the system's output and its input. As a result, it's good for simple, repetitive tasks but useless for ones requiring adaptation.

Control Theory—The Bare Minimum

Control theory is a way of defining the behavior of dynamical systems. We want the system to be in a certain state; for example, we want the speed of a vehicle to be some particular value. This, the desired output of the system, is the *reference*. The function of the control system is to ensure that the *output* of the system matches that reference. To achieve this, a controller manipulates the system inputs to obtain the required system output.[2] The difference between what we want (the reference) and what we actually get (the output) is the *control error*.

Consider an automatic pilot designed to steer a ship in a straight line—just fix the rudder in a central position and off you go. The problem is, this system can't adapt. What if there is a tidal flow or a strong wind pushing the ship off course? The rudder is locked in place, so the control system cannot compensate for changing circumstances, because there is no direct connection between the system output (the ship's heading) and the conditions encountered.

In a more flexible system the output (ship's heading) is monitored by some sensor, which feeds the data back to the controller, allowing it to continuously adjust the control input (the rudder position) to minimize the control error by keeping the actual heading as close to the desired heading as possible. A perfect feedback control system would cancel out all errors, allowing the system to adapt perfectly to changing circumstances and produce the exact outputs the user defined. A system that uses feedback like this is a *closed-loop controller*. Lashing the ship's rudder in a fixed position does not allow for feedback and so is an example of an *open-loop controller*. We encountered open- and closed-loop systems in Chapter 3, in the context of standing upright.

There are two basic flavors of feedback: positive and negative. Positive feedback, as the name suggests, provides positive reinforcement—the effect of the control loop is fed back as an input, acting to increase the magnitude of the effect and, generally, creating an unstable system. Hold an electric guitar too close to its amplifier and the result is positive feedback. So, too, is the initial mustering of termite workers to fix a breach in their fortress walls—workers release pheromones that attract more workers, who release more pheromones that attract more workers, and so on. Negative feedback, unsurprisingly, acts to decrease the magnitude of the effect. Our closed-loop ship-steering controller uses negative feedback, so the system acts to minimize the deviation from the set course. Thermoregulation in animals also involves negative feedback loops: the hotter the body gets, the more is done to cool it and vice versa. Note that "negative" and "positive" refer to the effect, not the action. A negative feedback loop may involve doing more of something to reduce the perturbation from the ideal state; a positive feedback loop may require doing less of it.

In a *feedforward* system the output is sent to a later stage of the process, so anticipating future activity. Before you picked up that book, your body had anticipated the shift in its center of gravity and started adjusting your muscles to compensate. Figure 23 shows a typical control loop.

Most control systems use conventional Boolean logic: the statement "Output equals Reference" is either true or false. Set a room's thermostat to 22°C, and if the temperature falls to 21.5°C, the heating system pumps out energy. When the temperature reaches 22°C, the system will stop, but if it falls again, it starts up once more. On—off—on—off all day long: not very efficient and very wearing on the hardware. We can reduce this problem by setting a *deadband* with upper and lower limits. For example, when the temperature falls below 22°C, we pump out heat until it gets to 24°C, then stop until it again falls below 22°C. We have a comfortable range of 22–24°C, and wear and tear is reduced. We could further improve this system by using feedback to create

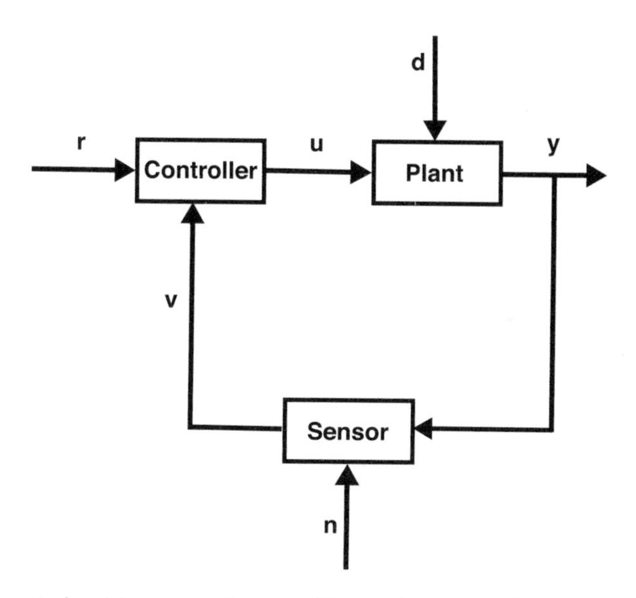

FIGURE 23 A closed-loop control system. Plant is the object to be controlled. Sensor measures the output from Plant. Controller generates the input. In addition, r is the reference or command input; u is the generated control signal; d is the external disturbance affecting the plant; y is the plant output signal; n is external noise, which affects Sensor's ability to accurately measure y; and v is the sensor output, which is feed back into Controller. In an open-loop control system, there is no Sensor loop feeding back into Controller.

a proportional control system in which, as the difference between the desired and actual room temperatures decreases, the amount of heat being input to the room is decreased in proportion. But wouldn't it be nice to be able to just define our ideal temperature as *comfortable*?

This idea leads us into the realms of fuzzy logic.[3] Here, things can still be true or false, but there are intermediate stages—degrees of true-ness and falseness. Try sorting people into Tall or Not-Tall (Short, if you prefer). Set your threshold at 4′ 6″, and anyone measuring this or more gets to ride the rollercoaster; anyone smaller gets to stand and watch. Fuzzy logic replaces this threshold by a value that quantifies the degree of tallness by giving Tall a value between 0 and 1. For an especially scary fairground ride, we might want an age restriction as well, so we'll set a threshold of 16 years and assign Old a fuzzy value between 0 and 1. So far, so obvious, perhaps, as all we're really doing is quantifying height and age, but there is a point to this example. Boolean logic allows us to ask simple questions about Tall and Old: Are they both true? Is at least one of them true? Is one of them not true? Fuzzy logic permits the same questions, but because Tall and Old now have values between 0 and 1, the answers to the questions also have values between 0 and 1.

If we apply these ideas to the thermostat, instead of fixed thresholds at which things happen, we now have a series of conditions, as shown in Figure 24. The system is no longer hot or cold but can belong to a number of subsets to different degrees. Thus between temperatures T1 and T2, it will belong, to different extents, to the subsets of Cold and Cool. We can do the same thing for the current Rate of Change of temperature in the room, giving it a series of subsets from High Negative through No Change to High Positive. The Heat Output from the system can be similarly divided into subsets ranging from Static to Greatly Increased.

We can now control Heat Output by building a *fuzzy expert system,* using a series of simple rules to make decisions. We can say "If Temperature is Cold and Rate of Change is High Negative, then set Heat Output to High Positive," but at the same time we can also say "If Temperature is Comfortable and Rate of Change is No Change, then set Heat Output to Static." All possible rule combinations involving all possible subsets are implemented, and fuzzy logic is used to assign values to the various outcomes. These are then mathematically *de-fuzzified* to produce a *crisp* value, and the control is set to this. Defuzzification can be as simple as taking the largest value from all rules or can involve quite complex sums, but the result is the same—a system in which the

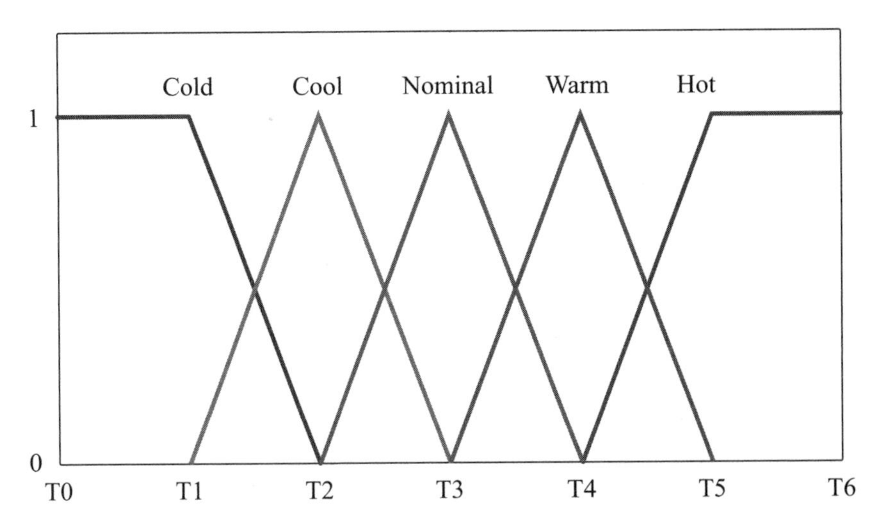

FIGURE 24 Fuzzy sets for use in a temperature-control system.

reference value varies smoothly with the inputs in a way that, to the casual observer, simply seems more sensible.

Toward a Better Robot

A sophisticated closed-loop control system uses feedback to adjust the control as required, while employing fuzzy logic to avoid "hard" thresholds. Our robotic arm can incorporate sensors that tell it the position and speed of its components at any time. We can add pressure sensors to its fingers; feedback from these enables the system to adjust its gripping force sufficiently to hold the book. (Minor modifications of the algorithms and their parameters would also allow us to adapt the robotic arm to lift other, more fragile things, such as eggs.) In fact, very sensitive pressure sensors would allow the arm to find the table and then feel its way toward the book. To achieve this, we need to build in three capabilities: to provide the feedback, to use that feedback to make a decision, and to adjust our control to implement that decision. All these capabilities require sensors, of course, but they also involve algorithms—ways of calculating how things are and what they should be.

We can build robotic arms that can do all this, including ones that can automatically adjust their gripping force to pick up delicate objects, but they are nothing like as adaptive and sensitive as animal limbs. However, they are often stronger, faster, and relatively tireless. Comparing animals with robotic machines can be illustrative and reasonably illuminating, but we must always remember that they are not the same thing. There are probably more differences than there are similarities. Nature and engineering may be trying to solve similar problems, but they often do so in totally different ways. Only the underlying logic remains in common.

You may be wondering what all the fuss is about. We all know that we think with our brains, so it follows that all this sensing, deciding, and doing is somehow a product of the brain. Well, it's not quite that simple. Not all creatures have a brain that's quite as sophisticated as yours, yet they still seem to function quite happily. So, does this mean control is not all in the brain? Or should we come at this from another direction and ask: What does a brain do? And who needs one, anyway?

The Brain's the Thing

In fact, to control a machine or an animal, you don't really need a brain as we would understand the term. All you need is a control system that can sense its environment, make simple decisions, and use feedback control to implement them. Even plants can achieve this in a very limited form—they grow toward light, for example. In animals this control is provided by the nervous system. This system does the core computing (processing data, making decisions, and issuing orders) and the core communication (transmitting the data and the orders to where they are needed as well as storing any information that needs to be retained). Basically, it serves as processor, data bus, and hard drive.

The building blocks of the nervous system are the neurons or nerve cells.[4] The general chemical structure and workings of neurons are pretty much the same in almost all animals—worms, insects, fish, reptiles, birds, mammals, and humans all have nervous systems assembled from essentially the same materials. However, the detailed structure and behavior of neurons can be very different for different types of animals. In fact, in a single creature, there will be many different types of neurons performing different functions, and their physical structure and operation will also vary. So there is no such thing as a typical neuron or even a typical neuron structure. Undaunted, however, we sketch the workings of a representative neuron in Figure 25.

Neurons share many features in common with other animal cells. What makes them different, and special, is that they can carry out long-distance communication with other cells by generating and propagating unique electrochemical signals. With reference to Figure 25, the important things we need to know about neurons are:

- They receive input from sensory cells or other neurons and integrate them until a threshold is reached, at which point they send a signal to other neurons or muscles.
- Inputs arrive via *dendrites* and are integrated in the *axon hillock,* giving rise to an *action potential* that propagates down the *axon* to the *synapse.* Here the action potential triggers the release of *neurotransmitters*—special chemicals that carry the signal across the *synaptic gaps* to other cells.
- There are three basic types. Sensory neurons have sensory inputs; Motor neurons output to muscles; inter-neurons connect only with other neurons.
- Depending on the type of synaptic gaps they have and the particular neurotransmitters that cross them, the effects of individual neurons can be either excitatory or inhibitory, two clumsy words you will never see used anywhere else. In case you're not a neurologist, "excitatory" means turning things on and "inhibitory" means turning them off.
- The output from a neuron is binary, either on or off. Keep stimulating the neuron and you get a higher rate of signals, not a stronger one. Overdo it and the neurotransmitters run out, leaving the neuron desensitized until it gets a chance to stock up again.
- Depending on what connections are made at either end, there are neurons that receive inputs from multiple sources, ones that send the same

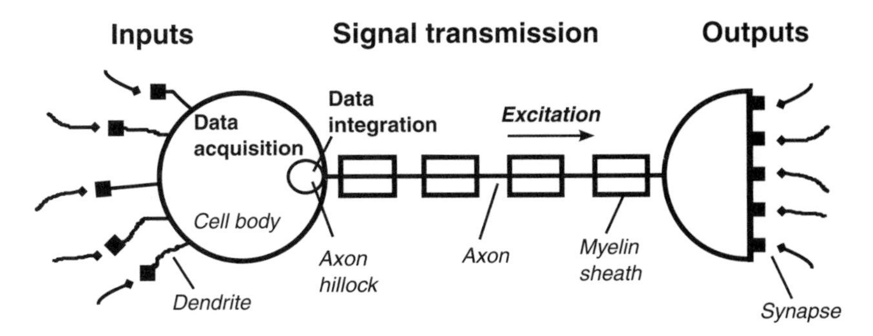

FIGURE 25 Abstraction of the motor neuron in a vertebrate animal, showing data inputs, integration, and signal output. One neuron can receive inputs from many sources and can output to many others.

output to multiple receivers, and others that can do both. The exact configuration depends on the neuron's job.
- Vertebrate neurons have insulated axons that transmit faster than invertebrate ones. Invertebrates compensate by using (relatively) thick axons where a quick response is required, for instance, in neurons that trigger escape behavior.[5]

Neurons form the building blocks of a control system, with the three basic types linked together to provide enhanced functionality. We use the term "ganglion" (plural: ganglia) to indicate a dense cluster of interconnected neurons that may be used, for example, to process sensory information or control motor outputs. Gather enough neurons together in the same place, wire them up in the right way, and we call the resultant structure a "brain." In fact, the most primitive brains we know of are really just collections of ganglia. The vast number of simple invertebrates getting along with such a system shows that, for uncomplicated creatures, it's a solution that works. As we move up in complexity through the vertebrates, we find that fish and amphibians have—relative to their body sizes—small but clearly defined brains, with those of reptiles and birds being even more intricate. In all cases we can identify areas devoted to specific functions, especially those associated with senses. Thus the brain of a bird that hunts by sight will—again, relative to its body size—have more neurons devoted to processing visual senses than to, say, processing scents, whereas the opposite would be true for a nocturnal mammal that primarily hunts by smell. It's a simple rule of thumb: the more data you want to handle, the more processing power you need, and thus the more knowledge you can acquire and so the better choices you can make.

Like all vertebrates, our nervous systems can be split into a central and a peripheral nervous system, as shown in Figure 26. The central nervous system consists of the brain and the spinal cord. The spinal cord is the body's information highway, carrying sensory data from the peripheral nervous system to the brain and motor data from the brain to the rest of the body. (It does make some decisions by itself—your reflexes are a reaction to threats that can't wait for the brain to mull things over. In this case the spinal cord is processing, deciding, and acting, with the brain being informed after the event.) The brain itself receives sensory data from both the spinal cord and other nerves directly connected to it (e.g., the optic nerves from the eyes). It processes these data and initiates appropriate, coordinated action by outputting motor data. Because the central nervous system is awash with data, one of its tasks is to select what really matters at any given moment and to ignore (or at least de-prioritize) what doesn't. Unless it's dangerously tight or uncommonly scratchy, how aware are you, right now, of your underwear?

The peripheral nervous system is further split into two functional subcomponents: the somatic nervous system and the autonomic nervous system. The

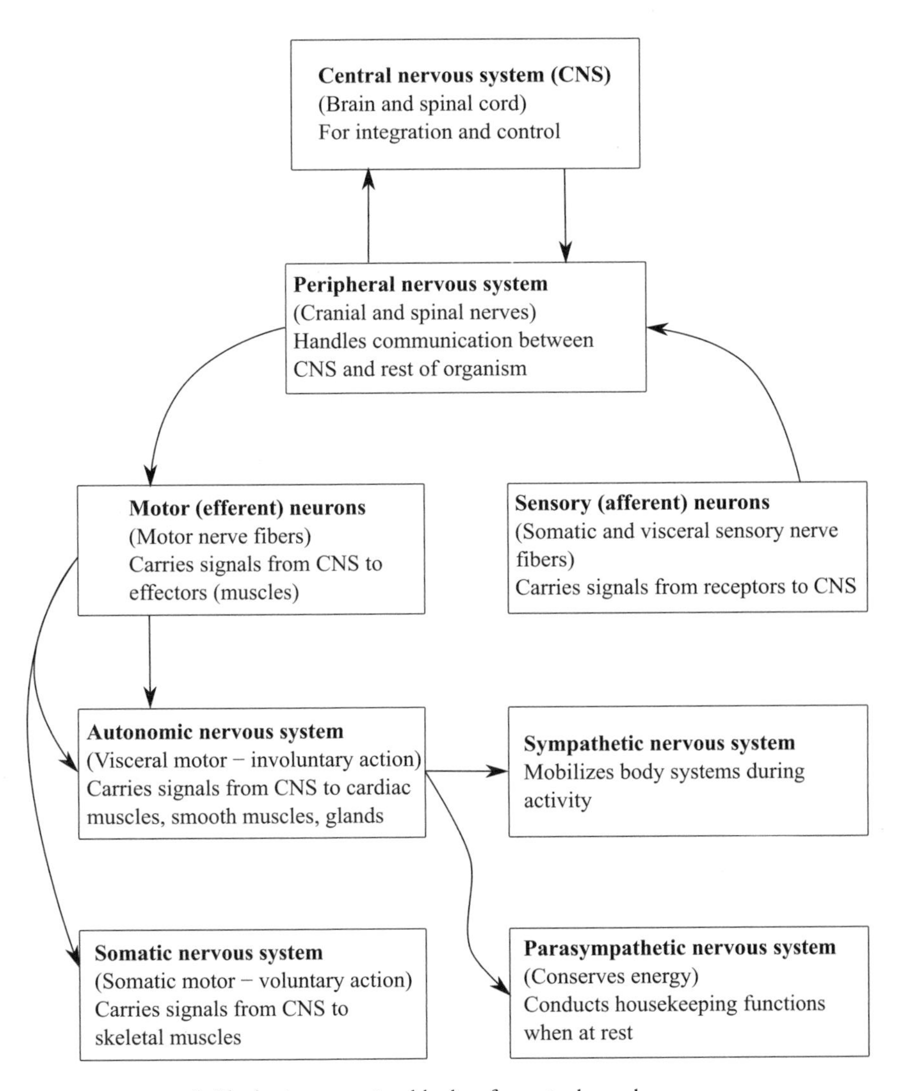

FIGURE 26 The basic processing blocks of a typical vertebrate nervous system.

somatic nervous system deals with voluntary control of movement through the action of skeletal muscles as well as the reception of those external senses that keep the body in touch with the outside world (e.g., hearing, touch, and sight). It includes all the neurons that connect with muscles, skin, and sensory organs and sends the signals that cause muscles to contract.

The autonomic nervous system maintains the body in a "steady state" by controlling the various completely unconscious processes used to keep the body operating at optimum conditions.[6] In other words, it controls such things as heart rate, breathing, digestion, perspiration, micturition (the discharge of urine), the diameters of the pupils of the eyes and, especially in warm-blooded creatures, body temperature—all without deliberate control or even any conscious registering of sensation. Some functions are connected to the conscious mind (you can, to some extent, control your breathing and micturition), but these processes are mostly wholly automated. Basically, system stability results from all the "housekeeping" functions that keep things ticking.

Invertebrates don't have quite the same set-up—the central "brain" is of less importance, and there is no spinal cord. The simplest nematode worms are so simple that we can count every neuron. Arthropods are more complicated, so we can, to a degree, identify central and peripheral nervous systems, with the latter being especially important. In fact it has been said that insects think in the periphery. An engineer might phrase it differently and say that insects employ a distributed control system. Among other things, this configuration makes them robust but individually less versatile.

So we can wire up our neurons and get on with processing. What we need now is some way of broadcasting information throughout the entire system. No matter how many connections a neuron may have to other neurons and cells, it is still essentially fulfilling the same function as an electrical cable, which may also get input from multiple sources and output it to multiple nodes. A common analogy is that of a well-designed telephone system in an office building. A conference call facility allows multiple handsets to input data and multiple handsets to receive data, but they are still all physically connected by wires or optical cables. Is there an equivalent of a wireless system? Some sort of radio transmitter that will broadcast information that can be received by anything tuned to the right frequencies, no matter where it is in the building? Like all analogies, this one is less than perfect, but in animals the equivalent of the radio transmitter is the endocrine system. Endocrine glands (and some specialized neurons—nothing is ever clear-cut in animal design) release chemical messengers straight into the bloodstream, allowing them to be pumped around the body or into the fluid between cells, where they can affect a large number of local cells at once. The right chemical detected in the right amounts by the appropriate parts of the body either triggers or suppresses an appropriate activity. The process is entirely involuntary: no animal can choose to ignore these messengers. They are usually referred to as "hormones," and as any parent knows, they play a vital role in determining animal behavior.

Figure 27 shows an animal brain at work. Is it really just a kind of computer? Probably not. There are similarities and there are differences, and the differences might be more important. Thus, a computer's electronic signals travel down wires at nearly the speed of light, more than a million times faster than the fastest of the brain's sluggish electrochemical signals. Perhaps as a consequence, brains exhibit massively parallel processing—they do lots of processing at the same time in lots of different processors. If they didn't, they probably couldn't cope. But consider this: the entire nervous system of the

FIGURE 27 What's going on in there? Blind instinct or slow reasoning? We thank Anita McFadzean for this image.

minute nematode worm (*Caenorhabditis elegans*) contains fewer than 400 neurons, and we're not entirely sure how it works. The human brain contains about 100 billion cells, with 3.2 million kilometers of "electrical wiring" and 1 quadrillion synaptic connections, yet it occupies a volume of less than 1.5 liters, weighs less than 1.5 kg, and consumes about 10 W of power. It can program itself, repair itself, and shift processing and memory about in response to damage. We can't even conceive of how to build a computer to rival that.[7]

Start Making Sense

Inputs to the sensory neurons can come from a variety of sources, both internal and external. In Part Two we look at the primary sensor systems animals employ to gather information about their external environment, but information on the animals' internal states is also important. Here all we need to consider is that all animals use a variety of organs to sense different things. These organs act as transducers, converting energy or chemicals collected from the environment into action potentials propagating down the axons of the neurons. Thus, the energy contained in a passing pressure wave in the air causes movement in tiny hairs buried deep in the ears, leading to the detection of a sound. Airborne chemicals dock with specific chemical receptors in the nasal tissues, triggering a cascade of action potentials that allows the chemicals to be identified and classified as a particular smell. The range of sensory inputs available will depend on animal type. Animals are usually most sensitive to the information that is most important to their lifestyle. Thus cave-dwelling creatures tend not to concentrate resources on perfecting their sight, in direct contrast with, say, predatory birds, for whom sight may be the primary sense.

Ask people how many senses we possess and the traditional answer is five—sight, sound, taste, smell, and touch. Ask a neurologist and you could get an answer between 9 and 33.[8] Ask an engineer, however, and you might get a more limited result, because animals' sensors fall into three basic categories:

- Electromagnetic receptors—photosensitive cells detect the presence of electromagnetic waves, usually in the form of light (vision) or infrared (heat); other cells are sensitive to the presence of electric and magnetic fields (see Chapter 10).
- Chemical receptors—molecular detectors respond to specific chemical structures to produce sensations of both taste and smell (see Chapter 7).

- Mechanical receptors—various mechanoreceptors convert pressure (or more usually, pressure changes) into the sensations of sound and touch (see Chapter 8).

Other senses are either a variation on these receptors or a combination of two or more of them. For example, we may think there is a "gravity sensor," because animals have an innate sense of up and down. But this is primarily the result of mechanoreceptors responding to the downward tug of gravity, which really makes it just a variation of pressure detectors. It is not a foolproof sense. Terrestrial animals, such as you, can become confused if they find themselves suddenly plunged deep into water. Unlike aquatic creatures, their sense of gravity isn't used to making an allowance for the support the water provides.

So far we have looked at the hardware employed to meet the computational requirements of the living machine. But as any engineer will tell you, the hardware may determine speed and throughput, but without the right algorithms, it's not going to do anything very useful. So what are the algorithms used to control an animal's functions? In electronic computers, algorithms are generally expressed in a mathematical form. Inputs are measured, more or less precisely, those inputs are compared with the preset reference values, calculations are made, and equally precise outputs are transmitted to the effectors that implement them, more or less exactly. (There has to be an allowance for uncertainties. No machine ever achieves perfect precision: there will always be errors in the measurements of both the inputs and the outputs.)

Think of an animal as a machine—an autonomous, self-powered, self-repairing, learning machine intended to operate entirely on its own. Human engineers are starting to build such machines, although they are nowhere near as sophisticated or versatile as actual animals. But what kind of functions would such a machine's control system have to fulfill? At the most basic level, the machine has to be kept operational. It requires subsystems to, for example, monitor the health status of its components and maintain them at optimum operating conditions. It requires other subsystems to monitor fuel status and control power generation and distribution to ensure that it maintains sufficient fuel reserves and is able to supply power as required to the components that need it, with minimum wastage. In a machine, everything from built-in test to fuel and engine management and thermal control falls into this most elementary category.

In an animal the nervous system is used to keep the creature alive by monitoring and controlling its internal organs. Details vary among animals, but much of the effort goes into system stability—maintaining the status of the organism. This involves a whole suite of mainly internal sensors and associated control subsystems working to check for and repair damage, control the intake and general processing of food and water, ensure that vital chemicals like oxygen are available where and when needed and that other waste products are removed, and keep the body temperature right—just like any well-built machine.

Another part of internal body monitoring produces what is known as *proprioception.* This is an animal's inner "body sense," a system mainly based on mechanoreceptors in the muscles, tendons, and joints. Proprioception enables an animal to monitor the position of the various components of its own body. This is how you knew where your arm and fingers were when you were reaching for the book back at the start of this chapter. Without it, any kind of controlled, coordinated muscular activity would be extremely difficult. You can see this for yourself—try picking up the book with an arm that has "gone to sleep."

Now that we know our machine is functional, we can move up a level of control and processing and get it to do something—say, sense its environment and move around with purpose. (Even animals that don't move about much still have body parts that move.) From a control point of view, movement is pretty straightforward. The decision is made; neurons fire; signals are sent to muscles; muscles contract; and all the flesh, bone, chitin, or whatever is attached to the muscles also moves. Appropriate low-level internal subsystems are engaged to keep track of where the various body parts are in time and space and how they are all relating to one another. Others check that no damage is being incurred as a result of the movement, or if there is damage, react accordingly. External sensor systems check that the subsequent relationship of the body parts to the outside world is as expected for the planned movement. Balance is maintained, and the appropriate parts move in the appropriate way. At the same time, energy is supplied to the components that need it, energy generation from fuel is initiated if required, cooling systems are triggered to bleed off excess heat, and overall system stability is maintained. The bigger the creature and the more parts it is moving, the more of these things are made to happen.

But what about this sensing of the external environment hinted at above? Surely that's more involved than just stimulating muscles and activating glands?

How does all that work? We look at animal sensor systems in the second half of the book. Here, we are interested in how the nervous system processes information to build a world-view. We've already done a quick fly-by of sensory neurons, outlining how they send data to other parts of the nervous system—how, for example, appropriate neurons can react to changes in pressure or temperature or to the detection of specific molecules by chemical sensors.

Or consider the marvel that is the sense of sight. In general terms we can understand how light falling on optical receptors in the eyes is converted to action potentials that stimulate visual processing centers in the brain. But that's just raw data; how does the brain make sense of the incoming signals? Our sense of vision is such a fundamental part of our lives that we rarely think about this aspect of it. It just happens. We look, we see images, and we make sense of them. We investigate visual capabilities more fully in Chapter 10, but consider this as a starting point: we are born with two eyes, which give us the ability to form stereoscopic images, which, in turn, allows us to estimate distance. If you look at a cat sitting on a wall 2 m away, each eye collects a separate—and marginally different—two-dimensional image of that cat, just as two cameras located side by side would take two pictures of it. Yet you see only one cat. The visual processing centers in the brain combine these two distinct sets of data to create one three-dimensional image of a cat, from which you can infer that the cat is about 2 m away.[9]

What more can we get from this image?[10] The retina of a human eye contains more than 100 million photoreceptors, generates images made up of about a million individual points of light (a million pixels) and does so about 10 times per second. These images aren't like photographs—the eye is not a camera but an extension of the brain, carrying out some local data processing before passing information down the million neuron axons that form the optic nerve for further processing deep in the brain. If, between one image and the next, something in that image has changed, then the brain is alerted to analyze that change. That's one reason why alarm lights work best if they flash. What the brain is really interested in is movement. To an animal, moving things matter, because they represent the most immediate dangers and opportunities. If the cat swishes its tail, you notice the movement first. The next thing to do with the image is search for edges—lines and boundaries that define shapes. Then you compare the shapes with a mental library of known shapes and make a judgment. Unfamiliar shapes barely glimpsed can cause difficulties. As more images become available, the identification becomes more reliable. (This process is one of the many reasons eyewitnesses to sudden, traumatic events can be very unreliable. Cameras merely record images; brains continuously interpret them and record the interpretation, not the image.) Once you've sorted out what's changed in the image and what shapes it contains, other information, such as the color of the shapes, can be added to the mix to help make sense of what you're seeing.[11]

Getting Stuff Done

A machine usually has a purpose—a mission it has to fulfill. We need to provide it with control algorithms specialized for the tasks with which it is faced. Ultimately, all animals are driven by three core motivations: to find sustenance (in the form of food and water), to avoid being killed, and to reproduce. The process of "not getting killed" means providing security for the organism, which includes finding shelter from possible dangers. The motivators are thus sustenance, security, and sex. (These various considerations can lead to different behaviors, which are discussed in Chapter 6.)

Let's think about lunch. From the animal's point of view, now that we're up and moving, we need to find some sustenance. If we're a predator that means catching something: detect prey, track it, and target it. You can try this for yourself, because we instinctively track and subconsciously target objects all around us all the time. Watch a car driving down the street. Follow it with your eyes, as it passes behind trees, bushes, other vehicles—stationary and moving—or even buildings. Watch it until it is out of sight. Easy, wasn't it?

Now try building a machine to do the same thing. We can do it, but our machines achieve this result by measuring specific data about the target and the surrounding environment. Thus, we would measure range and angles to all the objects in the field of view, from which we could locate them all in three-dimensional space. We would also want to measure the rates of change of these parameters, from which we could calculate their three-dimensional velocities through the environment. If we can't directly measure these quantities, we can estimate them by seeing how quickly the range and angles change with time. Similarly, we can measure or calculate the accelerations associated with these parameters—the rates of change of the rates of change. But all this gives us is raw data about objects in space. A machine has to be able to recognize the individual objects in it. That involves mathematics, and quite complex mathematics at that, in which all these individual data items are mixed together to sort out what's what, where it's going, and where it might be a few seconds from now. Modern sensor systems can do this operation—it's how the radars in Airborne Warning and Control System (AWACS) aircraft sort out and follow multiple targets, and it pushes their processors to the limits—but then they need to pick out the targets of interest from the more or less stationary background environment.[12]

If you have a whole series of different sensors all sampling different aspects of the external world, it would be stupid not to try to maximize the available information by merging all the data together. Engineers call this process "sensor fusion," and entire careers have been built around it, with a moderate degree of success.[13] Again, it's all done with math and it all depends on our ability to identify and isolate the individual objects in a scene. Until we know what the objects are, we can't be sure what different pieces of information pertain to each object. As animals, however, we do it all the time, every day, without even thinking about it—as suggested in Figure 28.

It seems unlikely that animal senses are used to precisely measure the range and angle to a target, let alone its velocity.[14] You can see this for yourself by simply looking at things. You subconsciously assess the range on an abstract scale from very close to far away. (Are we entering fuzzy logic territory? Very probably.) Beyond the limit of stereoscopic vision, you make this assessment using visual cues, like the object's size relative to other recognized objects, the way light reflects from it, or the way it moves against a fixed background landscape. Only after you've classified the distance in this fashion do you consciously try to put a number on it. Learning to accurately estimate distance in terms of meters or yards is much harder than you might think.

If animals don't have accurate measurements of target properties, can we even begin to guess what kind of algorithms they use to track and target their prey? Does the term "algorithm" even make sense here? Well, perhaps. When scientists study the way animals pursue their prey, they measure things. These things are then used to build mathematical models of what the animal is doing. That's how science works. Models provide a mathematical description of what is happening, and we start calling that description an "algorithm." So we can talk, meaningfully, about the algorithms animals employ to achieve certain tasks. But this terminology does not mean that they are processing data through those algorithms as a computer would. It just means that their behavior can be described using these specific input data and these algorithms to achieve these specific observed output data. Any animal (including us) is no

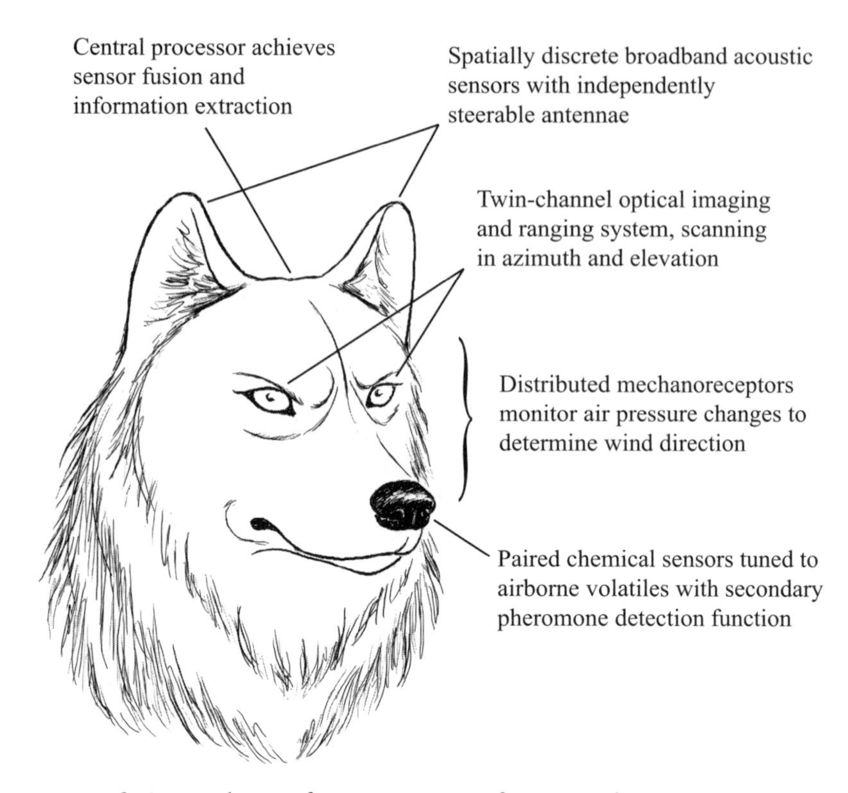

FIGURE 28 A natural sensor-fusion system. Data from many disparate sensors are collated, integrated, and perhaps most importantly, interpreted to form a coherent assessment of the overall situation. Decisions are made and appropriate actions are initiated: knowledge → decision → action.

more aware of the "algorithm" it is using than an apple falling from a tree is aware that its plunge to earth follows an algorithm defined by Newton's Law of Gravity.

Let's take an engineering example dear to the hearts of human missile engineers: how does an animal pursue its prey? The actual algorithms will probably never be known exactly, but for comparison purposes, we can do some interesting calculations. The first, simplest assumption is that, just like an early air-to-air missile, an animal maneuvers so that it is always pointing directly at the target, which we will call a Direct Bearing (DB) strategy. If the prey fails to maneuver, then a tail chase results, and if the pursuer is faster, capture follows. A better solution might be to predict ahead, using a mental model to estimate where the target will be at some point in the future and aiming for that point. We know we can do this, because we can throw things and hit a moving target. What we don't know is quite how we do it—are we predicting one second ahead? Ten seconds? Do we predict farther ahead for targets at longer range, reducing our forward thinking as we get closer? Marksmen give a faster or more distant target greater "lead" when they aim, but you can experiment for yourself simply by playing ball.[15]

A slightly more sophisticated approach, and one used for hundreds of years by human sailors, is to adopt a Constant Bearing (CB) strategy, in which the pursuer moves in such a way as to keep constant the angle between its heading —the direction it is moving—and the target. This technique is actually seen in many different types of animal, from spiders to fish to dogs and humans and is much easier to achieve than it sounds. Animals don't need math or maps on plotting tables, they just move in such a way that minimizes the rate of change in their visual angle to the target. Mathematically, the CB strategy can be shown to produce the minimum time to intercept, under the rather restrictive condition that the target moves in a simple and predictable fashion. (If the target stays put and the pursuer sticks with a constant speed, his path is a spiral. Mathematically speaking, this is a logarithmic spiral, and we shall see it again in later chapters.) Not surprisingly, prey tries to evade predators, and even if unaware that the predator is coming, many small flying creatures don't travel in a particularly predictable manner at any time. Is there a way of dealing with a maneuvering target? One solution is the Constant Absolute Target

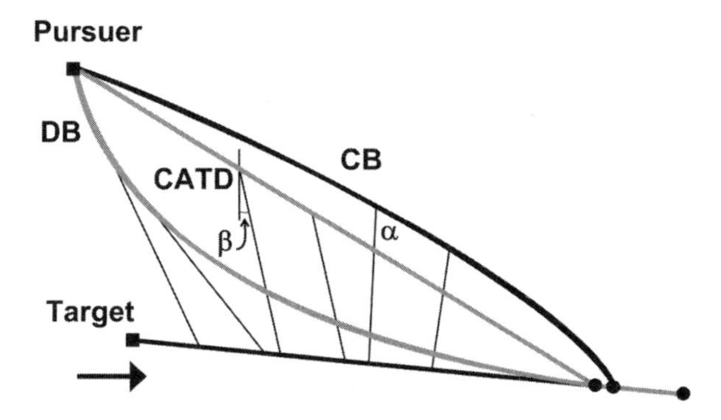

FIGURE 29 Pursuit strategies. DB points at the target at all times; CB moves to keep the angle α constant throughout the pursuit; and CATD keeps the angle β constant, where β is measured relative to some fixed direction.

Direction (CATD) strategy. This approach is a little more difficult to grasp, but Figure 29 should make it clear. The pursuer minimizes changes in the absolute direction to the target, relative to some fixed direction (e.g., the current angle to the sun) by maneuvering in such a way that it maintains the fixed optimum bearing at every stage of the pursuit. The result is that, if we compare any two moments in the pursuit, the lines between pursuer and pursued are parallel. Mathematically, this strategy minimizes the time taken to intercept an unpredictably maneuvering target.[16]

We have generated some simple computer models to examine these different strategies. We allowed for predators that predicted ahead to different extents and targets that moved in straight lines. Then we looked at targets that maneuvered and even targets that actively evaded when the predator got too close. To simplify the comparison of strategies, our predators moved at a constant speed; real critters are more variable. For a target moving in a straight line, there was no significant advantage of CB over CATD, and a DB strategy was almost as good (Figure 29). When the target started to maneuver, the CATD strategy really came into its own (Figure 30). However, for some targets with extreme maneuvering, DB was significantly better than the other options, so long as the pursuer had the speed advantage (Figure 31).

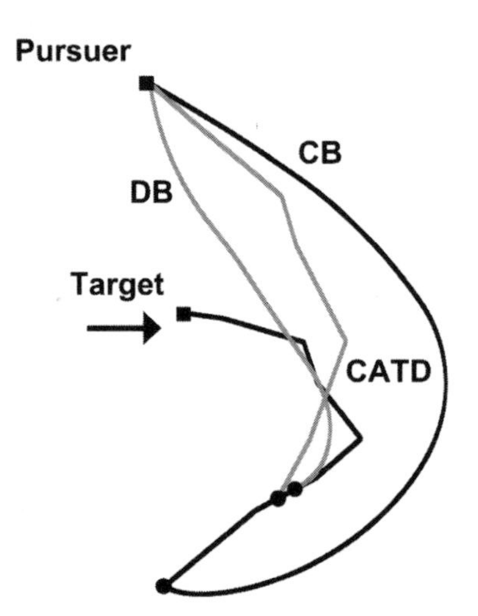

FIGURE 30 Pursuit simulations. In general, CATD does best against a maneuvering target.

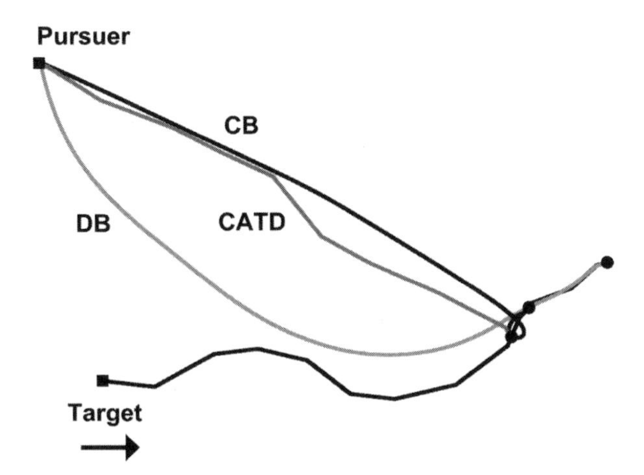

FIGURE 31 Pursuit simulations. At closer ranges, DB may provide a faster intercept if the target turns back on its trajectory.

So which strategies do animals use? It depends on the animal, the prey, and the stage of the pursuit. Bats can detect, localize, and capture rapidly maneuvering insects in less than 1 s. Experiments support the view that they use a CB strategy to start with, but as they close on the bug they change to CATD. Or to be more accurate, their movements are best described by a CATD algorithm —it seems unlikely that the bat's brain is selecting from a defined menu of available modes like a fighter plane's radar might do. Whatever it is doing, the bat does it instinctively.

There is another pursuit strategy, which at first glance, seems far too complicated for use by any animal. The pursuer moves in such a way that at every point in the pursuit, a line drawn from the prey to the pursuer passes through the same fixed point in space behind the pursuer, usually the point at which the pursuer started its chase—or maybe a distant fixed object. (In the extreme case, where this fixed point is at an infinite distance and the target doesn't maneuver, the geometry just becomes that of the CATD strategy.) The effect of this approach is that the pursuer has no visible sideways motion from the prey's point of view. The only sign that the pursuer is moving is that it will appear to get larger—to loom as it approaches. This is a Motion Camouflage strategy: the target literally doesn't see you coming (Figure 32). Using math, the pursuer's ideal trajectory can be readily calculated if you have perfect

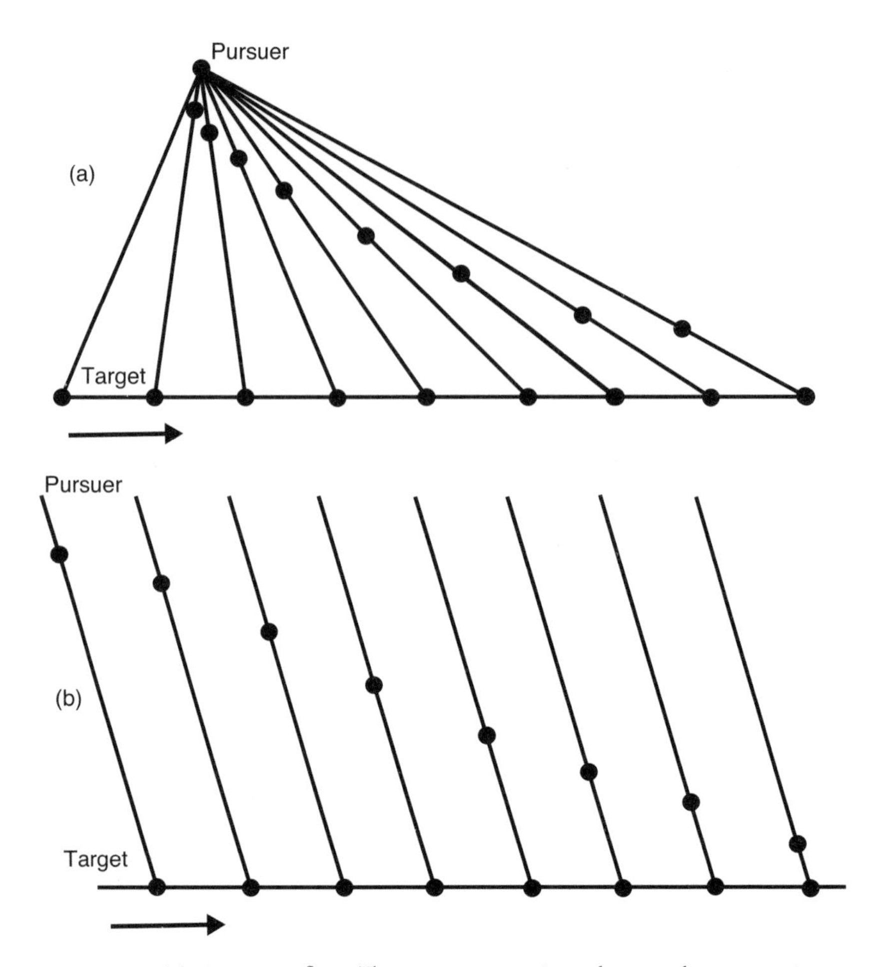

FIGURE 32 Motion camouflage. The pursuer moves in such a way that, at any time, a line between it and the target passes through (a) a fixed point in the distance or (b) a fixed point at infinite distance. The target sees no apparent movement.

knowledge of the prey's path. With the imperfect knowledge actually available to real animals, it's harder to see how they can work out an optimum approach path. Yet it can be done, because the remarkable thing is that many flying insects do it all the time.[17] Exactly how is still under debate.

We speculate that combinations of strategies are always used. Anecdotal evidence suggests that lions intercept zebras with the lion's head clearly pointed toward the selected prey but the body moving on an intercept path. This looks like a CB approach, which seems sensible, as the zebra is moving in a more or less straight line. As the range shortens, the lion's head and body line up on the target, suggesting a switch to a DB strategy in the last seconds before capture.

You can conduct your own experiments on animal pursuits.[18] All you need is two people, a flat area of ground, a ball, and a reasonably well-trained dog. The ball wrangler throws the ball across the dog's line of sight, and the dog wrangler lets the ball get a head start before releasing the dog. Watch how it runs to intercept the ball. Which way is it looking? Which way is it heading? See? Science really can be fun.

5

Built for Life

This chapter is about the structures that animals build. That covers beehives and termite mounds, ant nests and bird nests, beaver dams and gopher tunnels, and even spiderwebs, but there is much more to the story of animal construction engineering than the things they construct.

Grow Your Own

We all know that nests, dams, and tunnels are built by animals but are not part of the animals. In many cases, such as spiderwebs, the structures are even built *from* the animals: the building materials are made in their bodies. But we still think of the web as an external item, built by, and from, the spider, but not actually a part of it. There is a unique connection between the structure and the builder, but we clearly recognize them as being separate entities. Are we right to do so? How can we define what is, or is not, a structure built by an animal?

Snails have shells, spiders have webs. The spider inputs foodstuffs, processes them in its body, and extrudes building materials that it manipulates to form the web. The snail does the same thing, but on a much smaller scale: it manipulates microscopic particles to generate a shell. Although the shell is not composed of living material, there is an interface between the living tissue of the animal and the inanimate minerals of the shell. A spider can leave its web; a snail cannot leave its shell. Is that the difference?

An alternative view holds that building is a direct result of purposeful behavior, but growth is merely a function of body chemistry.[1] That's not a very useful definition. It just delegates the problem to a different department: now we must define "behavior." If a thousand termites process and manipulate mud to create a solid mound, that's construction behavior. If a thousand cells process and manipulate silicate particles to build a shell, why is that different? Above a certain scale why is it behavior when, below that, it is merely growth? We draw a line between *grown* and *built,* but it is a blurred line—precision in classification is a human obsession. So, for our purposes, a structure may be composed of organic materials, but it is not itself alive. In addition we shall exclude things that are part of or inseparably attached to the body. See Figure 33 and decide for yourself.

Why Build?

We can see that there are advantages to growing a shell, but why go to all the trouble of building structures? What are the benefits? We can identify three main reasons:

- to provide protection from environmental and biological threats,
- to find food and water, and
- to communicate with others.

FIGURE 33 Is it built or is it grown?

Protection, whether from environmental extremes or the predations of other creatures, is all about security. Digging holes or building traps to obtain food and water is the pursuit of sustenance. And for animals, structures built for communication are mostly about finding a mate or scaring off sexual rivals. So we're back to security, sustenance, and sex again—motive enough for any animal. But why do some need to build to satisfy these drives, whereas others, often closely related, do not? There is a common theme to all animal construction, however small and simple: construction extends a creature's control over its environment. If your physical body gives you all the control you need to survive and breed, then you won't waste time and energy on building something external to it. Animals don't build for fun. We are no different: if we weren't susceptible to extremes of heat or cold, damp or dryness, and the nocturnal attentions of predators, we would never have built houses. If we could catch fish with our bare hands, we would never have built traps. And once we started building, we had to build other things to support our new lifestyles. Our ability to dig was adapted into the mining of stone and flint for the tools required to build. We needed dry, animal-proof storage for surplus food, so we adapted our house-building skills. In time this process gave rise to workshops, factories, and offices designed to provide protection from weather and theft. To communicate our strength, vigor, or suitability as a mate, we built otherwise unnecessary structures and imbued them with sacred or ritual powers. The pyramids of Egypt stand as declarations of the status and virility of those who ordered them built.

Like us, animals have found that the process of construction generates fresh problems to be solved. Perhaps we build a home that is secure and dry, then discover it quickly becomes hot and airless: we have to build in a solution to keep the air cool and fresh. Animal builders have had to adapt to cope with the same problems. And like us, they have adapted their structures for other purposes. A single building can fulfill multiple functions and display adaptations for those functions. Thus, a single structure may protect from both environmental and biological threats; serve as a place to gather, store, or even grow food; and at the same time provide a clear signal of the occupier's strength and status. After all, that's what our houses do.

So who are they, these animal engineers? Who are the planet's master builders? Well, actually, we are. In terms of scale, scope, function, and sheer adaptability, nothing comes close to human constructions. But it's not a family trait: our closest relatives, the great apes, only manage some crude "nests" for sleeping in and show little inclination for a life in construction. In fact, mammals in general are pretty indifferent to the idea. The most prolific mammalian builders are the rodents who specialize in building tunnels or dams. Some do so for predator protection, others to find food. Many combine the two and even expand into food storage. Birds, of course, build nests. There are 10,000 bird species, and they build a lot of different types of nests, from the unbelievably crude to engineering masterpieces. The bower bird builds an intricate structure that is neither a nest nor a trap. It doesn't provide any protection, and it doesn't generate any meals. The bower attracts a mate—a striking example of a structure purely intended for communication. Reptiles and amphibians are, in general, mediocre builders. Fish, too, have their burrowers and nest builders, but their results are hardly inspiring.

So, in terms of lifestyle complexity and available processing power, we started at the top, with ourselves, and we've run all the way through the vertebrates looking for top-quality builders. Birds come out best,[2] and a few rodents get an honorable mention, but for the most part, mammalian construction activity is restricted to some minor tunneling or nest building. So much for a direct correlation between raw intelligence and sophisticated construction techniques.

Invertebrates also do a lot of burrowing, with assorted worms even taking the trouble to line their tunnels in various ways. Some go further: the octopus —definitely at the smarter end of the mollusk spectrum—will arrange stones and even coconut shells to form a shelter. However, it is the arthropods, the all-pervasive bugs with their exoskeletons and jointed legs, that really take the prize. In particular we have the spiders, most of whom build something, and the insects. And at the top of the class—after us, of course—are the eusocial insects: ants, termites, wasps, and bees. They may be small and not very smart, but they are many and they are *organized.*

Form and Function

Protective structures can provide both a life support function and a defensive one. The defensive function covers security against both predator and parasite threats. The best way to be predator-proof is to build in inaccessible places: up trees, on cliffs, and so on. If that's not an option, any kind of tunnel that hides your presence or restricts predator access will help, and if there is more than one exit that's a definite boon. As an extreme example, consider the Brants' Whistling Rats (*Parotomys brantsii*) of southern Africa. These creatures dig convoluted burrows that truly deserve the title of "tunnel complex" with dozens, if not hundreds, of separate openings, yet they are used by typically only one, and at most three or four rats. Is this rat especially paranoid, or is there a more mundane explanation for this excessive digging? In fact these rats live in arid regions where the plants they eat are spaced quite far apart, but the soil is sandy and relatively easy to dig through. The safest way to get from plant to plant is to go underground, creating exits as you need them, as close to lunch as possible.

Fleas, ticks, and other external parasites are a problem for many creatures. Many burrowing mammals, like badgers or the meerkats of Figure 34, try to control them by building multiple sleeping chambers and moving between them every few days, hopefully leaving most of the pests behind to die. An alternative approach exploits plants that produce natural insecticides—some birds build these into their nests, presumably to control parasites.

Life support is about maintaining ideal environmental conditions for the owner/occupier to thrive. Primarily this means keeping the right temperature and humidity while maintaining the oxygen supply, but it can also include adaptations for cleaning house by removing food and body wastes. In other words, the structure is used as an external aid to maintaining the body's steady state conditions. If you burrow beneath the surface, then you may run short of oxygen. You need to get oxygen-rich fluids—air or water, as appropriate—flowing through your home. Human miners have faced this problem for millennia and have used all sorts of ingenious means to pump fresh air to the mine-face. Animals living in burrows face similar problems. The human solution is to install a pump. Not having access to our technology, other animals may themselves become the pump; for example, the aquatic lugworm uses its own body contractions to drag oxygenated water down into its simple burrow. This approach is fine for a basic tunnel, but it does require the inhabitant to do all the work. However, with a working knowledge of fluid dynamics, a more sophisticated builder can get free ventilation.

Enter the humble mud shrimp. These diminutive creatures live in elaborate personal tunnels dug down into the sediment of the ocean floor. Each tunnel

FIGURE 34 They dig extensive, well-ventilated tunnel networks with a variety of specialized chambers, so they deserve a photograph. Besides, everyone loves meerkats. We thank Anita McFadzean for this image.

has several openings, some of them in low mounds that the shrimp has raised above the surrounding sediment, while others are either at sediment level or below, in little craters dug into it. As ocean currents waft across these openings, two effects can come into play. The first is Bernoulli's principle, which states that, along a streamline of moving fluid, the pressure decreases as fluid speed increases. Thus, as water flows across the raised mound, its speed increases and the pressure drops relative to the other openings, so that water is pulled in through these and forced out through the mound. The second effect is *viscous entrainment,*[3] which results from the drag-induced velocity gradient of water flowing over the mound. Fluid near the bottom of the flow is slowed by its interaction with the seafloor. From these two effects, you can see that water will flow up an open tube placed more or less vertically in the flow. (The same effects explain why tall chimneys draw air upward more effectively than do short chimneys.) In the case of the mud-shrimp tunnel, the net effect is a ventilating flow through the burrow. Anywhere you see an animal burrow with some entrances raised while others are at or below surface level, you can be sure this principle is at work (see Figure 35). Worms, termites, fish, and prairie dogs all exploit these same basic processes to ventilate their homes without having to do any work other than the initial construction.[4]

Truly subterranean animals find their food underground, either hunting underground prey or searching out roots and tubers. Others, like Brants'

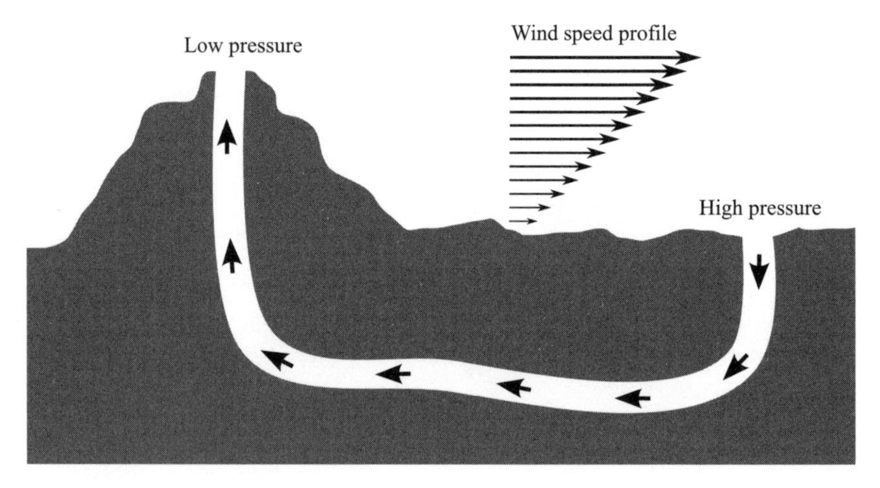

FIGURE 35 Free air conditioning. Wind direction is irrelevant to the operation, so long as there are higher and lower tunnel openings. This set-up even works under water.

Whistling Rats, use their tunnels to get close to food supplies without too much exposure to predators. Once you've got the space, if you're the sort of creature that hoards for the future, using your tunnels to store spare food seems a sensible thing to do. Some creatures dig special chambers for this purpose, whereas others just adapt their tunnels. Still others, especially various termites and ants, go even further and build chambers in which they grow their own food, thus freeing them from the vagaries of the surface seasons. Termites maintain gardens where tough grasses, indigestible to tiny termite guts, are turned into edible fungus; leaf cutter ants do something similar, but their fungus prospers on the toxic leaves of tropical trees that the ants couldn't possibly eat. There's no conscious planning here, but this *is* animal agriculture.

Apart from humans, trap building appears to be an exclusively invertebrate trait. Traps fall into two general categories: snares and pits. The latter are the least interesting: dig a pit in loose, dusty ground and bury yourself at the bottom of it. When something falls in, eat it. Two unrelated insect larvae, the ant lion and the ant worm, employ this technique, which relies for its success on the predator staying in the pit and spraying potential prey with traction-defeating dust to help it down that slippery slope. Snares are more elaborate affairs, constructed of self-extruded silk or mucus, often in the form of nets or webs. Mucus nets are terribly fragile things and so only work for aquatic creatures—some worms and free-swimming snails use them to trap tiny plants and animals drifting in the current. Silk, however, is an amazing material, much used by arthropods. The best known of these builders are spiders, but insects also make silken traps for the unwary. The simplest are just sticky lines dangling from above. Prey snags on the lines and is either reeled in or the predator scurries down to fetch it. At the other end of the complexity scale, the most elaborate silken traps are the orb-webs made by spiders.

Structures built primarily to aid communication also fall into two categories that we can distinguish as enhancers and displays. Enhancers take an existing signal and boost it. A stronger signal carries farther and is more detectable. This boost is most useful for acoustic signals, and a prime example is the mole cricket, who builds a burrow that acts like the horn of a brass instrument to amplify and direct his mating call. This burrow is no haphazard construction—the specifics of acoustical physics require that it be built to exacting specifications. We revisit these burrows in Chapter 12. Although they're not as proficient at it, some frogs use burrows in a similar way. One frog species even turns this concept around: the burrow is used to detect incoming signals, acting like a satellite dish to focus signals onto the receiving frog sitting at the acoustical center. The burrow is even optimized to receive sounds at the peak received frequency, which is less than the peak transmitted frequency, because the damp air of the local environment absorbs the higher frequency sounds.[5]

Displays, however, *are* the signal. A nest or burrow built to attract a mate will have evolved to send out the best possible signal. Most remarkable, as already mentioned, are the various species of bower birds whose elaborate constructions serve no other purpose.

Building a Better Builder

Physical Adaptations

All creatures that build do so with their mouths, their limbs, or some combination of the two. To date we know of no limbs utilized solely for building, even among the multi-limbed arthropods. Nor is there any known animal whose mouth has evolved into something used only for construction. Indeed, there are few cases where the mouth parts of even the most successful animal engineers show any adaptations that are clearly specific to building. Birds may be top-class builders, but their beaks are optimized for feeding, not building.

We do, however, see body parts adapted to better suit the needs of the builder. Consider the mole: an expert mammalian tunneler, it is characterized by its short powerful legs and broad clawed feet. The legs act as levers, converting the muscles' work into a strong force, whereas the feet are basically shovels. The mole looks exactly like we would expect a tunneling mammal to look. Now consider another prolific mammalian tunneler: the rabbit. Rabbits do not display the same degree of adaptation to a tunneling lifestyle; instead, like their close relative, the hare, they look like they are built for running. In some minor ways rabbit legs are adapted for digging, but not nearly so much as the mole's legs. The reason is that, to survive and breed, the entire creature must be adapted to support the whole lifestyle. Rabbit and hare legs are primarily used for locomotion on the surface, with a strong focus on rapid evasion of predators. In contrast, the mole's legs are mostly used for comparatively slow perambulation beneath the surface. The degree of adaptation possible is constrained by what else you need those parts for. The mole's legs, along with the rest of it, are more suited to the underground life, because that's how it lives. We see the same principle at work in arthropods. The more time they spend underground, the more they can afford specialist adaptations for the lifestyle. You can see some examples in Figure 36: they all dig burrows but they don't all show it.

But what about spiders? Although they build with self-produced silk, the actual building is done with limbs and mouth parts. The evolutionary adaptation to manufacture silk predates its use in construction. We're not quite sure what it was initially used for, but it probably involved reproduction.[6]

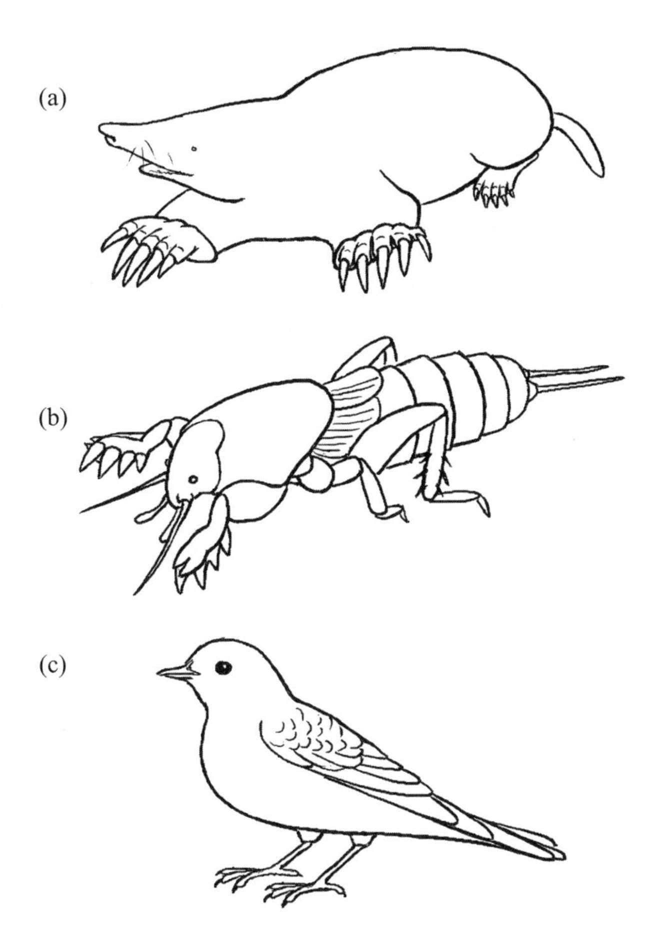

FIGURE 36 Adaptations to a tunneling life. (a) Moles and (b) mole crickets dig with their legs. (c) Sand martins need their forelimbs for flying, so they haven't invested in specialized digging equipment. This failure to specialize limits what they can excavate, but their burrows still serve their purpose.

Behavioral Adaptations

At the start of this chapter we frowned skeptically at the notion that building could be defined purely in terms of behavior. But observation and reflection may oblige us to reconsider. Consider this: two species of insects may be virtually indistinguishable to the untrained eyes; or two bird species may have the same physical structure and even exhibit similar feeding, flocking, and mating behavior. Yet one species might be an accomplished builder whereas the other is not. The difference between them is behavior, and behavior is the result of control algorithms.

We wander into dangerous territory when we start questioning how much of this building behavior is instinct, programmed at a genetic level, and how much is cultural, learned from experience and observation. Put 10,000 human babies in a sealed room, provide them with basic care, food, water, a generally benign environment, and absolutely no adult instruction, and we are willing to bet they will not grow up to spontaneously build a 200-story air-conditioned skyscraper.[7] Ten thousand baby termites would do so, if one of them was a queen.[8] They are not taught, they do not learn, they simply *do*. We examine how in Chapter 6.

Some young birds build nests badly.[9] With practice, they get better. Compared to termites, these birds have a more sophisticated control mechanism that allows them as individuals to fine tune their instinctive behavior, like a human infant figuring out how to walk. The control algorithms are all there, but a little trial and error is needed before we find the best values for their parameters. We describe this behavior as "practice" and not "learning": practice makes perfect, but learning is discovering how to do something you didn't know was possible. (Once you've learned how to make your body perform the actions, you may still need a lot of practice before you're any good at it.)

During an individual's lifetime, the control algorithms themselves can be modified as a result of experience—we call this "learning," and because as a species we're quite good at it, we think it makes us very special. But it happens to all animals. It's just that, in most cases, these modifications take place at a much slower pace, over the lifetime of species, not individuals—what we call "evolution." This is why we haven't really contradicted our earlier viewpoint of building: animal behavior has evolved to assist construction, not to define it.

Building Blocks

We can think of building materials as being substances that have been processed. Here we specifically mean the processing required to produce the building materials, not the processing involved in actually building with them. Thus, wooden planks are natural timber that has been cut and planed to shape; house bricks are molded and baked clay. The same is true for animals' building materials. The easiest to acquire are naturally abundant substances that need little or no processing, but these will come in all manner of shapes and sizes. The easiest to build with are standardized in both shape and physical properties—you know what you've got and what it can do, and you don't have to put much effort into working out how to use it.

The most basic processing involves simply collecting a naturally occurring substance, such as stones, twigs, or grass. Going one better you could sort it by size, shape, or other useful properties like strength or weight. The next step is to alter it in some way, by cutting, molding, or otherwise shaping it. The larvae of caddis flies build little houses around their bodies. Some species laboriously sort through available materials to find just the right sand grains; others cut standard-sized building panels from nearby vegetation—a little more processing but a much easier build. Processing can also be used to change a material's basic properties, as when, for example, liquid is added to soil to create a malleable mud.

That's about as far as animals go in processing the things they collect from their environment. They don't have factories to bake mud into standard bricks or cut timber into standard lengths. Instead, if they need to manufacture something, they have to use their own bodies and secrete what they need, which has the huge advantage of providing standardized materials that don't need to be collected. Broadly speaking, animals work with what's already available: body wastes (feces and urine), saliva, mucus, assorted sealant waxes, and (if they have the right equipment) silk.

Animal dung is a very useful building material. Mixed with mud, it acts as a binding and stabilizing agent. Add fibrous materials like straw or hair and let it dry in the sun, and you can build quite elaborate structures with it. Why does the fiber help? Because its different mechanical properties make it more resistant to crack propagation: when the crack reaches a length of fiber its force is dissipated or even stopped, making the material tougher and structures stronger.[10]

We've built with dung for millennia and still do in many parts of the world.[11] So do animals: the enormous termites mounds that dot the arid plains of Africa and Australia are built largely of a mixture of local soil and termite feces, with some chewed up wood or vegetable matter thrown in as available. Dung is such a marvelous material that some animals actually use it to produce their own standardized bricks: the caterpillars of some moth species (who, as leaf eaters, produce prodigious amounts of fibrous dung) build themselves little houses from uniform dried fecal bricks and beams, held together by their own silk.[12]

For gluing materials together, many creatures use their own body fluids. Thus, many wasp species chew up wood pulp to create paper from which to build their nests. Birds mix mud and saliva to cement cup-shaped nests to rocks, trees, or houses. Saliva usually has to be thoroughly mixed with another substance and allowed to dry if you want to build with it; however, mucus can be used on its own. Many worm species, aquatic and otherwise, line their burrows with it. (This is one reason the worm casts on your lawn are so persistent.) Other aquatic worms use mucus to glue sand and mud into protective tubes. Fish, such as sticklebacks, build crude nests from plant material held together by mucus secreted from their kidneys. Water-insoluble wax (produced by many terrestrial arthropods to prevent themselves from drying out) has also been adopted as a building material. Bees and wasps take its production to extremes and use wax to build the honeycombs in which they store their food and raise their young. It has the great advantage of being malleable, especially when warm. Honeybees construct wax cylinders around themselves, using their own bodies as a size template. Stacked vertically, the heat of the hive softens these cylinders and gravity drags them down, forcing the original circular cross-section to distort into the shape that most efficiently fills the space. The result is a comb of hexagonal cells, just the right size for developing larvae. No planning, no forethought, just simple physics.

The Spiral of Doom

But when it comes to self-extruded building materials, silk beats them all. It's such a useful material that many other creatures use second-hand silk in their building work, either as linings and insulation or to hold things together, in which case it may be used as thread for stitching or as a sort of natural Velcro loop. As a building material, silk is strong and flexible and can be spun and worked under water. In fact, "silk" is really a generic term for a range of similar materials secreted by a variety of invertebrates. Chemically, silks are polypeptides—the long chains of amino acids that form the basis of proteins. A typical spider thread combines loose, disordered coils of glycine that stretch when pulled with highly ordered crystals of alanine that give it stiffness and strength. The specific structural properties of different silks depend on the exact mixture and arrangement of the contributing proteins.[13]

Consider a typical garden spider of the sort that builds an orb-web to catch flying insects, like that shown in Figure 33. The web is stretched across a gap, anchored to suitable rocks, vegetation, and the like. The spider's first design problem: it needs to maximize the prey capture area while minimizing its use of precious resources, specifically, energy and the self-produced silk. Thus, we get a web, because a net uses less material than a solid sheet. (It also offers less resistance to potentially damaging breezes.) The trick is to ensure that the web mesh is not larger than the prey. To conserve resources, many spiders recycle amino acids by eating their old webs. While balancing budgets, the construction must also balance stiffness and strength against flexibility and resilience, so that when a bug hits it, it doesn't break the web or get thrown back out by elastic recoil. It also has to ensure that, once stopped, the bug can't just fly away. A typical web has five basic structural components all based on spider silk and all doing very different jobs.

First there are the attachment disks, a series of sticky anchors that fix the web to its environment. Next is the basic outer frame of supporting threads that runs between these disks and then, within it, the straight-line radii, running from center to frame. Between them, these define the shape of the web. This combination of anchors, frame, and radii gives us the mechanical structure to absorb the shock of impact. Laid down over this structure is the actual net—the capture spiral that will trap the prey. But this capture spiral needs something extra as its final component, some sort of restraint to hold the prey in place. Sticky droplets (absent on the frame) glue the victim to the capture spiral; these have to stay sticky, which means they must not dry out too fast.

Our garden spider, like all its kin, has a series of spinnerets, each extruding a different silk through a variety of spigots. The thread diameter is determined by the spigot's aperture, and the thread's properties result from its precise chemical composition, itself determined by the particular silk-generating gland into which the spinneret taps. The process of squeezing silk through the spigot aligns the polypeptides, rearranging them into a flexible thread that hardens as it leaves the spider's body. So the web is really built from a series of different threads, each specialized for its particular task. Attachment disks have a multitude of short, gripping threads. Frame and radii threads are tough and break-resistant, but the capture spiral is stretchy. The capture droplets themselves contain sticky centers of glycoprotein and are hygroscopic, which means they attract water, helping them stay damp.

To build its web, the spider spins a silk dragline, baited with an attachment disk, and uses air currents to waft it across a suitable gap to catch on leaves, twigs, or any other support. The spider then goes back and forward across this line, laying down more silk to strengthen it before dropping from the centre of the bridge to lay a vertical line to some suitable anchor point. To this Y-shaped beginning the spider adds more frame lines for support before moving on to the radial spokes. Now, from the center of the web it spirals outward, laying down a temporary, nonsticky scaffolding thread. The spider joins this to each radius in turn, ensuring that the threads always intersect at the same angle, creating a spiral in which the turns get looser toward the outside. This stabilizes the structure, allowing the more closely spaced capture spiral with its sticky droplets to be laid, this time starting at the rim and spiraling in to the centre. All turns of the spiral are spaced the same distance apart, an effect achieved by keeping the outside front leg always in touch with the previous spiral turn—Simple, but effective.[14] As the spider builds the spiral, the scaffolding web is rolled up and eaten.

But will a web stop a fly? The trick to stopping a flying bug lies in a combination of resilience and aerodynamic damping. Spider silk is not especially strong—you can snap it easily—but it stretches a long way before it breaks. And it has low resilience . . . very low resilience. The web brings the bug to a halt, and the energy available for "bounceback" is insufficient to overcome the sticky droplets on the capture spiral. Dinner is served.

Aerodynamic damping is simpler to grasp: when the flying bug hits, the whole interconnected web is forced to move through the air, which naturally resists that movement, leading to impact energy being lost as heat. Loss of energy means loss of speed: the web slows down, and so the bug slows down.

Optimum Webs

Is there such a thing as an optimum web? Theoretically, perhaps. Consider the hypothetical Invented Spider, *Araneus commentus.* This figment of our imagi-

LIVING IN THE MATERIAL WORLD

When a material is subjected to a stress (defined as force per unit cross-sectional area), it experiences a strain (a measure of how much it is stretched, given by the change in length divided by the original length). Put under too much stress, it is overstrained and breaks. Its tensile strength is the maximum stress it can take without breaking, and its extensibility is the maximum strain (see Figure 37). Figure 37a shows a loading curve—a plot of stress versus strain as the material is stretched. The initial slope of the curve gives the material's stiffness, a measure of its resistance to being stretched. Energy must be applied to stretch it, and the material stores this energy internally. The shaded area under the curve is the work of extension, a measure of the energy stored in the strain.

When the stress is removed in a perfectly elastic material, the material relaxes back along the loading curve to its original configuration, releasing all its stored

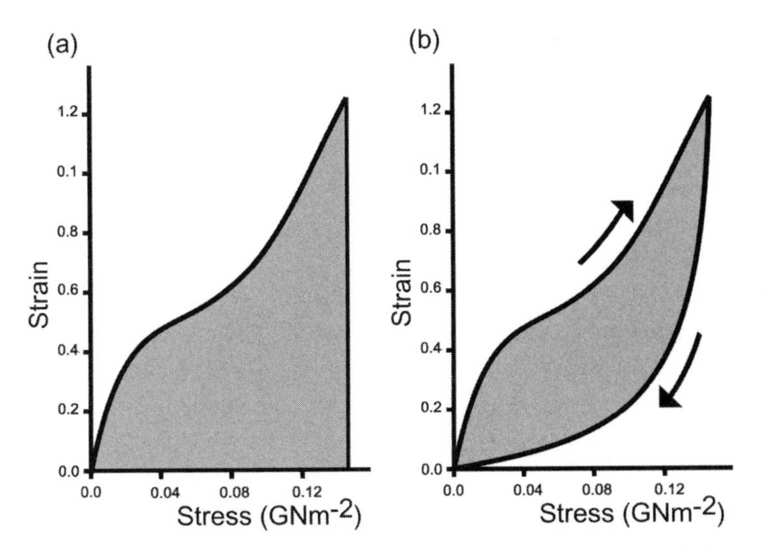

FIGURE 37 Stopping flying bugs. (a) A perfectly elastic material is stretched. When released, it relaxes by following exactly the same curve on the stress versus strain plot. The energy stored in the material at maximum stretch is represented by the shaded area under the curve. (b) Theoretical spider silk is stretched in exactly the same way to exactly the same point. But when released, it follows a different route back to the origin. The shaded area between the two curves is the energy that the silk dissipates. The data are from Hansell (2005).

(*continued*)

energy. But no real material is perfectly elastic; instead, they lose energy as heat and follow an unloading curve back to the start, as shown in Figure 37b. The shaded area between the two curves is a measure of the energy dissipated by the material. The ratio of energy out from unloading to energy in through loading is the material's resilience: the lower the resilience, the more energy is dissipated when the material is stretched and released. Finally, the energy per unit cross-sectional area that must be input to break the material is its toughness.

nation builds an idealized orb-web to catch imaginary insects. However, to simplify our calculations, the Invented Spider does not spin a spiral structure into its web. Instead, as shown in Figure 38, our idealized web consists of seven polygons and five radii forming a rather too-perfect geometric shape for which we can calculate the total surface area (allowing for the gaps between the polygons) and the total length of silk required. By generalizing the geometry, we can allow for any web with N polygons and k radii and calculate a dimension-

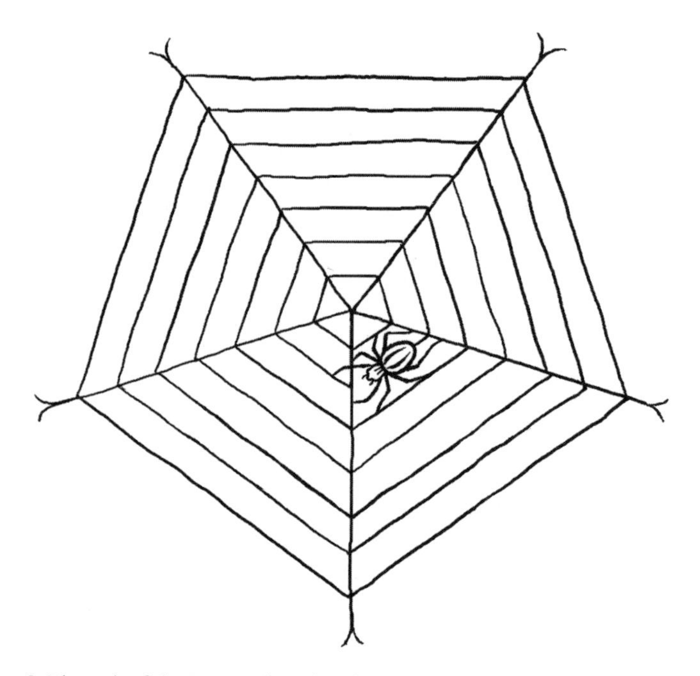

FIGURE 38 The web of the Invented Spider, showing its basic components as described in the text.

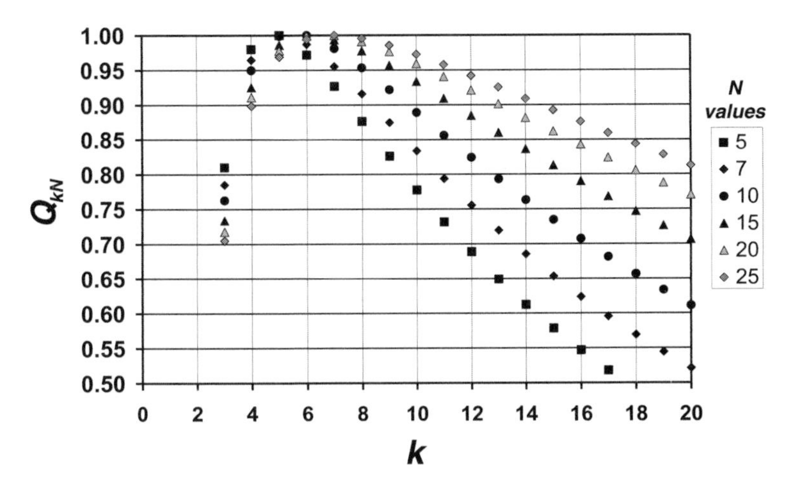

FIGURE 39 Optimum spider webs. The optimum web is the one with the maximum value of Q (a measure of quality). For various values of N (the number of polygons), the normalized Qs are plotted versus k (the number of radials). For small webs, the optimum number of radials is $k = 5$; for larger webs it is up to $k = 8$. The Invented Spider's web has $N = 7$, $k = 5$.

less "figure of merit," Q_{kN}, which is just the ratio of area to $(\text{length})^2$. Or, put another way, it is a measure of capture area relative to silk production. By plotting Q_{kN} against k for different values of N, as shown in Figure 39, we predict that a small web will tend to have about five radii, whereas a larger one may have about eight. For the webs of real spiders there are all sorts of physical considerations other than maximizing capture area while minimizing silk production, but we would still expect to see the number of radii fall within a limited range of preferred values, unless in response to some more pressing constraint. If you have a persistent neighborhood spider, you can check for yourself.

Changing the World

Human construction engineering can drastically alter the environment, often in ways we hadn't anticipated and don't always want. Animal engineering is no different. The best-known environmental engineers are the beavers. These giant rodents cut down trees, dam rivers, and build lodges in the resultant ponds. Trees are also required for food (specifically, the foliage, branches, and inner bark, although they really prefer wetland plants). Beavers dam streams because they like to live in the deep, slow-moving water of a pond or lake. There is no design committee: if the water is moving and conditions are right, dam building starts. Although experiments suggest that the sound of running water plays a large part in triggering this behavior, it is clearly not the only stimulus. Beavers seem to respond to the water flow, whether heard, seen, or felt.[15] Damming allows them to keep the lodge entrance under water, hidden from predators, but it also provides them with a place to store water-logged trees as a winter larder. Beavers stick close to the safety of the water and so dig canals that allow them to swim to new stands of trees and make it easier to bring felled timber back to their store.[16]

Clearly, turning a woodland stream into a pond represents a significant change to the local environment, which can be good news for animal and plant species that favor slowly moving water, but bad news for those that were there before. However, the beaver's influence doesn't stop at that. Beavers prefer deciduous trees, so we might expect to see those kinds becoming rare near their dams, but flooding kills conifers faster.[17] Cropped deciduous trees can grow back from surviving root systems, or new, fresh growth might come into the area, so long as it can cope with the wet ground and the beavers' browsing. The dam and pond system acts as a buffer to water flow, reducing the downstream flooding danger during heavy rains while ensuring a relatively steady flow even when there is no rain. The pond and its associated wetlands also play a role in filtering and cleaning the water that passes through them.

However things play out, the net effect is that the local environment is drastically changed. Some species benefit, others lose out. One of the losers might even be the beavers if, eventually, there's nothing left to eat. At that point they move on. Eventually—and it can take years—the abandoned dam gives way, with nutrient-rich sediment flooding out to create an open meadow, which favors flowering plants, which attract different species of insect, which bring in birds, and so on. In time, perhaps, the forest can reestablish itself, often with a tendency to favor the beaver-approved deciduous species, and so the beavers come back. Sometimes, however, it doesn't work out like that and the area may persist as scrub or bog. Perhaps unsurprisingly, beaver-friendly landscape regeneration occurs more often where the beaver population density is low. Arguments rage among biologists over whether beavers encourage or suppress local biodiversity and how sustainable their browsing is.[18] So the beavers' selfish pursuits cause drastic environmental changes, deplete the natural resources, and if their population density is too high, leave a landscape incapable of supporting them. Sound familiar?

A less obvious, but hugely effective, environmental engineer is the earthworm. As a builder, it's unimpressive, spending its days burrowing through soil, sifting out scraps of food, and dumping the processed earth out on the surface as nutrient-rich worm casts. Gardeners recognize the earthworm's importance for plowing the soil, aerating it, breaking up clumps and clods, and helping fertilize it by dragging leaf litter down below the surface, all of which assists other subterranean life from bacteria upward.[19] That's enough to make them important ecosystems engineers, but what is less readily appreciated is this simple invertebrate's impressive ability to bury whole civilizations, given enough earthworms and sufficient time. Charles Darwin figured this out when he examined earthworm activity at an archaeological dig and calculated that the depth at which ruined Roman walls lay below the surface could be explained by 1,500 patient years of accumulated worm casts.[20]

Air Con 101

Animals require air and generate heat, so the homes they construct may need ventilation. Solitary animals (e.g., mud shrimps) or small communities (e.g., those of prairie dogs) make do with a simple pressure-driven ventilation system, because they're not using much oxygen or generating a lot of heat, but a colony of thousands or millions of individuals must construct a more sophisticated solution.

Wild honeybees build nests in natural holes in tree trunks. They seal the cavity with self-secreted wax, leaving just one opening. Any exchange of warm stale air for cool fresh air has to be achieved through that entrance, which is usually near the bottom of the hive. The problem is that warm air is lighter than cool air and so rises to the top of the hive, where heat and carbon dioxide concentrations build up. The bees have to find a way to pump it out and get fresh air in. They achieve this by doing work (which consumes oxygen and generates heat, but these effects occur just outside the hive entrance). Bees face the hive entrance and fan their wings to pull air out, which has the effect of pulling down the warm moist air and reducing air pressure in the hive. All bees engaged in the job fan together for a time, and then they all stop and rest, letting fresh air flow back in. The bees seem to respond instinctively to some combination of oxygen or carbon dioxide levels and hive temperature. The more ventilation that is needed, the more bees take part in fanning.[21]

To heat the hive, say, on a cold morning, bees huddle in a tight cluster. The warmest bees are in the center, surrounded by a middle ring of "shivering" bees, who work their bodies to generate heat. Finally the bees in the coldest (outer) ring lock their chitinous hairs together to act as insulation. Again, the warmer they need to get, the more bees will participate in shivering and insulating tasks.[22]

Many species of ants and termites ventilate their nests using the same pressure-driven principles that we see elsewhere. They just do it on a much larger scale. A nest, which can house millions of individuals, is built under a raised mound. Air channels are provided at ground level, and there are one or more vents leading up to the top of the mound. Cross tunnels through the mound allow air to circulate freely, and the natural tendency of warm air to rise helps things along.[23]

The structures of other termites are a bit more problematic, because their mounds don't seem to have vents in the top. Found primarily in hot, arid regions of the world, these are some of the largest structures in the insect world: vast, near-vertical towers that, relative to the size of the builders, dwarf most human constructions. Unlike the open pressure-driven ventilation systems of their cousins, these towers—"mounds" seems inappropriate—are closed to the outside world. What's going on? Are these truly sealed systems, like earth-bound spaceships from which fragile termite work parties cautiously emerge into the comparative cool of the tropical night? If so, how do they work?

Details vary among species, but closed mounds seem to be characterized by a large enclosed air space at the top and an even larger cellar beneath the actual nest (see Figure 40). This cellar is not just a quarry for building materials but seems to play a vital role in the working of the mound. A quick analysis identifies the following major components of the mound:

- the nest, where the termites live;
- the fungus gardens, where they grow their food;
- the cellar beneath the nest;
- the central vents or chimneys;
- the air-space or chimney cap to which they lead;
- the surface conduits, running up the outside of the mound walls;

- the transverse passages connecting surface conduits with chimneys; and
- other passages connecting all this structure to the nest and fungus gardens.

It seems that the mound is interacting with the external environment and acting as a giant lung. The surface conduits are porous to air. On the windward (or high-pressure) side, air pressure forces fresh air in; on the low-pressure leeward side, mound air is drawn out. Natural wind is chaotic, changing speed, strength, and direction all the time, which gives rise to a tidal effect with ever-changing pressure fields around the mound.[24]

We know that tidal ventilation in the vertebrate lung exchanges respiratory gases in three phases. The termite mound seems to work in a similar way. Around the surface conduits and the chimney cap, the wind creates a region of forced convection, as in the upper bronchial branches of the lung. Farther down the chimney, near the nest, density variations due to the heat and oxygen consumption give rise to a natural convection zone where gas exchange is dominated by diffusion, as happens in the alveoli of the lung. In the middle chimney and the lateral passages, gas exchange occurs in a mixed forced/natural convection region, as found in the intermediate zones of the lung. The mound architecture combines the metabolic energy in the nest with wind energy to drive a colony-level ventilation system. Moreover, it can adapt to changing circumstances. The mound itself is continuously modified, adapted, and repaired by termites in response to natural erosion and damage caused by

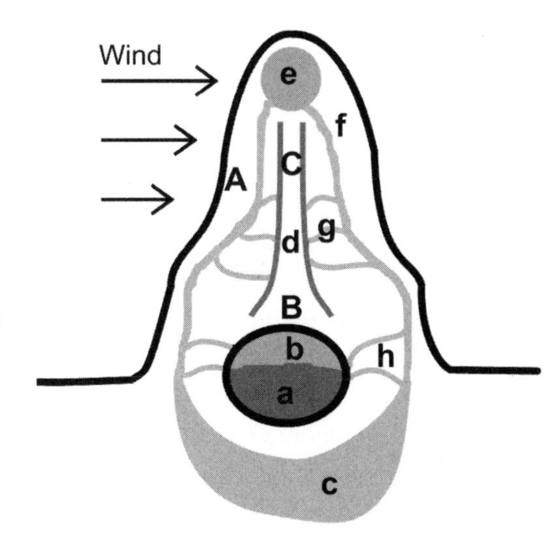

FIGURE 40 Schematic of a closed termite mound. Shown are a, nest; b, fungus gardens; c, cellar; d, central chimney; e, air space; f, surface conduits; g, transverse passages; and h, connecting passages. Also shown are A, the forced convection zone; B, the natural convection zone; and C, the mixing zone.

predators or natural disasters. A mound grows with the colony, reaching up into stronger winds that produce more vigorous ventilation. Simple modeling[25] suggests that termite mounds grow and adapt in line with the carbon dioxide gradient in the nest, with structures built to reduce these carbon dioxide concentrations. Building material is carried perpendicularly across the lines of constant carbon dioxide concentration—the CO_2 isobars. Species that don't farm fungus produce less heat and less carbon dioxide, so the isobars, and hence the mounds, are almost hemispherical. Add a fungus farm sucking up oxygen and pumping out heat, and the resultant buoyant forces distort the carbon dioxide concentration, stretching it upward and forcing the termites to produce the chimney. Because the fungus, not the termites, produces most heat and carbon dioxide, there may be grounds for regarding the termite tower as a homeostatic structure built by fungus, using termites as its agents.

Tool Time

We have been looking at things that animals create to extend the capabilities of their bodies. It would be remiss of us not to insert a few words about tools. This huge and often controversial subject is approached through a minefield of difficult definitions. What is a tool? What behavior constitutes tool use? In the creation of an apparent tool, where do we draw the line between accident and design?[26] Using a stone to crack open a shell seems less impressive than shaping a stick to prize grubs from their burrows. Do animals actually make tools?

In fact many do, for example, by tearing off twigs or blades of grass. Fewer actively shape them for the task at hand. One species of Galápagos finch uses twigs or cactus spines to prize tasty grubs from holes in branches. If the tool is too thick, too slender, or too weak, they try another one, or sometimes they carefully modify the one they've got until it works. In a similar way the New Caledonian crow makes two distinct and different tools for getting at wood-burrowing insects. One of these involves the careful tearing of serrated palm leaves to create a very precise extraction tool. Chimps use sticks to fish for termites—they poke the stick into a termite mound, termites rush to defend the nest, the chimp pulls out the termite-covered stick, scoops up the insects, and eats them. Sometimes they deliberately flatten the ends of sticks by chewing them, creating more space for doomed termites to congregate. Is any of this evidence of intelligent tool design or is it just learning from serendipitous experience?

Using tools, making tools, and adapting tools—none of these behaviors seem to be uniquely human and none seem to require an abundance of intellect. Like anything made or used by animals, tools are just a means to extend capabilities beyond the limitations of the available body. Perhaps we consider humans as special because we seem to be the only animal that can consider an entirely theoretical situation, imagine a wholly new tool that could help us deal with it, and then explain the concept to others without actually making it. No other animal has thought of the wheel or the International Space Station. It's that leap of imagination coupled with the ability to communicate the idea to others that makes us the true Master Builders.[27]

6

Simple Complexity

Emergent Behavior

Most of this chapter will be concerned with animal behavior. You will not learn how to tell if a ferret is friendly or an iguana irritated. Instead we take a look at some of the control algorithms that make animals behave the way they do. In particular, we ask the question: how simple can they be?

Numbers Count

We've already looked at animal brains and the mechanisms they use to control the body. From an engineering viewpoint we've seen that, like any machine, an animal has an information processor for making decisions and a control system to implement those decisions. However, a quick glance at the world around you will show certain fundamental differences between human-designed machinery and living animals. Machines do what they were designed to do. Automated factory production lines can carry out a series of intricate steps over and over again with impeccable precision to churn out identical copies of the same prototype. Animal behavior is more erratic in its execution—no two beaver dams are exactly the same, for example—yet, if we watch an animal, we soon realize that its behavior is both complex (it does a lot of different things) and adaptive (what it does changes with circumstances).

When we first examined animal brains we noted that size isn't everything, but that, as a first approximation, neurons count—quite literally. As a general rule of thumb, the more neurons a brain has, the more information it can both store and process. It stands to reason that bigger, smarter animals display the most intricate behavior as the result of their hugely complex and sophisticated control algorithms. Doesn't it? Well, we will see.

A Roach for All Seasons

Consider the intellectual capacity of a cockroach. We know that roaches, like most insects, employ a sort of distributed computing to provide the necessary control of the body. We know that they can react—sometimes with astonishing speed—to external stimuli. They can communicate with one another by touch and by means of chemical pheromones (more on these in Chapter 12). Some even care for their young, albeit not in a very exemplary fashion. We also know that they can get everywhere. But with such limited processing power to drive the machine, how do they achieve so much? The answer is that, like most creatures, *a cockroach is just as smart as it needs to be.* Remember the mantra: sustenance, security, and sex.

People have studied cockroaches for a long time (usually as a prelude to finding better ways to kill them); recently there has been a lot of interest in the cockroach as a useful analog for autonomous robots.[1] Cockroaches can navigate round a room, finding things that interest them and avoiding things that could harm them. If we could find way to reproduce that behavior using software algorithms, we could program simple robots to, for example, explore Mars on their own. Unfortunately, trying to reproduce cockroach behavior in a computer model has proved to be extremely difficult. There are just too many rules to be programmed in, too much information to be processed, too many things that the creature has to think about. Building a model cockroach would appear to be an extremely difficult thing to do.

Except that it isn't.

Initially, scientists had adopted a top-down approach: look at roaches and try to program in their behavior. Such an approach led to complex, unwieldy models that didn't behave like a real roach. The answer was to come at the problem from the other direction—to adopt a bottom-up approach. Stop thinking about what cockroaches *achieve* and instead look at what they *do.* And the reality is that they don't do much, but they do it all the time. If they're awake and next to a wall, they follow it. If they sense food, they head toward it, then eat it until they are satiated. If they're ready to mate and they sense a potential partner, they head toward it. If they're threatened, they run away. They avoid bright lights, which means they seek shelter in the day and come out at night. And if they have no more pressing business elsewhere, they wander randomly until something interesting turns up. These basic rules can be used to program a computer model that uses a simple set of prioritized commands, asking "Is this the current situation? If so, do this, otherwise check the next possible situation and act accordingly." By linking these rules together, the result is a model that behaves in a remarkably cockroach-like manner. It turns out that such behavior can be quite involved. Thus, a series of individually simple responses, when chained together in the right way, can produce remarkably complex behavior. This behavior emerges spontaneously because of the way the different responses mesh together. Or to put it another way, complexity arises from the interactions of simple rules.[2] Take a look at Figure 41. Can you tell them apart?

The essence here is that the whole is greater than the sum of its parts. In the natural world, emergent behavior is an important mechanism for solving practical problems. At heart, nature is inherently simple, and natural solutions are based on simple principles. Humans design things based on our ability to project into the future. We can plan ahead, consider the consequences, and build mental models, constantly saying to ourselves, "If I do *this,* the result will be *that.*" In our smarter moments, we solve problems by thinking about

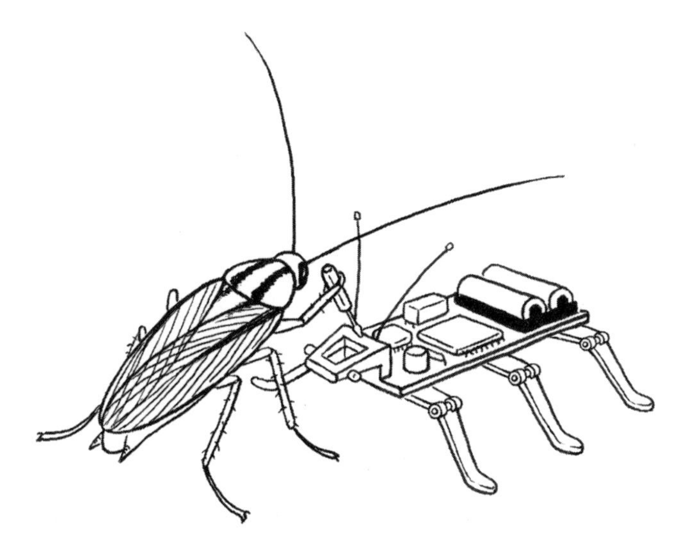

FIGURE 41 Living cockroaches behave remarkably like robo-roaches. Or is it the other way around?

them, running through theoretical scenarios, and making decisions based on our projections and our experience. Some big-brained animals may be able to do similar things on an individual level, but in general, nature doesn't work like that. Instead, nature evolves solutions by letting things act. On evolutionary timescales, the things that work are kept, and the things that don't work get dumped (Darwin's natural selection). The solutions may not be the best possible ones—in scientific terminology they are often "suboptimal"—but they do the job. They are, in other words, fit for purpose. Humans design solutions through a combination of brainpower and hard work, but nature just lets them evolve. We will leave it to the reader to decide which approach is more effective.[3]

For members of a single cockroach species it is reasonable to assume that they all want the same things out of life. Yes, we're back to sustenance, security, and sex. It is perhaps not surprising, therefore, that they are often found grouped together in what scientists call an "aggregation."

Creatures of the same type, constrained by the same physical laws and driven by identical basic requirements tend to form aggregations.[4] When cockroaches meet, they interact, but they don't really cooperate. Cockroaches aren't famed for working together toward a common goal. In the end, the

THE ROACH APPROACH

You can do this experiment for yourself. All you need are some cockroaches (the ubiquitous German Cockroach [*Blattella germanica*] will do just fine), some slightly spoiled food, water, a simple shelter (e.g., a wooden board perched on low rocks), and a spare room you'll never want to enter again. Put the food and the shelter an equal distance from each other and from the room walls. Now free your cockroaches and watch them go. Wait until dark when they are most active, then hit the room lights and watch them scurry for cover. Try marking individuals with, say, a nontoxic paint, and track their movement and behavior throughout a day. See for yourself how simple their lives really are. Yes, Roach Manor provides hours of harmless, scientific fun for all the family! You'll never need to watch TV again.

cockroach is just a rugged individualist, doing what it has to do to survive.[5] We can see that, for a creature with very limited processing power, a few simple, hard-wired rules might work well enough to get by, but it stands to reason that more intricate behavior must require correspondingly more sophisticated algorithms.

E Pluribus Unum

Termites, often erroneously called "white ants," are in fact very closely related to cockroaches; they separated at some point during the Cretaceous Era. However, unlike the cockroach, the termite is a social insect. This doesn't mean that it goes to cocktail parties or hangs out on Facebook, but rather that it lives in colonies in which the colony members deliberately cooperate with one another for the benefit of the colony as a whole.[6] Biologists define them more precisely as *eusocial* insects, which are characterized by three principal traits:

1. They cooperate to care for their young, which are incapable of surviving on their own.
2. The lifespans of the generations overlap (without which they couldn't care for their young).
3. They exhibit a reproductive division of labor, being divided into distinct types or *castes,* of which only a restricted subset will ever breed, while the

> others work. Different castes thus perform different functions and, in some cases, may actually have physically distinct forms.

That last point is the most important one, so we shall repeat it. To be classed as eusocial, a species must exhibit a clear division into different castes, some of which breed and some of which work. Most mammals, for example, care for their young and have overlapping generations. Many of them are gregarious creatures, living in packs, herds, and prides with intricate social relationships. However, with the notable exceptions of two species of African naked mole rats, they are not eusocial, because they do not divide themselves into distinct castes with distinct functions.[7]

It is among the insects that we find the most eusocial animals. (Ants, bees, and wasps are, of course, very closely related, all being part of the order Hymenoptera, with termites forming the separate order Isoptera.) So, in a typical eusocial insect colony there will be a reproductive caste, which produces all the young, and one (or more) productive castes, which do everything else. In some species of ants and termites the productive castes may consist of both workers and soldiers, and even within these groups there may be specialized roles. In the Hymenoptera the productive caste are all sterile females, but among termites they may be male or female. The reproductive caste typically involves one dominant female (the queen) and one or more males. In the Hymenoptra the reproductive males' only functions are to breed and then die, whereas among termites they fulfill other important roles in the running of the colony.[8] The most important point is that the reproductive caste produces all other colony members, which means it's not really a colony, it's a family. It therefore makes perfect sense for the nonbreeding workers to devote themselves so fully to the care of their siblings. In fact, ants are genetically closer to one another than they would be to their own young, which would share only 50% of an individual parent's genes.

So we have a colony of eusocial insects, living and working together to form a kind of super-organism, whose individual components organize themselves into a collective, working for the common good. The results they achieve can be quite extraordinary. They build, expand, and maintain nests, sometimes complete with air-conditioning (see Chapter 5); they scout territory, communicating information on threats and food sources and then find the quickest route to and from the latter. In some cases they farm livestock and fungus, they wage wars and conduct slave raids, and they solve sophisticated geometrical problems. But individually, termites, ants, and bees are no smarter than a cockroach. So how do they achieve this remarkable level of organization and cooperation? Are there insect planners, supervisors, and middle managers? Of course not. As a general rule, insects don't know what they're doing. They just react to circumstances in instinctive ways that have evolved to be effective. For the eusocial insects the key to this phenomenon is communication, driving hard-wired, instinctive behavior. We shall discuss animal communication in Chapter 12 but for the moment we just need to note that, in the eusocial insects, most communication seems to be achieved via chemical signals.

In a typical colony—whether ant, termite, bee, or wasp—the queen does not spend her time issuing detailed orders. Instead she keeps busy laying eggs while every colonist, including her, reacts to a series of chemical signals from the eggs, the young, one another, the environment, outsiders, food sources, and so on. Based on these input signals, each colonist then reacts according to a set of simple behavioral rules encoded in its genes. There is no central decision making. There isn't even local control. There is just an incoming signal and a hard-wired response, with no requirement for intelligence and in most situations, little requirement for memory.

And yet these colonies seem to display all manner of complex behavior. So how do they do it? Can it really just be down to simple interactions between lots of individuals? Let's take a look at what they achieve and see if it is possible to work out how they do it.

Ants are foragers. When you see a solitary ant ambling across your kitchen floor she (worker ants are always female) is looking for something (food, water, or building materials) that is of use to the colony. Actually that last sentence suggests a little too much deliberation on the part of the ant. She, like the solitary cockroach we admired at the start of this chapter, is just wandering at random until something interesting turns up. If she finds something good, such as your sugar bowl, she will pick up a sample and start heading home, but now, as she walks, she will secrete pheromones from her abdomen. These pheromones encode information on what she's found, including its quality and quantity. (How ants find home depends on the species but may involve contributions from one or more of an internal pedometer, the position of the sun, the polarization of sunlight, recognition of local landmarks, and even Earth's magnetic field. In fact some desert ants have such good on-board navigation that they go straight back to base.[9] We discuss much more about ant and other animal navigation in Chapter 11.) However she does it, she will return to the nest, having marked out a trail back to the food source.

This trail will fade with time, which prevents the area around the nest being so overwritten with trails that one cannot be distinguished from another. More importantly, it means that any of her wandering sisters who stumble across a pheromone trail can follow the fading signal gradient until they, too, find your sugar bowl. They then pick some up and head for home, laying a pheromone trail as they go. Once they get back to the nest, they drop the food in piles and follow the pheromone trails back for more. The more ants that get back from the food source, the stronger and higher quality the pheromone trail. Being a scent trail, it has a small but finite width, which means overlapping trails tend to merge together and become straighter. And because individual ants don't follow these trails with perfect precision—they always have a tendency to wander a bit—they will, over time, create shortcuts in the meandering path. Thus, as more and more ants drop more and more pheromones, the trail between nest and food gets smoothed and refined into the shortest possible path. Once the food source is exhausted, the ants that get there will find nothing to carry back to the nest, and so the pheromone trail will fade (Figure 42).

What the ant colony achieves by this activity is to first find food and then to establish the most efficient path between the nest and the food. But by examining what the individual ants actually do, we find that we can achieve this result with five simple rules:[10]

1. Wander randomly, avoiding obstacles. Ants don't bang their heads against walls.
2. If you sense pheromones nearby, favor a direction of movement that tends to take you toward them. (You don't have to head straight for the pheromones, just show a preference for going that way.)
3. If you're holding food, head back to base and drop pheromones as you go.
4. If you find food and you're not already holding some, pick it up.
5. If you're holding food and you're standing next to a pile of other food, drop the food you're holding. (This rule ensures that food returned to the nest gets stockpiled and not just dumped in the corridors.)

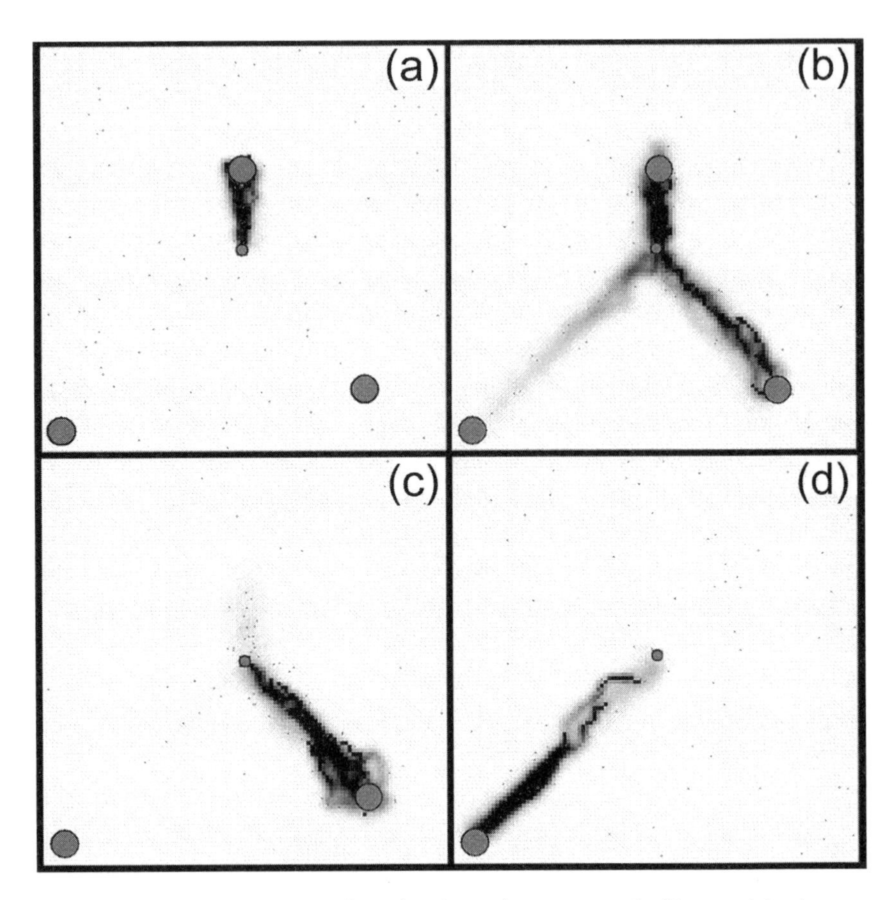

FIGURE 42 A time sequence of simulated ant pheromone trails. The nest is in the center; three food sources are located in different directions and at different distances. (a) The closest food source is exploited first. (b–d) Once the nearest source is exhausted, the pheromone trails leading to it will fade away as the ants favor the other sources.

These rules can be captured in a computer simulation that shows that, although an individual forager may have little chance of finding food and getting it back to the nest, if you send out 20 or 200 or 2,000 foragers, you will eventually find any accessible food source within the range of your scouts. (The ant's theoretical range is determined by how long it can go without starving to death. In practice, it's more likely to be eaten by something before this happens.) As the successful foragers return, you then evolve the most efficient route to that food source. If that route is then blocked for some reason, the pheromone-following ants will revert to random wandering until they pick up the trail again, and the process will continue with a new efficient route being created. Ant colonies can maintain several different routes to several different food sources at the same time, with each route being the best possible at any given moment. That the simple rules given above should lead to that result is not intuitively obvious on first inspection. Hence, the behavior is emergent. Of equal importance is that the rules allow the ants to adapt to changing circumstances. Hence, it is adaptive behavior. It's emergent and adaptive: complexity from simplicity. Clever, isn't it?

All eusocial insects forage in much the same way, except bees and eusocial wasps; they fly and so, instead of laying down pheromone trails, they dance. Successful foraging bees return to the hive and then do their famous waggle dance to show their fellow workers the direction, distance, and quality of a food source, usually nectar (more on this in Chapter 12). Other bees then follow the scout back to the source; the more attractive the location, as promised by the dance, the more bees will follow the scout. Chemical signals may also be involved, with the bees picking up the nectar scent from the dancer and using that for terminal guidance as they close in on the source.

The computer simulation we outlined earlier could be extended to study other aspects of these creatures' behavior. For example, in the nest ants are constantly sorting through all eggs, larvae, and food to arrange them in chambers by categories. Thus, when an egg hatches, the larva gets moved to the nursery with the other larvae. This behavior is just a variation on foraging: here, the ant encounters a larva surrounded by eggs and so picks up the larva and wanders through the nest until it finds a place where there are other larvae. It then drops its charge and wanders off. In the same way, if it finds a dead ant, it will carry it until it finds more dead ants. As a result, ant colonies routinely dispose of their own dead by dumping the bodies in a spot that represents the maximum distance from all colony entrances. Modify these sorting and moving rules, apply them to termites, add a modifier for carbon dioxide concentrations, and we can understand how Chapter 5's termite mounds got built. The behavior of bees and wasps can be analyzed in a similar way.[11]

There are many other examples we could cite of the apparently complex, yet essentially simple, behavior of eusocial insects, from the ways in which bees build hexagonal honeycombs to the evolution of their caste system. Careful study of these behaviors often leads to novel engineering solutions to practical human problems; in some instances biological solutions are better than those obtained by more conventional mathematics. Your car's satellite navigation system probably uses an ACO algorithm, where "ACO" stands for "ant colony optimization." Studying ants helps us find our own way home.

So it would appear that having very limited individual processing power is no real barrier to the development of apparently complex behavior, at least not if there are enough of you and you can cope with a solution that may be inefficient in terms of the individual—who may expend much energy achieving very little—but is efficient in terms of the colony.

Ant Bridge

Many ant species are remarkably adept at other aspects of civil engineering, especially bridge building. When faced with a river (or rather, given their size, a small stream), they cross it by creating a bridge from living ants.[12] With no central decision making and no forward planning, everything we know of ants suggests this behavior must be yet another example of simple rules interacting. So we built a simple simulation to examine it. The ant rules are:

1. As ever, avoid obstacles.
2. Once again, wander randomly but show a preference for following pheromone trails.
3. If you encounter an edge, stop with half your body over the edge. Wait for a period of time. If nothing happens go back to moving.
4. If you encounter a stationary ant, climb on top of it and keep moving forward.
5. If there's another ant hanging off you, don't move, but emit a pheromone call for help.

The results are shown in Figure 43, where we can see that ants can indeed build bridges. The tree-dwelling weaver ants of the African, Australian, and southeastern Asian forests go one better: they use much the same algorithms to bridge high-altitude gaps between branches and trees.

Bird-Brained Genius

But what of bigger animals with more processing power—surely they must be following much more complex rules? Birds, as a rule, are bigger than insects[13] and are warm-blooded as well, which means they can afford to install a more

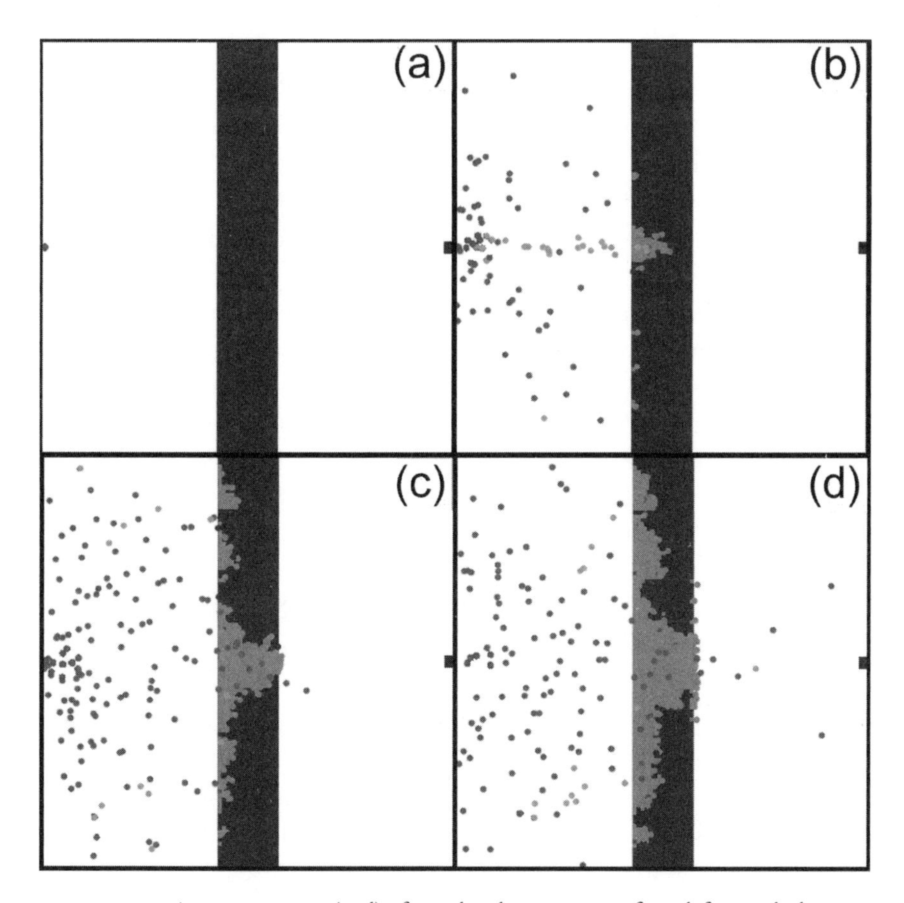

FIGURE 43 A time sequence (a–d) of simulated ants moving from left to right by bridging a stream. The ants enter from the left and are drawn to the small rectangular food source on the right hand edge. The black strip is the stream; gray dots are the ants. Our simplistic cyber-ants tend to maintain their false starts: real ones usually produce a single bridge that is as narrow as the water flow allows.

complex, energy-hungry brain and invest in more sophisticated control algorithms. Consider, for instance, the European Starling (*Sturnus vulgaris*). This bird is common throughout Europe, western Asia, and North America, and most people, especially city dwellers, would regard it as little more than a nuisance. Starlings are gregarious—they like to be with other starlings. In fall and winter they like to be with lots of other starlings, forming vast flocks as they head for their evening roost. Flocks of between 1 million and 1.5 million birds have been reported from many parts of the world. Starling flocks are tightly bounded and fast-moving formations, expanding, contracting, and changing shape and direction in an instant. All without filing a flight plan or holding a rehearsal and all achieved without the benefit of air-traffic control or instantaneous radio communication. How do they do it? How do they know which way to turn? Why don't they ever collide with one another or with trees? Why aren't the evening skies filled with a shimmering rain of stunned birds tumbling from the sky?

By now you've probably got a pretty good idea what the answer is going to be. As for computer models of cockroach behavior, researchers spent years trying to simulate bird flocking on a computer but found it too difficult until someone had the good sense to actually look at what the birds were doing and realized they were following just three simple rules:

1. Separation—avoid crowding neighbors by maintaining a minimum separation from them.
2. Alignment—steer toward the average heading of neighbors.
3. Cohesion—steer toward the average position of neighbors.

Add in a tendency to avoid flying into fixed obstacles, and there you have it.[14] No leaders and no all-seeing, all-knowing control function. Just three simple rules applied in the right way can reproduce the most important aspects of flocking behavior. An individual bird doesn't need to know that it's a member of a million-strong flock, nor does it need to know where the center of that flock is or even where the whole flock is heading. Instead it only needs to be aware of the positions and movements of its immediate neighbors, and because every member is responding to local conditions, the flock as a whole displays impressive large-scale coordination. Each bird need only be aware of what's happening in a small, local region of space, defined by a distance from its current position and an angle from its direction of flight. Neighbors outside this region don't influence its steering and so can be ignored. The nuts and bolts are shown in Figure 44.

Perhaps we shouldn't be surprised that the flocking algorithm is so straightforward. Why waste energy and resources on a more complicated algorithm when you don't need to? Nature seeks simplicity. By implementing these simple rules in computer software, it is possible to produce impressive simulations of animal flocking behavior. As always, of course, the devil is in the detail. How close is too close? How far apart is uncomfortably far? If a bird wants to get closer to another flock member, how quickly does it turn toward it? Does it constantly change speed to maintain its ideal position or does it try to stick to a steady, energy-saving rate? How many neighboring birds does it take into account when deciding what to do? And so on. The rules are simple, but implementation of the algorithm involves many parameters whose values may change with circumstances, with individual birds, and across species. The basic structure and behavior of a flock will be determined by these parameters. Indeed, many experienced birdwatchers have learned to identify bird species purely from the shape of their flock—at least if the flock is large enough. Flocks occur throughout the natural world, with other names, such as herds, shoals, schools, swarms, or in humans, crowds. The names may change, but the basic rules remain the same.

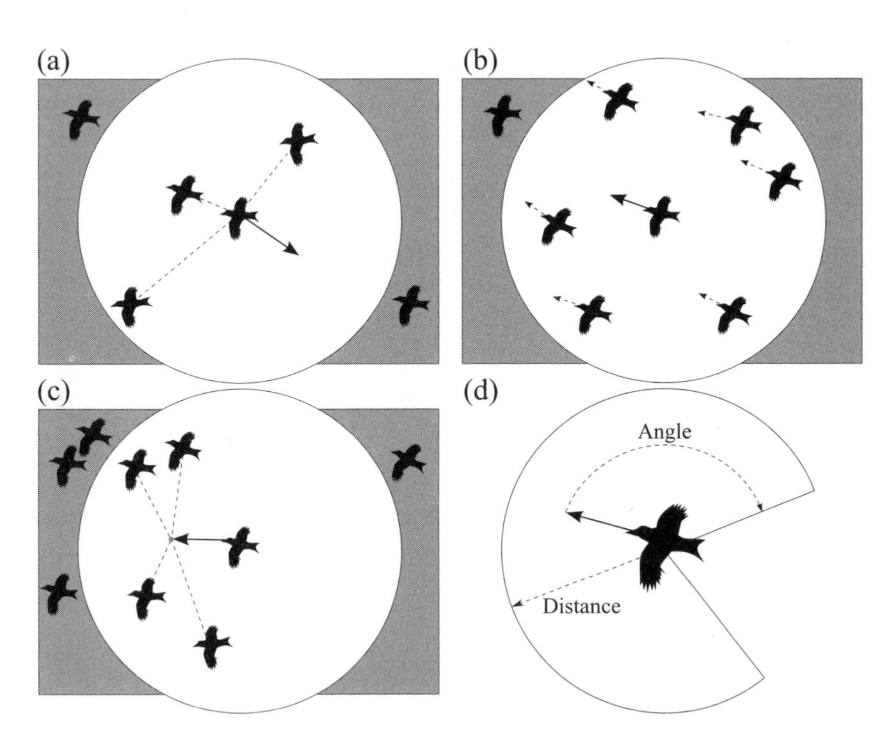

FIGURE 44 The basic rules of flocking: (a) separation, (b) alignment, and (c) cohesion relative to neighbors in a localized perception space (d) defined by an angle and distance within which neighbors influence steering. Watch a group of friends moving through a crowded mall. It's the same process.

We can easily understand that individual animals will form aggregations based on the location and availability of food, shelter, and mates, but what are the benefits to the individual animals of coagulating into a vast cohesive flock? The most obvious benefit is protection. A solitary animal pursued by a predator will either be eaten or escape; if it flocks with others, there's a third possibility—someone else gets eaten instead. Statistically, the individual has a much better chance of survival if it shares its risk with others. In many cases, the one that gets eaten is the one that has been separated from the flock. Forming a flock has other advantages in the arms race between predator and prey: a cohesive flock is what the military would call a "target-rich environment," and this can confuse a predator, which simply can't decide which individual to go for. Furthermore, a flock benefits from a distributed sensor system with massively parallel data processing. Or, to put it another way, it has a lot of eyes, ears, and noses; when they detect something of interest, they react instinctively, and the rest of the flock follows. This collective action gives the flock tremendous advantages in spotting both threats and food.[15] It can also provide protection against the cold. For example, in the depths of the Antarctic winter, Emperor Penguins (*Aptenodytes forsteri*) incubating their eggs on the bare ice will form a compact huddle with each bird leaning forward onto a neighbor. The birds on the outside of the huddle, where it is colder, want to move to the inside, where it is warmer. Their gentle but persistent shuffling for position produces a slow churning of the flock, which sees each bird moving from outside to inside and back again. As a result, everyone stays warm enough to ensure that they, and their eggs, survive.

So, if flocks offer all these amazing benefits, and you don't have to be especially smart to engage in flocking behavior, why don't all animals form flocks? The most obvious disadvantage is that if a large flock finds a small food supply, not everyone gets to eat. (The number of individuals in a flock may itself be determined by the interactions of a series of simple rules that balance the advantages and disadvantages of different flock sizes.) Another down-side to flocking is that some predatory species have learned to take advantage of the flocking behavior of their prey by herding them into positions where they can be more easily caught. In most cases the actual capture involves breaking the integrity of the flock to isolate the weaker, slower individuals. Once separated, they become much more vulnerable. However, species as varied as cetaceans and sea birds can use a variety of—often cooperative—techniques to herd prey like small fish or krill into clusters that are so tightly packed that the predators simply cannot miss. And for a tiny fish faced by a predator the size of a humpback whale, sticking really close to your buddies isn't going to be much help at all.

Animal species that flock are ones that benefit from flocking behavior; for them, advantages outweigh disadvantages. Most flocking creatures are ones preyed on by something else. Starlings flock; sparrowhawks don't. Predators that form flocks, like those shown in Figure 45, do so for reasons other than protection (most probably enhanced sensory capability and more effective hunting). So, sheep form flocks, but wolves, well—wolves are a little different.

Like a Wolf on the Fold

Wolves are carnivorous mammals. As such, they are generally opportunists when it comes to dinner time, eating anything they can, from a beetle to a bison. Given that an adult generally needs somewhere between 3 and 10 pounds of meat a day, it makes sense for a wolf to go after prey at the bigger end of the scale—a larger animal costs more energy to catch but provides much more energy if it is caught (recall Chapter 1). You need to chase and eat a lot of rabbits to get as much food as a single deer will provide. But what

FIGURE 45 Cape (or Painted) Hunting Dogs. Highly socialized and mutually supporting, they may be the most successful pack predator. More than 75% of hunts end with a kill; North American wolves manage about 25%. We thank Anita McFadzean for this image.

ultimately matters to the wolf is the ratio of energy expended to energy gained, which makes that large energy-packed herbivore a very attractive target. Of course, for a lone wolf to bring down a bison would be quite an achievement,[16] but wolves have developed a social solution to the problem—the wolf is a pack animal. It lives in packs, and it hunts in packs, and a pack is just a flock with teeth.

How wolves hunt seems to vary with what they are hunting. Consider cooperative hunting, where a pack rather than an individual is involved. But therein lies a problem: when hunting, how does each individual wolf know what to do? Like the flocking of birds and the behavior of cockroaches, this problem puzzled scientists for a long time. Most suggestions involved rather implausible mechanisms in which wolves either engaged in some undetectable communication during the hunt or, equally improbably, had participated in a round-table briefing session beforehand. Whatever the truth was, it had to involve both the wolves and the prey only sensing what they could really sense, knowing what they could really know, and doing what they can really do.

Clues emerged from studying the animals. Like most predators, wolves don't preferentially eat the weak, the sick, or the young: they eat the ones they can catch. Statistically, those individuals tend to be weak, sick, or young, but the wolf doesn't need to be an expert in veterinary medicine to pick its target. It just needs to spook the flock and see who drops behind.[17]

Moose Chess

And now we come to another experiment. You will need six wolves, a reasonably healthy moose, and a large open space. Ideally find a space far from inquisitive eyes, as the results of your experiment may cause distress and disruption in your community. Or substitute a chessboard, as in Figure 46, with half a dozen pawns as the wolves and a knight as the moose.

The rules are simple. You can only have one animal per square. The moose can move one square, the wolves can move up to two in one turn. In each turn the moose moves to the square that is farthest away from the nearest wolf. If it can't do this, it is trapped and stays put. Now the wolves have to do a bit of math. Each wolf moves to the neighboring square for which the Korf algorithm[18] gives the biggest number:

$$S = (\text{distance to moose}) - k(\text{distance to nearest neighbor wolf}).$$

What happens depends on the value of k, which determines how close the wolves will get to one another. Start with a value of $k = 0.5$ and six wolves, and see what happens. Try using fewer wolves and different values of k, and see what effect these changes have. The wolves are drawn to the moose but try to maintain a degree of separation from one another. Balance the forces of attraction and

FIGURE 46 Moose chess. This time, our money's on the wolves. . . .

repulsion, and they will eventually surround the moose, without any discussion or communication. Make the moose faster than them, and it will escape.

It's not much of a game, because the moose usually loses. In real life, of course, animal movements aren't constrained by regular grids, and there are lots of other factors that come into play, making the game more biased in the moose's favor. You could, for example, allow everyone's speed to vary, depending on the ground cover or slope of the terrain. You could allow for them to tire as the chase goes on or for individual wolves to become dispirited and give up, if their distance to the moose passes some threshold value. You could even allow the moose to choose a more sensible tactic than just running from the nearest wolf. (In reality, if the moose is healthy and stands its ground, the wolves will usually look for easier prey.[19] You may want to bear this in mind when out alone in the woods in winter.[20])

Wolfettes and Sheeplings

Once again, the rules of cooperative hunting lend themselves to fairly elementary computer simulation. Because real wolves are reasonably intelligent creatures (sporting a goodly number of neurons) with a range of sophisticated sensory systems (sight, smell, and hearing), creating a simulation of a whole wolf pack would involve guessing at too many unknowns, so we restricted ourselves to examining the cooperative hunting behavior of the little known Wolfette, a cyber-beast that preys on the equally obscure Sheepling.

The Sheepling is a docile herbivore that gathers in small flocks of 20–30 creatures to gently graze the inside of the computer screen. With eyes on the sides of their virtual heads, they have a wide field of vision but indifferent eyesight. They exhibit all the behaviors we expect from flocks, and an individual tends to ignore threats like the Wolfette until they get too close. Then the Sheepling will simply run away until it no longer feels threatened. Flocking behavior encourages other Sheeplings to tag along, even if they haven't seen the threat for themselves. Surviving on low-nutritional pixels, Sheeplings must spend most of their time eating, and so, when a Wolfette is out of sight or far away, they ignore it and revert to grazing.

Wolfettes, for their part, live in small packs of 6–12 individuals, and in the absence of visible prey, they exhibit a loose flocking behavior not unlike that of the Sheepling. However, when they spot a Sheepling that is within a critical attack distance, they will head toward it, and their loose flocking behavior will tend to turn the whole pack toward the Sheepling. Blessed with forward-facing predator's eyes, the Wolfette compensates for its more limited field of vision by the simple expedient of moving its head from side to side when hunting. As it closes on prey, it stops scanning and its visual field shrinks. The Wolfette's keen eyes can also identify a Sheepling at greater range than the Sheepling can identify the Wolfette. If an individual Wolfette can see both Sheeplings and other Wolfettes, it will, deep in its simple brain, execute a revised version of the Korf algorithm (see the previous subsection), aiming to close on the nearest sheep while maintaining separation from its packmates. The repulsion factor in the Korf algorithm decreases as the Wolfette gets closer to a Sheepling, meaning that Wolfettes start their attack by spreading out to surround the prey and then close in for the kill. Wolfettes are no brighter than Sheeplings in that they forget about potential prey and packmates as soon as they are out of sight. In other words, they only react to what is currently in their field of view. Unlike most real predators, they also lack the ability to predict the future movements of other creatures—they react to where things are now, not where they might be heading in the next few seconds. Unlike real wolves, for example, they will never attempt to intercept their prey, they just see and chase.

The creatures in our pack and in our flock are all individuals, in that they each have slightly different characteristics. Most importantly, some of them will tend to be faster, and some slower, than the others. We also allowed for the occasional chance that a Sheepling, when threatened by Wolfettes, would not run but would stand its ground defiantly. So sometimes the flock runs and the pack tries to catch one, and sometimes, as the flock runs, one tougher Sheepling turns at bay, and the pack has to gang up on it. The Wolfette algorithm remains the same:

1. Form a loose "flock," so no one gets lost and you have more eyes looking for lunch.
2. If you see a Sheepling, head toward it.
3. If you see a Sheepling and other Wolfettes, execute a Korf algorithm.
4. If you contact a Sheepling, try to bring it down.

Having fabricated our virtual sandpit, we set our creations free to play (or, more accurately, kill). The resultant behavior depends on what the Sheeplings do. For example, most Sheeplings succumb to a single Wolfette, but the larger, more dangerous, ones have to be surrounded and confused just like a bull moose. Using the same basic rules, the Wolfettes' behavior adapts to different circumstances. What we observe does not really reproduce the exact behavior of any actual species, but it does demonstrate the central point that simple rules can interact to produce apparently complex and adaptive behavior. A time-lapsed image of a typical simulation run is shown in Figure 47.

Once again, each individual in the predator-prey system both influences and is influenced by the entire system. A behavior emerges from the interplay of simple rules governing attraction and repulsion, and the hunt proceeds with no need for either communication or pregame tactical conferences.[21] Nature in the raw is simple at heart.

Problems can arise when relying on simple rules that have evolved to meet special circumstances: in a human-dominated world, those circumstances can

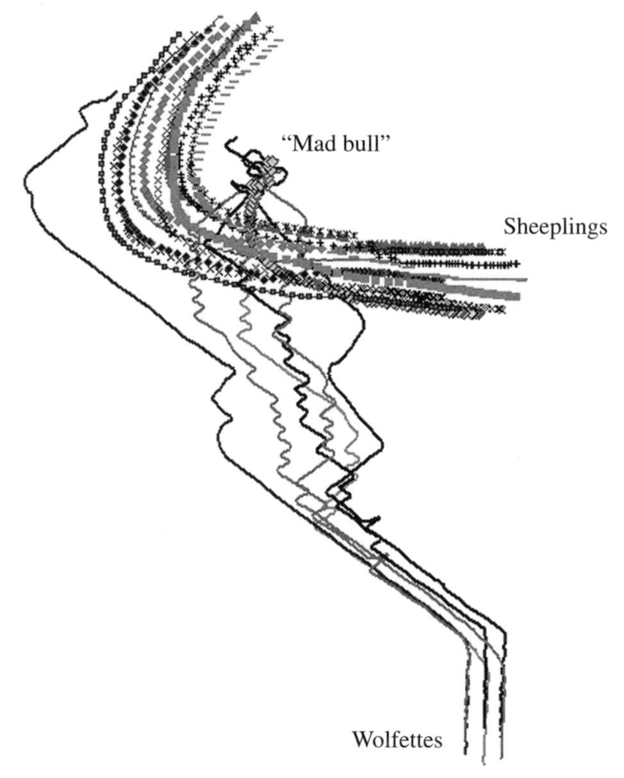

FIGURE 47 A typical Wolfette hunt, showing the paths taken by all involved. Note how the Wolfettes spread out once they start to close on the flock. The slower "mad bull" Sheepling turns at bay but proves too weak to face down his pursuers.

change too fast. Thus, flocking is a poor defense against gun-armed humans, as the Passenger Pigeon never got a chance to learn.

All Things Have a Pattern

So far we have looked only at how apparently complex animal behavior can be driven by remarkably simple rules. A final thought: can we extend this search for simplicity from preprogrammed behavior to other characteristics of animals? Can equally simple rules be used to explain their appearance?

How did the leopard get its spots, or the zebra get its stripes? Why do some spotted animals have striped tails, but striped animals never have spotted tails? How many different ways are there for a snail to develop its shell? Can any of these problems be understood as an emergent property of a simple rule set? Well, yes.

In 1952 mathematician Alan Turing proposed that patterns in nature arise due to reactions between certain chemicals, dubbed "morphogens."[22] Over short distances, these reactions promote one another, so their chemical products form concentrations; over longer distances, the reactions inhibit one another, so the chemical products are scarce. The overall effect is that individual small disturbances grow larger but create spaces between themselves. Given the right geometry and conditions, if the chemical product is a skin pigment, this effect could produce spots, stripes, and bars. Later work showed that the markings on a spotted animal's tail would always change from spots at the thick base to stripes at the thin tip, as shown in Figure 48. You just can't get thin spotted tails. Related work has looked at stripes on creatures as varied as zebras and fish. The mathematics is certainly compelling.[23]

And the snail shells? Well, we can grasp the math—the shells are almost always based on a logarithmic spiral, so it's a constant bearing thing again, as we saw in Chapter 5.[24] We're just not entirely sure why (although algorithmic simplicity might have something to do with it), and that's the weakness here—these are studies of the underlying mathematics. Knowing that nature can be described mathematically is not the same as proving that it uses that mathematics. Pinning down the chemical and biological processes is proving a little harder than describing them. But that doesn't make the math any less valid—we knew the math of planetary motions long before Einstein invoked curved space-time to explain *why* they moved that way. In our case, the math illuminates but doesn't quite explain.

FIGURE 48 Constrained by mathematics, the leopard is forced to change his spots into bars. (a, c) This animal is actually a jaguar. Note how the markings on the lower legs also become more bar-like. (b) For this leopard, the tail markings merge into bar-like blotches rather than spots. We thank Anita McFadzean for these images.

Emergent, self-organizing behavior is seen everywhere in the natural world. Should this really surprise us? Perhaps not. The closer we look at life, the more we discover that everywhere, from molecules to flocks, the individual rules of behavior are really quite simple. It's just that, even in the most basic creatures, there are a lot of rules with many parameters, and they are all interacting constantly. And that's what makes life complex.

PART TWO

Remote Sensing

7

A Chemical Universe

Human beings are highly visual animals. We experience the world through our eyes. Sight so dominates our lives that visual metaphors dominate our language—if you see what we mean. If not, we'll cast more light by stating it bluntly: humans *see* the world around us.

Other animals are different; they rely primarily on a chemical sense that we call smell. They need to find the food chemicals that provide the energy and raw materials required for survival and growth; they need to identify and avoid those chemicals that represent threats to their survival, in the form of predators or poisons; they need to trace the source of those chemicals that promise an opportunity to reproduce. It should not surprise us, therefore, that chemical senses were the first to evolve and, for many creatures, are still the most important. Although there are animals that are naturally blind and animals that are naturally deaf, there are no known animals that lack some form of chemical sense.[1] That's how important smell is, which is why it seems so odd that, in the field of chemical sensing, human beings really are quite inept.

Things to Smell and Do

About 3.5 billion years ago life developed in the oceans.[2] These early life forms needed a simple means to detect chemicals dissolved in that water—"in solution," as science puts it. We see this process at work today in simple organisms, like bacteria. Despite being single-celled creatures with limited sensory capabilities, bacteria display a behavior known as *chemotaxis,* which allows them to move toward food and away from toxins. Arranged around the outer membrane of its single cell, a bacterium has a number of different types of receptors in the form of protein molecules that can bond with other chemicals in the external environment. Basically, it's like a series of locks and keys—a specific chemical key fits in a specific receptor lock and that sends a specific signal to the bacterium's flagellum, a hair-like spinning propeller mounted outside the membrane, by which the bacterium can move. Triggering a receptor tied to a chemical signal associated with food—usually in the form of sugar-like molecules—drives the bacterium toward the source; triggering one associated with toxins pushes it the other way. Bacteria can even sense chemical gradients, propelling themselves toward or away from the source of the chemical signal. For a single-celled creature with no obvious means of simultaneously comparing the chemical concentration at different parts of its body, this ability implies something akin to memory—the bacterium knows that the concentration here is different from what it was there and so keeps heading in the right direction.[3]

As we will see, this lock and key mechanism forms the basis of all animal chemical sensing. From the simple bacterium to the hypersensitive nose of a dog, the underlying physical principles are the same, it's just that Fido has a lot more receptors and he does a lot more with the chemical signal.

Taste the Wind

We smell odors carried in the air and we taste flavors in our mouths, right? Wrong: we don't really taste our food, we smell it. The whole culinary experience is confused by the fact that the mechanisms of tasting and smelling overlap, in both function (both systems may detect the same chemicals) and interpretation (smell and taste pathways converge in the same part of the brain). Consequently, when you put food in your mouth what you experience as "flavor" is a result of both smell and taste. In fact, most of food's flavor is actually its aroma, in the form of volatile molecules carried up through the nasal passages.[4] Some of these waft up directly via the nostrils before the food goes into your mouth, but others take a back-door route via the pharynx, which connects the back of the mouth to the nose. That's one reason why food has more flavor when it's hot: more molecules pass into a vapor state and so more pass from the mouth into the nasal passage to be detected as smell. Similarly, although to a lesser extent, when we smell certain odors we may be tasting them as well, because vapors entering the nostrils go down the pharynx into the mouth and stimulate the taste buds.[5]

Because of this merging of the experiences, scientists don't usually talk about smell or taste, but refer to *olfaction* and *gustation*. Gustation happens in the mouth and is a contact sense, whereas olfaction happens in the nose and is a ranged sense, which detects airborne chemicals from more distant sources. Right? Well, not really. Examples: fish smell under water, some taste with their outer skin, and invertebrates don't even have noses. Take a look at Figure 49. Do they all taste or smell the same?

A more useful definition is based on purpose, because olfaction and gustation play very different roles. Gustation is all about food—is it good to eat or is it poisonous? Olfaction is all about sampling the external environment. (What's out there? Threats, opportunities, friends, enemies, or potential mates?) A salmon can smell out the river of its birth. A dog can smell cancer.[6] For vertebrates the mechanisms of the two senses are quite different: gustation happens at the taste buds, but olfaction is the product of the olfactory receptor neurons (ORNs), which are both neurons (as discussed in Chapter 4) and sensory receptors. In fact, they are the only sensory neurons that receive direct stimuli from the outside world—presumably a throwback to their origins in those earliest single-celled creatures. All other senses use intermediary cells or chemical signals to provide a link between the external detectors and the actual

FIGURE 49 (a) Snail, (b) lobster, (c) shark, and (d) otter: how do they really taste? We thank Anita McFadzean for these images.

receptor neurons that process the information. This includes the sense of gustation—vertebrate taste buds aren't neurons but just specially adapted skin cells that can generate electric pulses, causing their associated neurons to fire and relay the taste signals to the brain. Here, for vertebrates at least, we have a physical difference between olfaction (direct stimulus of the receptor neurons) and gustation (skin cells react to chemicals, triggering a neurological response farther down the pipeline).

Because invertebrate neurons follow the same general design as vertebrate ones, we might expect the basic olfactory mechanisms to be broadly similar. This is almost true, in that there is a receptor neuron whose axon communicates directly with the nervous system.[7] The problem lies with gustation: in invertebrates the gustation sensor cells are also neurons, communicating directly with the nervous system. It's really only in the arthropods—insects, spiders, crustaceans, and the like—that we can clearly distinguish different mechanisms for olfaction and gustation, and even so they are more alike than they are in vertebrates. So this ORN versus Taste Bud definition really only works for vertebrates. Because olfaction and gustation mean different things for different types of animal, and because these words are unfamiliar, we here refer to smell and taste, and only occasionally to olfaction and gustation. Just remember that when we say "taste," we mean the chemicals to which your taste buds respond. We follow convention here and call these chemicals "tastants." Similarly the chemicals that tickle your ORNs, which you experience as odors, we call "odorants."

We know that different animals experience taste and smell differently. What repels one type of creature may attract another. Even in the same species the experience differs among individuals—we don't all like the same things to the same extent.[8] In part this is because we are being driven toward the things we need, which, among individuals, will vary with time and circumstances. The smell of home cooking is less enticing after a heavy meal; if you're suffering from a stomach bug, the chances are that any food-related smell will repel you. Satiated or not, we are hopelessly drawn to the fuel-cells of life—sweet, high-energy foods—so you can always manage dessert. There may be other advantages to this preference for different tastes: if everything craved identical tastes, all life would be competing for the same foods.

The major chemical sensors for typical vertebrates are identified in Figure 50. The chemical senses detect things that actually exist at the time and place of the detection. Sight and hearing can be deceived by reflections and echoes, but a smell or a taste indicates chemicals that are physically present here and now. This may mean the source is here, or was here, or that there is an odor trail leading to the source; only a little further investigation is required to figure out which is the case. As a result, chemical senses are good for detecting sources, because the rate of false alarms is low.

Remember that sensory neurons, like all neurons, are essentially digital: the action potential is either on or off. The strength of a sensation is determined by the rate at which action potentials are generated (greater for stronger stimuli). The downside is that most neurons become less sensitive when repeatedly stimulated, as we saw in Chapter 4. This is why we often stop noticing smells when exposed to them for a long time. We "get used to it," which simply means our ORNs are flagging.

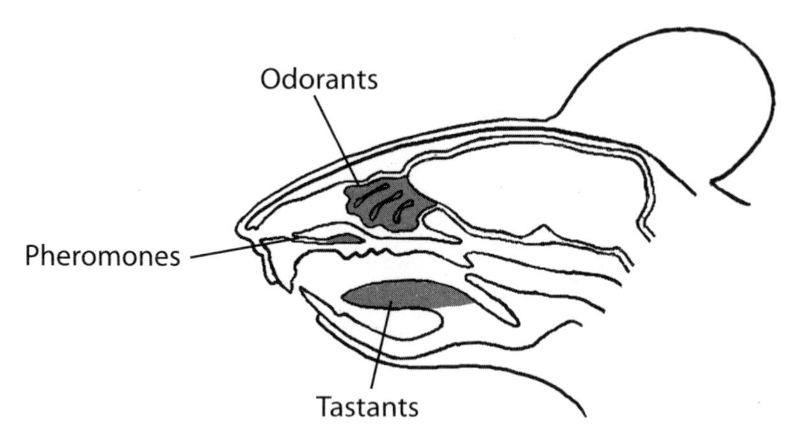

FIGURE 50 Chemical sensors (shaded areas) in a vertebrate head. A rodent is shown, but the principles are much the same for all vertebrates. Shown are the olfactory epithelium (where it smells odorants), the taste buds (where it tastes tastants), and the vomeronasal organ (where, among other things, it detects pheromones). Some vertebrates, especially among the fish, also have external taste buds.

Smell Never Sleeps

It might seem strange to say that smell is the king of the senses, because we humans seem rather useless at it, and it plays a remarkably minor role in our lives. However (and at the risk of generalizing, because the animal kingdom is a pretty big place) for most other animals, smell is their most important sense, used to explore the world and to communicate with other creatures. It works in the dark, at a distance, and requires only the most basic of hardware with no moving parts. In science-speak, olfaction simply involves a receptor protein at a cell boundary chemically binding with an external molecule with which it has an affinity, and then generating an action potential in a neuron. Stated more simply, we have a key that fits a specific lock and triggers a response, as for bacteria.[9]

In the vertebrates, smell *is* what happens in the nose (more accurately, in the nasal cavity). Showing its primitive origins, the olfactory system detects chemicals that are in solution (i.e., dissolved in water). For aquatic animals those chemicals are already dissolved in the water in which the creature lives. For air breathers things are a little more difficult—the molecules to be detected must first be volatile, so that they can be breathed in with the air, but they then also have to be dissolved in liquid before the olfactory receptors can bind with them. And that's one reason why air-breathing vertebrates tend to have runny noses: it's the mucus in the nasal passages that dissolves the odorant chemicals and prepares them for detection. (It has other functions, such as filtering out microbes and other debris, which is why fish also have nasal mucus).

In the nasal cavity, ORNs cluster on the olfactory epithelium, a fancy term for a patch of skin covered with these neurons. Their dendrites stick out into the nasal cavity, usually in the upper part, and these dendrites are covered with large numbers of tiny, hair-like cilia projecting into the olfactory epithelium's mucus covering. These cilia are themselves covered with specialized proteins; these are the olfactory receptors that bind with the odorants. The binding of the odorant to the receptor generates an action potential in the ORN, although the exact details depend on the type of organism. Individual ORNs get replaced every couple of months from a layer of stem cells in the olfactory epithelium. (This is one of the few examples where an adult animal can regrow neurons.)

It seems that each ORN only contains a single type of odor receptor—although it has a lot of them—which only responds to a single chemical key. So each ORN has many locks, all of the same specific type and responding to one specific type of key. When fit with keys, these locks send signals through the dendrites to be integrated in the neuron, leading to a series of rapid firings of that ORN's action potential. In any given nose there will be a huge number of each type of ORN distributed across the olfactory epithelium. (Evidence suggests that they are not randomly distributed, but that ORNs of the same type tend to be grouped together. We emphasise the word "tend"—there is still a fair degree of mingling.) Different smells are thus the result of different odor molecules stimulating different types of odor receptors. Does this mean we need a set of ORNs specific to every possible chemical we might need to smell? Not quite. It seems that individual olfactory receptor proteins don't necessarily respond to a unique molecule, but rather to specific arrangements of atoms forming a substructure within it. Thus the unique keys that fit the specific locks are actually only parts of the whole odorant molecule. There are two consequences: different molecules can trigger the same receptor if they have the same key, and a single molecule may have different keys that fit the locks of different ORNs.[10]

On first glance the substructure response of smell receptors seems to complicate things, because it means that the firing of a specific type of ORN will not give us a unique identification of the molecule that triggered it. A single ORN firing once as a result of bonding with a specific key in that molecule is certainly a detection, but it doesn't really convey any useful information. Imagine a sensitive digital camera taking a picture in a darkened room. A single photon of light may trigger a response in a single pixel. That constitutes a detection, and it tells you there's light out there, but it's a long way from being a picture of the light source. For your nose to create an olfactory picture of the environment, you need a whole suite of ORNs firing in response to lots of molecules and you need some clever processing to make sense of it all.

We now know, for example, that mice have more than a thousand different types of ORN, whereas humans have fewer than half that, yet we can distinguish tens of thousands of different odors. It seems that having a large array of different ORN types, each of which may respond to several different molecules (and may overlap with other ORN types responding differently to those molecules), allows the combination of detections to identify a huge range of different molecules using a relatively small number of ORN types.[11]

Here's how it works: ORNs fire in reaction to their molecular keys, driving action potentials down their axons into a region of the vertebrate brain known as the olfactory bulb. In this bulb are a large number of structures called "glomeruli"—bundles of neurons. Each axon connects with a single glomerulus, but one glomerulus may receive inputs from many thousands of ORNs. These ORNs are all of the same type—they are sensitive to the same key—and so each glomerulus is also specific to a unique key. The olfactory bulb contains many examples of each key-specific glomerulus. Each glomerulus then connects with a number of other structures called "mitral cells," which pass signals into the olfactory cortex. Olfactory structure is illustrated in Figure 51.

Many signals from identical detectors are combined in the glomerulus, which suggests some kind of integration process. The use of multiple glomeruli handling the same type of signal suggests a distributed or parallel processing arrangement. These signals are passed to multiple mitral cells—which may simply indicate a degree of redundancy to protect against component failure—and are then sent for further processing in the olfactory cortex, where each cortical neuron ends up receiving multiple signals that originated in multiple ORNs of different types. What happens thereafter is still not exactly clear. It seems that the brain is making use of not just the number and rate of ORN detections, but also of the pattern of firing of different types of ORN

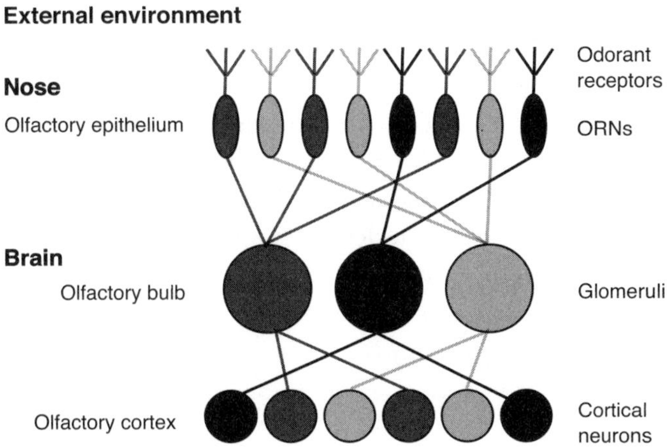

FIGURE 51 Olfactory pathways in vertebrates. Odorants from the outside enter the nose. On the olfactory epithelium, specific chemical keys in the odorants interact with specific olfactory receptors and trigger associated olfactory receptor neurons (ORNs). Signals from these converge on matching glomeruli in the olfactory bulb of the brain. From there they are passed deeper into the olfactory cortex, where each glomeruli may connect to several cortical neurons. The general structure for invertebrates is broadly similar.

responding to different keys on the same type of molecule. It's not just the signal strength that counts, but the structure within it.[12]

To illustrate with a simplified example, consider two different molecules, A and B. Molecule A has chemical keys that fit olfactory receptors R1, R2, and R3, whereas molecule B has keys tied to receptors R1, R2, and R4. Signals from each detector are fed to an associated glomerulus (G1, G2, G3, and G4) for further processing. Detecting signals in G1 and G2 does not allow us to unambiguously choose between the molecules, but if we also get a response from G4 and none from G3, then we know it is molecule B that is being detected. But what if both types of molecule are present in some quantity? In this case the ratios of the number of R1 to R3 responses and of R2 to R3 responses will be wrong for molecule A; similarly, the R1/R4 and R2/R4 response ratios will be wrong for molecule B. However, if the olfactory processing can distinguish these ratio differences, it may be able to recognize that both molecules are present and quantify their proportions. (Experiments show mixing odors can produce strange results: sometimes a new odor is detected, or the odors may be sensed successively, or one masks the other, or nothing is detected.) Now imagine a third molecule, C, which has keys for all four receptors R1, R2, R3, and R4. If we sniff a gas made up entirely of molecule C, will it have a different odor compared to one that is a mixture of A and B? In principle, it might, and once again, the answer lies in the ratios. Compared to the gas of molecule C, the mixture of molecules A and B produces a higher proportion of R1 and R2 responses relative to R3 and R4. Figure 52 illustrates some possible combinations of activated glomeruli.

In reality, differentiation of odorants won't be quite this straightforward. For one thing, a single molecule of any of our substances will bind with a single receptor, generating a single response. So if we only have two molecules

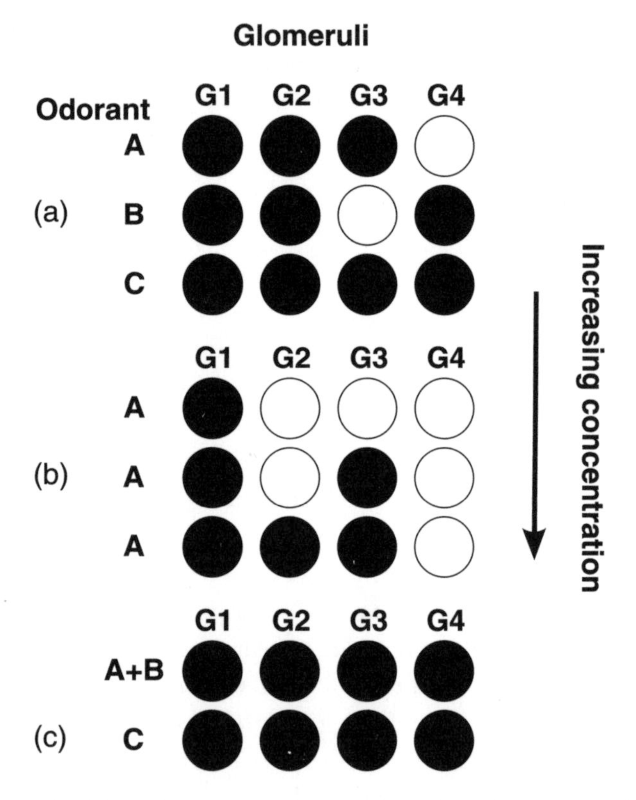

FIGURE 52 Simplified olfactory processing. Open circles indicate activated glomeruli; filled circles are inactive ones. (a) Different chemicals activate different glomeruli. (b) Increasing concentration causes more glomeruli to be activated. (c) A mixture of Molecules A and B gives the same response as Molecule C.

of substance A, we may get an R1 and an R2 response, or two R1s, or any other combination of R1, R2, and R3. For another, the system has to be capable of responding to a lot more than four chemical keys.

Does the smell experienced depend on molecular concentration? To an extent it must. A stronger concentration means more molecules at the olfactory epithelium, which means more ORNs being stimulated more often (until they get fatigued). But once we have a strong enough concentration to get those ORNs firing, the ability to detect and identify odors becomes less dependent on concentration, because the relative proportions of the different components stay constant even if the concentration alters. Experiments show this to be the case.[13] Although concentration (beyond a certain minimum) may not be important for detection, it does seem to play a role in determining how an animal responds to an odor: a hint of something cooking far away won't produce the same behavioral response as an overwhelming smell of dinner freshly served in the next room.

Information about the detected odor is sent to the olfactory system in the brain, where multiple signals are further processed to form a synthesized olfactory perception.[14] Some of these areas control conscious perception of the odor, but others have more intriguing associations: in the human brain, the amygdala is involved in emotional and autonomic responses to odor, whereas the hippocampus is involved in motivation and memory. A direct consequence of this association is the very human way in which scents can trigger sudden memories of past events.

For humans "smell never sleeps," because the ORNs connect directly to those parts of the human brain associated with memory and emotions. Stimuli from other senses pass through the thalamus, which is switched off during sleep.[15]

For most invertebrates, smell is their principal sense, and they have evolved a high degree of sensitivity and specificity in their equipment. Invertebrates don't have noses, but the basic olfactory mechanism is almost the same as for vertebrates. Recent work on insects suggests that the biochemical pathway by which the odorant triggers the neuron is different (and faster), but from a sensory engineering perspective, the effect and much of the subsequent processing is essentially the same. In insects, ORNs are usually found on the antennae, in the sensilla. An antenna may not look much like a nose, but it does the same job: volatile molecules enter tiny pores, are dissolved by a thin liquid layer of mucus, and are detected by ORNs. The axons from the ORNs converge in glomeruli in the antennal lobe, which is analogous to the olfactory lobe. From here the signals are passed to more sophisticated parts of the insect brain. Our understanding is a little vague, but it seems unlikely that insects are doing anything very different with the signals than are vertebrates. They just tend to have less processing power and simpler needs.[16]

Good and Bad Smellers

Exactly how many ORNs does one animal need? We know that humans typically have a few million ORNs, whereas dogs have perhaps a couple of hundred million.

The sensitivity of the sense of smell, and thus its importance to a creature's existence, varies enormously among different animals, and even between closely related animals. Birds, with few exceptions, have a poor sense of smell. This may be because they spend most of their time up in the air, where scent trails are least likely to linger.[17] Most mammals have a very good sense of smell, particularly carnivores and ungulates (deer, cows, sheep, and goats): carnivores have to find their prey, and ungulates need to avoid the carnivores that eat them. Such creatures as moles, which dwell underground and hunt in the dark, have very good senses of smell, whereas primates, who took to the trees early in their evolution, are generally unimpressive smellers. Most cetaceans (whales and dolphins) can't smell—with the exception of Bowhead Whales, they don't even have the equipment, the genes that should produce the ORNs being nonfunctional. They compensate with a good sense of taste, which illustrates the difficulty air-breathing creatures have in adapting the sense of smell to an underwater environment. Pinnipeds (seals, sea lions, and walrus), not being so highly adapted to the aquatic life, use smell in air, not water, and seem to be on a par with primates. The noses in Figure 53 may look the same, but they don't smell the same things. Amphibians and reptiles show the same variations between species, with the amphibians, like pinnipeds, favoring airborne olfaction. Unusually for air-breathing water-dwelling vertebrates, some turtle species can smell underwater. Curiously, the same turtles seem to have very few taste buds, mostly on the tip of the tongue. For them, olfaction is probably more important for open sea navigation and nest site selection. In general, fish have a very good sense of smell, which is hardly surprising for totally aquatic creatures. Intriguingly we see evolution at work in aquatic creatures that have moved to the land—some air-breathing land-dwelling crabs

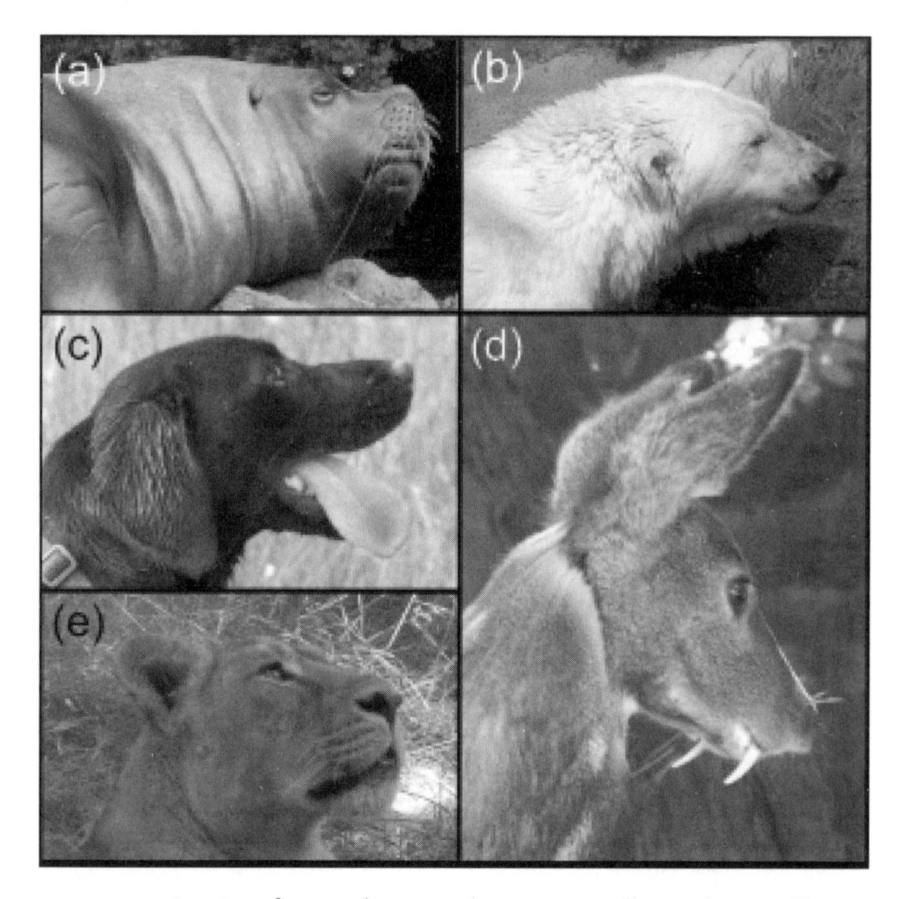

FIGURE 53 A variety of mammalian noses that may not smell quite the same. They are, in order of owner's affinity for water: (a) sea lion, (b) Polar Bear, (c) domestic dog, (d) musk deer, and (e) Asiatic lion. We thank Anita McFadzean for these images.

have a "terrestrial" olfactory system which is quite different from that of their oceanic cousins and seems to have evolved independently.[18]

How to Use a Nose

When humans encounter an interesting smell, we tend to explore it by taking a deep breath. This response seems sensible, because it will draw more of the odorant across the olfactory epithelium. But watch a dog that's found an interesting scent. It uses a series of shallow, rapid sniffs to suck odorants into its nasal cavity without whipping them down into its lungs, where they are no good to anyone. This strategy allows the odorants to linger in the nasal cavity, increasing the odorant concentration and improving the chances of identifying their source. What's more, this sniffing is interspersed with sudden, explosive snorts, which may serve to flush old odorants out of the nasal cavity, allowing a fresh sample to be taken. Other mammals display similar behavior, but dogs and cats have the advantage of the magnificently named subethmoidal shelf, a bony projection in the nasal cavity that serves to trap odor-rich air while it is being sampled.[19] Ungulates and rodents have these to a lesser extent; primates and humans barely at all. So if your dog can smell better than you, it's not just because it has more ORNs, but also because it benefits from some cunning structural adaptations in the skull.

A Spare Nose?

Pheromones are a special type of chemical dear to the hearts of many animals. They share many similarities with hormones: the difference is that an animal's hormones affect itself, whereas its pheromones affect other members of its species. Pheromones trigger behavioral responses in a more immediate, hardwired way than mere scent detection. Usually these are connected to sexual or social behavior.[20]

All animals (including humans) employ pheromones, although any commercial product claiming to use "human pheromones" to make you sexually irresistible is, perhaps literally, hogwash. We do know of at least one human pheromonal effect: women living in close proximity synchronize their menstrual cycles. The effect can be reproduced by taking swabs from one woman's armpit and transferring the chemicals to another's upper lip. And another experiment for you to try at home: teenage boys won't sit in seats that have been rubbed with male sweat. ("Who would?" you might ask, but, in part, this aversion may be due to human pheromones at work.)

In many insects the olfactory system is highly tuned to pheromone detection, because pheromones are their primary mechanism for communication among individuals. Vertebrates go even further—some of them have a whole secondary olfactory system that seems to be heavily involved with pheromone detection. These creatures are tetrapods (amphibians, reptiles, birds, and mammals). In many tetrapods, part of the olfactory epithelium has evolved to form the *vomeronasal organ* (VNO), or *Jacobson's organ.* The exact form, structure, and location of this organ varies considerably among species.[21] The receptor neurons in the VNO are superficially similar to ORNs, except that the receptor-laden cilia are replaced by distinct structures called "microvilli" and the electrochemical transduction mechanisms are a little different. These vomeronasal receptor neurons send signals down axons that terminate in the accessory olfactory bulb of the brain that, in turn, projects to areas of the brain different from those served by the main olfactory bulb. All of which suggests that the VNO does a different job.

However, it is a mistake to assume that the VNO is just a pheromone detection system. For example, the flickering tongue of a snake is actually sampling air-carried molecules and delivering them via the mouth to the snake's VNO; the molecules it is detecting have more to do with finding prey than finding a mate. Other reptiles, and even opossums, also use the VNO for food detection. Conversely, we know that some vertebrate pheromones are detected in the primary olfactory system.

If the VNO is not a dedicated pheromone detector, what is it for? One suggestion is that it handles unlearned responses: the VNO responds to chemicals to which an individual has not yet learned to react. Through experience those chemicals then become associated with the primary olfactory neurons. This idea is intriguing but as yet untested; it suggests that we have to learn how to interpret smells. Another possibility is that the VNO deals with low-volatility molecules that are too large to reach the dorsally located olfactory epithelium. Given that some of the known tetrapod pheromones are large, heavy, low-volatility molecules, this could explain why the VNO favors pheromone detection but not exclusively. In fact, the VNO may be specialized for the detection of larger, heavyweight molecules, whether they are volatile or not. Research continues. At present, all we can say with any confidence is that the VNO is definitely not just a dedicated pheromone detection system.[22]

Follow That Scent

Odorants always have a source that disperses them through the medium into which they are released. At the simplest level this dispersion is the result of chemical diffusion—molecules tend to move about, and so the odor slowly spreads out from its source. The olfactory system's ability to figure out concentration gradients allows an animal to move toward or away from the odor source, just like those early bacteria. This is simply a more sophisticated form of chemotaxis. But unless you're as small as a bacterium, or you live in a very still environment (as many small, soil-burrowing animals do), chemotaxis just isn't going to be good enough.

The problem is that most odorants are carried by a fluid medium (air or water) and so rarely get the chance to spread by simple diffusion.[23] Both air and water exhibit currents, which leads to an odor plume. That's not so bad. Visually we can follow a smoke trail to a burning house, so how hard can it be for a good nose to track an odor plume? Harder than you think. Unfortunately, air and water currents can vary considerably in strength and direction over both long and short timescales. As a result, an animal trying to head up an odor plume will tend to receive bursts of odor. At short range, bursts of high odor concentration may come fairly close together, but if the source is distant, there could be substantial intervals between them. To follow the gradient, the animal has to be able to average the odor concentration over both time and space. A much simpler way is to detect the odorant and just turn into the current.[24] The direction of the current is determined by mechanical sensors, which basically means the sense of touch. Moving upstream brings you closer to the odor source, and as more chemical and current data become available, you can correct your course. If the current is at all turbulent, you can improve your accuracy by zig-zagging upstream. Many creatures, like insects, show a basic behavioral response to odor detection: detect interesting chemical, turn into current, move forward, and start zig-zagging. It's a behavioral pattern now being programmed into small robots, initially as a means of simulating (and so investigating) insect behavior, but increasingly as a way of making the robots perform genuinely useful tasks, like sniffing out chemicals.[25] In Figure 54 we consider how odor plumes really form. Figure 55 presents the results of some simple simulations where differently equipped animals track them to their source.

A Matter of Taste

In contrast to the multifaceted sense of smell, gustation generally limits itself to determining whether something is good to eat. Taste's capabilities are remarkably limited, perhaps echoing its origins in the first, primitive, chemical sense. In the matter of taste, vertebrates and invertebrates do things very differently. We start with the vertebrates.

All vertebrates, except the hagfish, have taste buds.[26] We know that taste buds all work in much the same way (from their form and structure, and from

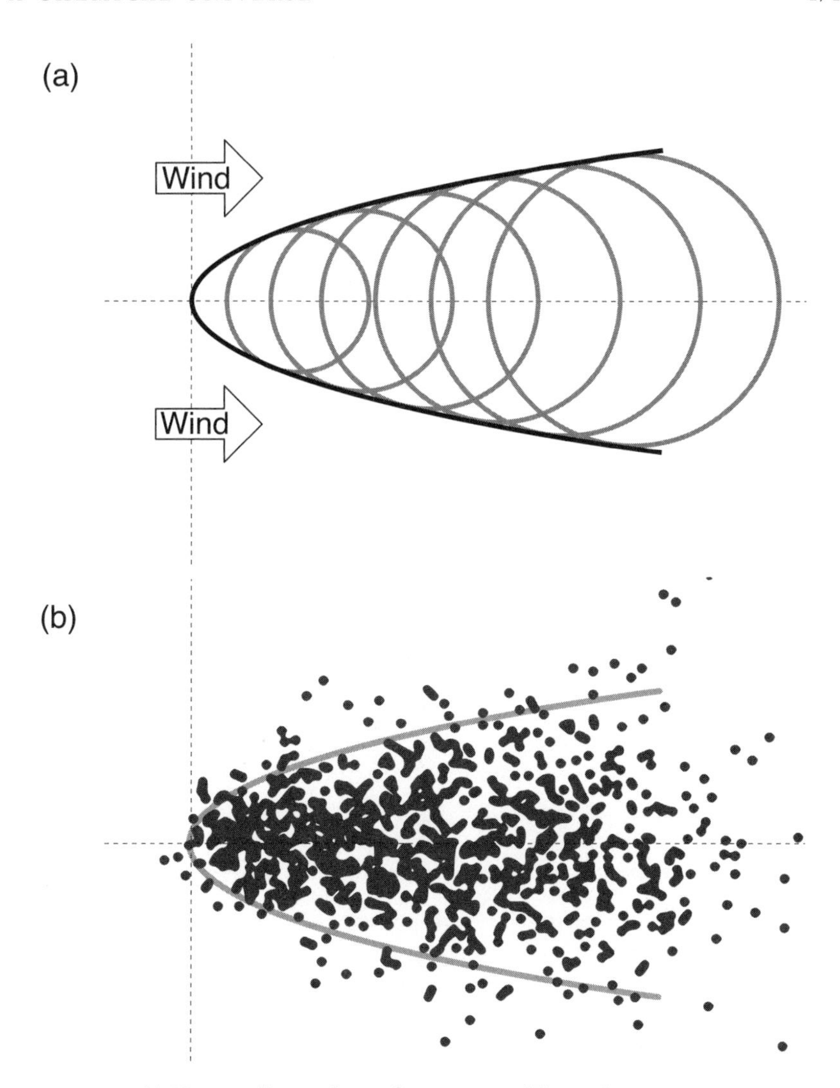

FIGURE 54 (a) How smell spreads out from a center. The circles are not concentric, because the wind moves the origin. (b) The results of 100 random walks starting from 10 different times (approximating a continuous smell).

experiments in which taste buds are stimulated by various chemicals and the resultant neurological responses are measured). In humans taste buds are usually found in the mouth (on bumps—*papillae*—on the tongue), but in other vertebrates they are more generally distributed around the tongue, palate, and pharynx. Some fish species also have external taste buds, frequently distrib-

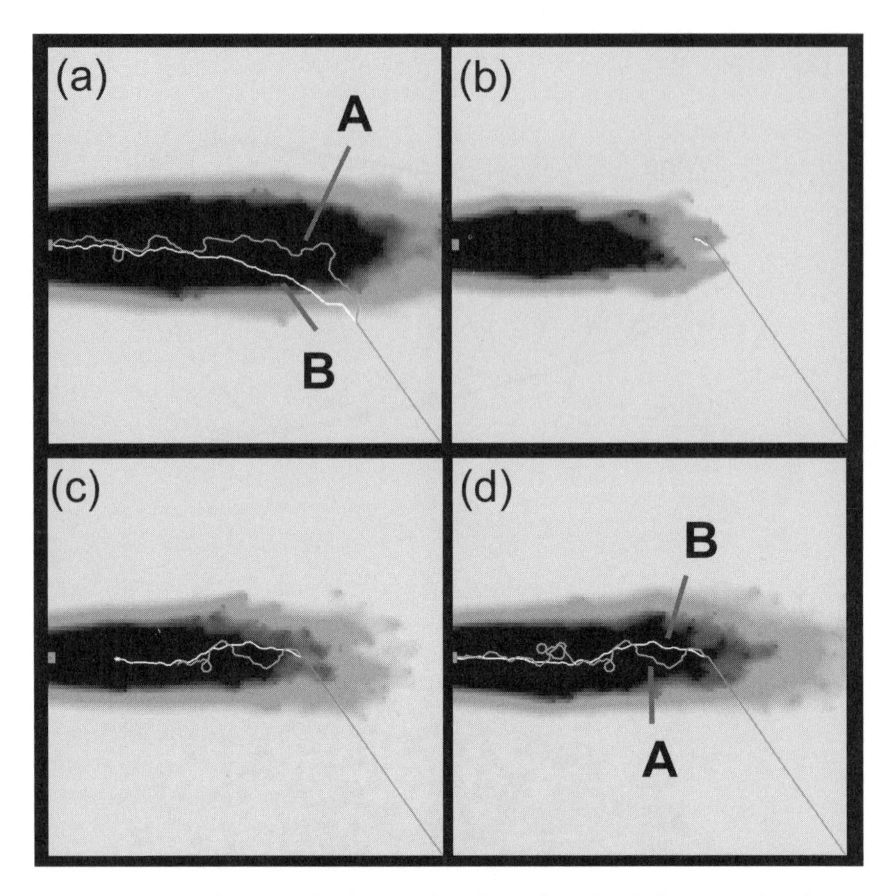

FIGURE 55 Simulation results showing the effects of wind turbulence on odor plume tracking as, starting from the same spot, two different scavengers, A and B, search for a meal. (a) Tracking of a well-developed plume. (b–d) A time sequence of a developing plume. In each case, both scavengers follow the same route into the plume. Thereafter, A heads toward the strongest smell, is mislead by the patchy nature of the windborne scent, but gets there in the end; the other scavenger, B, has an evolutionary advantage—pressure sensors enable it to factor in the average wind direction and so it gets there first.

uted around their heads or in trailing barbels; in some cases they can be found all over the body surface. External taste buds permit an aquatic animal (often living in an environment with little light but rich in dissolved chemicals) to assess an item's food value merely by swimming near it. Curiously, the fossil record suggests that vertebrate taste buds originated in the mouth; external taste buds evolved independently on several subsequent occasions.

Mammals tend to have the most taste buds of air-breathing vertebrates, whereas birds appear to have the fewest. For example, chickens may have only a few dozen taste buds, but a cow can have about 25,000. Humans typically have about 10,000 (which makes us wonder what a cow tastes in a mouthful of hay that we don't).[27] About 25% of people have an overabundance of taste buds, giving them much greater sensitivity in this area and presumably enhancing their job prospects in the wine-tasting profession. In contrast, regular smokers have fewer taste buds than normal.

In the vertebrates each taste bud is a modified skin cell made up a large number of taste receptor cells interspersed with supporting cells that hold everything together.[28] In a human tongue there are 50–150 receptor cells in each taste bud, which forms an onion-shaped organ. The tips of the receptor cells contain slender microvilli, which stick up into the surrounding fluids through a tiny taste pore on the tongue's surface. Dissolved chemicals in the fluids contact the microvilli and bind to specific receptor proteins on them, causing certain ion channels to open or close, thereby depolarizing the cell. Sounds a bit like an ORN? So it should. The difference is that the receptor is a skin cell, not a neuron—it doesn't connect directly to the higher processing centers. Instead, when a receptor cell is triggered, it releases neurotransmitters to the dendrites of the associated sensory neurons, generating an action potential in that sensory neuron, which then propagates to processing centers deeper in the nervous system. The underlying principles are the same in all vertebrate taste buds, although, as always, the details may differ (e.g., for aquatic animals the food chemicals are dissolved in sea water, whereas for terrestrial vertebrates they are dissolved in the animal's saliva).

Taste buds are among the shortest-lived cells in the animal body. In humans they are estimated to only last for a few days before being replaced, which means that the replacement taste bud must have a response to stimuli that is consistent with its predecessor. This continuity appears to be achieved by having the taste bud replacement process controlled by the downstream nerves to which the taste bud is attached. In fact, experiments show that cutting the nerve suppresses taste bud regeneration, but if the nerve is allowed to grow back, the taste bud does too, although the exact mechanism is still being researched.[29]

We know that our taste buds respond to a very limited number of tastants (the chemicals that actually cause taste). It has long been known that human taste buds respond to tastes characterized as sweet, salty, bitter, or sour. More recently a fifth has been identified (though not yet fully accepted), which is experienced as a savory taste, often referred to as "umami."[30]

Being able to distinguish only four or five tastes might seem rather limiting, especially compared to the capabilities of olfaction, but from an evolutionary perspective, this limited range is useful—it allows the sense of gustation to cut to the chase and answer the important questions. Olfaction may lead an animal to a potential foodstuff, but gustation tells it what expect from eating it. Each taste conveys important information about a potential foodstuff:

- Sweet—complex organic molecules like sugars and carbohydrates indicate a high-calorie foodstuff that will provide the organism with energy.
- Salty—indicates sodium chloride (table salt), which bodies need. It is especially important for nerve conduction and is essential for maintaining bodily equilibrium.
- Sour—acids (the taste is a response to hydrogen ions in the acid), which may indicate the presence of important vitamins or, equally likely, that food has spoiled or is unripe. It may also indicate foods that could be dangerous if eaten to excess. (You can eat one crab apple and get away with it but not a whole basket of them.)
- Bitter—indicates organic alkaloids or metallic salts, which are usually associated with poisons.
- Savory—associated with monosodium glutamate, but it probably also indicates the presence of vital amino acids the body needs to build proteins.

In the past it was believed that the taste receptor cells in an individual taste bud all reacted to the same single tastant and that all the taste buds of one type were concentrated in the same area. In fact each taste bud may contain several different types of receptor cell, and all the different tastes can be experienced in all locations of the tongue, although some of them are more dominant in one place than another (thus humans tend to experience sweet and salty on the front, bitter on the back, and sour on the sides of the tongue).[31] Interestingly, experiments in mice have shown that receptors for sweet and bitter tastants rarely occur in the same cells, although cells bearing various receptors are bundled together in taste buds. Again we seem to be seeing an evolutionary distinction between sweet (good) and bitter (bad).

To summarize: a few substances stimulate only one of the receptor types, though most stimulate two or more to varying degrees. Consequently, an individual taste bud may be capable of reacting to several tastes, but the form of the signals (intensity, duration, etc.) generated in the associated neurons will differ among them. In the downstream processing the central nervous system determines the actual taste from the discharge pattern across a sequence of nerves. Identifying taste is, like smell, a question of pattern recognition. As for olfaction, the concentration of the compound being tasted is determined from the cells' discharge rates, which increase with concentration. At too low a concentration, a definite identification may prove difficult; at too high a concentration, the sensors may be overloaded.

The exact biochemical mechanisms for detection of individual tastes vary considerably—sweet and bitter tastants are handled differently from salt and sour. (We don't yet know how umami's mechanism works.) Clearly, taste mechanisms are more diverse than olfactory ones, but the coding systems used to interpret the results are probably simpler because of the smaller number of tastes.

Compared to humans, animals tend to have a high detection threshold for salt, sweet, and sour tastes, but a low threshold for bitter ones. In other words, they detect bitter substances at much lower concentrations than salt, sweet, or sour ones. Mix equal quantities of all tastants in pure water, and only the bitter will be sensed. Given the association of bitter tastes with poisons, this discernment makes good evolutionary sense. Curiously, the ability to detect artificial sweeteners seems limited to Old World primates, with one endearing exception, shown in Figure 56.

All vertebrates follow this same model for gustation: tastants stimulate receptor cells, which send signals to the gustatory nerves (they are the same in all vertebrates), which carry them deeper into the gustatory cortex of the brain for further processing. Invertebrates do things differently. They have taste receptors, not taste buds, and the way these work seems to be much closer to olfaction—both taste and smell receptors are neurons with sensory dendrites and axons connecting to the nervous system. Inevitably, the details are very different in different types of invertebrate. We find taste receptors in the oddest places—on the heads of worms; on the tentacles of octopods; and on the legs, antennae, and mouthparts of many arthropods. The primary difference between these and the olfactory systems seems to be that the taste receptors are used to sample substances in solid or liquid form, leaving the olfactory receptors to handle gaseous volatiles.

In insects, for example, the taste receptors usually manifest as hairs, pegs, or pits where the dendrites of several sensory neurons are exposed to the environ-

FIGURE 56 The molecular structure of the Red Panda's sweet-taste receptor appears to be unique. Why is not known, but it does suggest that taste receptor mechanisms are more complex than previously thought. We thank Anita McFadzean for this image.

ment through a single opening in the surrounding cuticle. Each neuron seems to respond to a different range of compounds (sugar, salt, acid, etc.).[32]

Internal Messengers

All living things are essentially chemical machines, so it's not surprising that the state of the machine is monitored and regulated by internal chemical sensors. We touched on this idea in Chapter 4, when we considered the nervous system as a control mechanism. Basically, in any multicellular organism, communication and control between its component cells is essentially chemical and is achieved by a mix of hormones and neurotransmitters.[33]

Internal chemical sensors are also needed at a systems level to maintain the body's steady state conditions. In vertebrates, for example, chemoreceptors constantly check the oxygen and carbon dioxide levels in the bloodstream and brain. In part this check is done to maintain the blood's acid-alkali balance within its narrow operating limits, and in part it is done to tell us it's time to take another breath. (Depending on the animal, there can be several different systems acting in different ways to convince it to breathe more rapidly.) Other chemoreceptors react to an increased carbon dioxide level by driving up the heart rate and narrowing the smaller blood vessels to rush oxygen to where it is needed most. Some chemoreceptors monitor blood sugar levels, whereas others control the emission of those all-important messengers, the hormones. Still other chemoreceptors are attuned to detecting those hormones and stimulating an appropriate response. Chemoreceptors in the vertebrate stomach monitor acid-alkali levels to control the production of enzymes and acids needed for food digestion. Finally, some creatures, such as us, have chemoreceptors in the mucous membranes of the eyes, nose, and mouth, which detect potentially noxious chemicals and, in many cases, trigger a protective reaction to remove them. That's why your eyes water when you're chopping onions or breathing tear gas.

Similar systems exist in the invertebrates, although the details differ. Insects, for example, don't pump oxygen-rich blood around their bodies. Instead, air enters through spiracles, a series of small, muscle-controlled openings in their sides, and moves through the tracheae, a branching network of smaller and smaller tubes that deliver oxygen directly to the tissues that need it. Used air goes back out the same way. But insects still need to monitor oxygen and carbon dioxide levels and decide when to shut down or open up those breathing tubes.[34]

At the core of all these systems are sensory cells that react to chemicals outside the cell body—which all started with our simple bacterium.

8

Sound Ideas

Bioacoustic passive receivers ("ears") have evolved many shapes and sizes in many groups of animals—an elegant expression of the usefulness of hearing. In this chapter we share some of the engineering ideas associated with passive acoustical sensing (listening to what is out there); in the next chapter we delve into the more intricate world of active acoustical sensing (echolocation).

Here we restrict ourselves to sound that travels through fluids—air or water —and so say little about vibrations traveling through solid matter. Such vibrations are indeed sound waves, but the receptors developed by animals for sensing them are different. *Mechanoreceptors* in insects are often specially adapted structures, whereas birds and humans make use of feet or hands to sense vibrations. Spiders have specialized leg structures to pick up the minutest vibration of their webs, perhaps indicating prey. Monotremes, such as the platypus, have specialized mechanoreceptors; we return to these strange creatures and their mechanoreceptors in Chapter 10, but here we quickly move on to airborne and waterborne sound waves.[1]

The first few sections serve to review basic concepts, such as sound intensity, frequency, and wavelength, and how they impinge on animal hearing. Having mastered Acoustics 101, we will then be in a position to better appreciate the capabilities and adaptations that evolution has provided for animals—including ourselves.

Acoustics 101

Hear, Hear

Biologists measure the sensitivity of an animal's hearing and present their data in the very compact and useful form of an audiogram—a generic version is shown in Figure 57. From the figure you can see how the frequency range of hearing (and the frequency range of best hearing) is defined. There is an optimum hearing frequency for an animal, represented by the lowest point on the curve. Audiograms for most animals have the same basic shape, though the curve of Figure 57 may be stretched or squashed horizontally, or displaced horizontally or vertically, as we move from species to species. Thus, birds tend to hear less well than mammals (with the glorious exception of owls), and so their audiograms' curves are displaced upward from the mammal curves. These avian audiogram curves are also usually squashed, because birds' ears are sensitive to a narrower range of frequencies. Humans and elephants hear better than cats and dogs (i.e., we can detect quieter sounds at a given frequency), who hear better than rats, who hear better than horses and cows. Small animals tend to make use of higher frequencies than do large animals, for reasons that we will explore, and so their audiogram curves are displaced to the right. Their hearing, too, is often less sensitive than ours (compared with other animals, human hearing is pretty good—we will go into the reasons for this), and so their audiogram curve is also displaced upward. Thus, the range of hearing

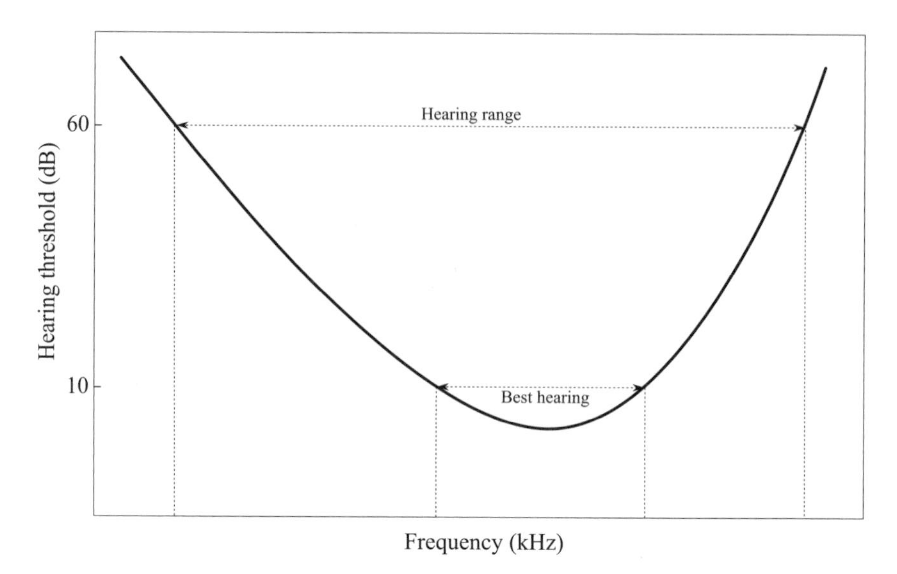

FIGURE 57 Audiogram of animal hearing capability. Most animal audiograms have this basic U-shape. The threshold at which a sound can be detected (minimum required acoustic power) is plotted against acoustic frequency. The threshold of hearing is set somewhat arbitrarily at a level of 60 dB: the range of frequencies that you can hear below this threshold is your hearing range. Your best hearing occurs over a frequency range below the 10 dB threshold—you are very sensitive to such sounds. The frequency that you are most acutely sensitive to lies at the lowest point of the audiogram curve.

for a young human is typically 30 Hz–20 kHz, whereas for a mouse the figures are 2–80 kHz; however, the range of best hearing is five octaves (from 250 Hz to 8 kHz) for humans and only a narrow 0.4 octave band around 16 kHz for mice.[2]

Making Waves

A sound wave is a wave of oscillating pressure that travels through the atmosphere (or through the oceans). In air, sound travels at about 340 $m \cdot s^{-1}$—say, 760 mph—give or take a little, depending on atmospheric water content, temperature, and other meteorological factors. In Figure 58 we show how a tuning fork sets up such a sinusoidal pressure wave by vibrating back and forth. The pressure wave is really a wave of overpressure, meaning areas of pressure that are above or below the background atmospheric level. So a sound wave is a stream of positive and negative overpressure, above and below the average air

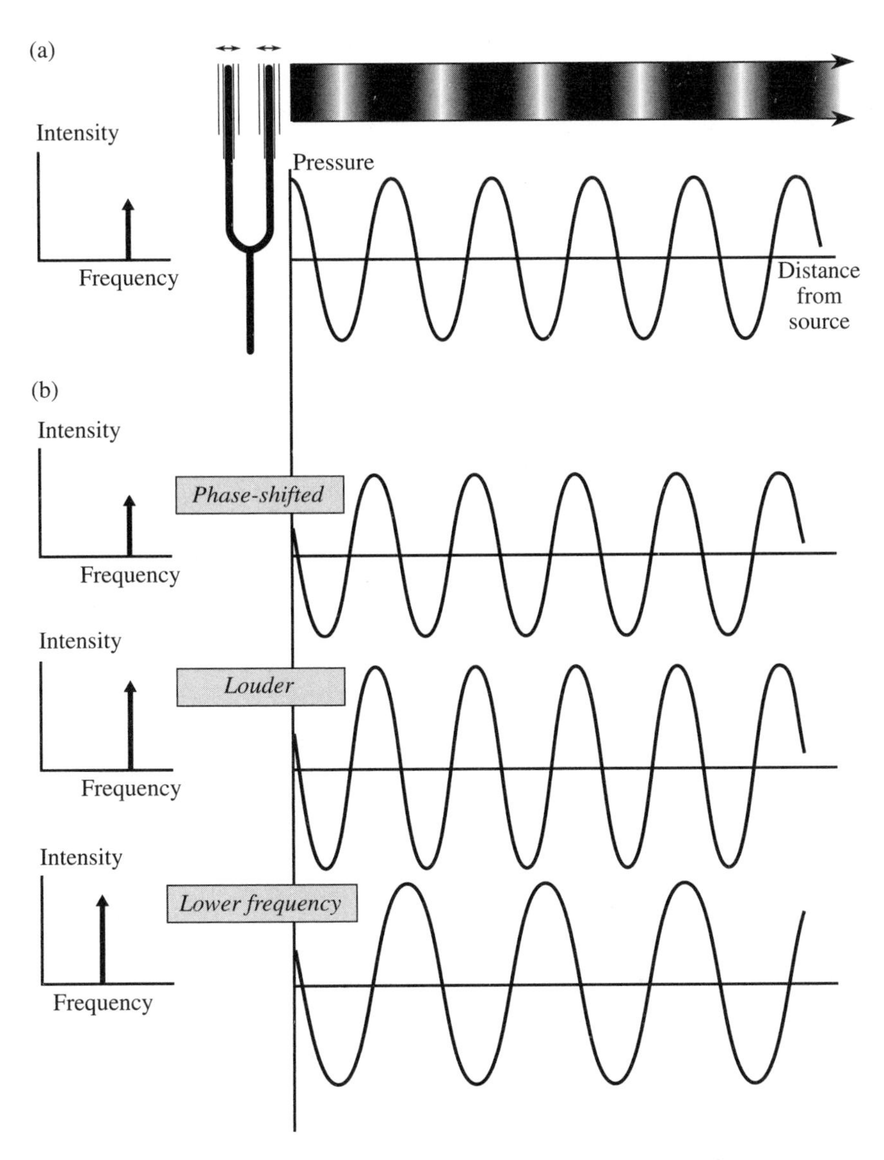

FIGURE 58 (a) Sound waves generated by a tuning fork. The increase/decrease in pressure is shown as dark/light regions of the sound wave that is sent out by the tuning fork. Engineers represent this fluctuating pressure by the oscillating wave shown. On the left is a plot of the sound *spectrum* (intensity plotted against frequency). For a tuning fork of pure pitch, the spectrum is a single frequency spike. (b) The same sound wave, but with a phase shift, which does not change the spectrum at all. Also shown is the same sound wave transmitted louder—more variation in pressure—and the corresponding increase in spectral intensity. Finally, we show the sound wave for a lower-frequency tuning fork—longer wavelength and with a spectrum concentrated at a lower frequency.

dB

Sound pressure and sound intensity are measured in decibels (dB). These strange units may be unfamiliar to many readers; hence this brief explanation. Stated mathematically, a dB is a *logarithmic ratio,* useful for expressing very large or very small differences. It is ten times the logarithm of the ratio of the measured sound to a reference sound level. Thus, a dime is ten cents: it is 10 dB greater than one cent. A dollar is 100 cents: it is 20 dB greater than one cent. Ten dollars is 1,000 cents: 30 dB greater than one cent, and so on. Confusion in decibels arises because people forget to state the reference level. Even worse, in acoustics we use two reference levels. Sound waves are measured in terms of the pressure (actually overpressure) required to send the wave through the medium. In air, that reference level is set at 20 μPa (micropascals) pressure. Underwater, we use 1 μPa, just to confuse things.

Sound intensity (loudness) is proportional to the average pressure squared: a sound 17 dB above the reference pressure level is 34 dB above the reference intensity level. Squaring a number means doubling its dB representation.

pressure. The *wavelength* of a sound wave is the distance between pressure peaks. Now imagine that you are standing still and watching sound pressure waves pass by; if 523 peaks pass by every second, then the frequency of the sound waves is 523 cycles per second, or 523 Hz. So, for example, if the tuning fork is pitched at the musical note of C_5 (which happens to be 523 Hz), then from the speed and frequency you can easily work out that the wavelength is 0.65 m, or about 25.5 in. The general rule is that the speed of sound c equals frequency f multiplied by wavelength λ, or $c = f\lambda$. (In fact, this rule applies to all waves, not just sound waves.) So, high frequency means short wavelength and low frequency means long wavelength.

The *phase* of a wave is a relative thing, describing the relative position of the peaks of two waves. If two waves with the same frequency are lined up, with the peaks all together, then the two waves have zero phase difference. If one of the waves is delayed a fraction of a cycle, then the peaks will not align, as in Figure 58, and we say that this wave is phase shifted. Phase shifting does not alter the wave intensity or frequency, but merely shifts the timing. We will see that phase shifts are important cues that animals use to localize the source of a sound.

Hear and Now

There are many reasons why hearing is a good idea—why evolving the gear to hear helps the hive to thrive.[3] Here is a list of auditory advantages (or "hearing aids"—some obvious, others more subtle) that we will unpack throughout this or later chapters.

- *Communication with animals of the same species.* The frequencies that an animal transmits when it sings, calls, howls, wails, hoots, neighs, grunts, barks, quacks, buzzes, trills, or screams overlap its frequency range of hearing, generally speaking. The obvious conclusion to draw from this evidence is that hearing evolved in most species to aid communication. There is more on communication in Chapter 12.
- *Prey detection.* Many predators detect their prey from the sounds that the prey emits, intentionally or unintentionally. We will see how owls have refined this process to an exquisite degree.
- *Predator detection.* Rabbits, deer, and many other prey species who evade their predators by running away have excellent hearing, which acts as an early warning system to give them a head start.
- *Sounding the alarm.* Social animals post sentries that can alert the rest of the herd or flock by emitting a cry.
- *Orientation.* It is clearly useful for an animal to be able to locate the source of many types of sound, be it from a potential predator, prey, or mate, or environmental noise (e.g., running water).
- *Navigation.* There is growing evidence that some birds (e.g., pigeons) can hear very low frequency infrasounds emitted by the environment, and this ability assists orientation during long distance flights, such as migration. We have more on this subject in Chapter 11.
- *Echolocation.* Animal sonar is the subject of Chapter 9. Clearly, for an animal to echolocate, it must first be able to hear.

There are broad conclusions that biologists have reached following many, many experimental audiogram tests on all kinds of crawling, flying, swimming, and slithering creatures. Mammals differ from almost all other vertebrates in that they can hear sounds with frequencies exceeding 10 kHz. Fish, amphibians, and reptiles generally are sensitive only to frequencies below 5 kHz. Frogs are the only amphibians that are well adapted to detecting airborne sounds. Birds hear quite well, though not as well as mammals, and over a more limited range of frequencies. Many insects have specialized hearing over one or more narrow bands of frequencies. (For example, cricket hearing is sensitive to the frequencies of their own songs but also to a much higher frequency band—quite separate from the first—that listens in on the echolocating call of their bat predators.) Smaller animals generally are sensitive to a higher band of frequencies than are larger animals. Many species use frequencies at the higher end of their hearing range to locate the direction of the sound source. The accuracy with which they can estimate source direction is variable. Humans hear relative pitch (we can tell the difference between frequencies), whereas birds hear the absolute pitch (they are sensitive to the number of cycles per second). Birds are much more capable (than we are) of detecting rapid changes of intensity of a sound wave. Mammals exhibit a huge range (nine octaves) in the lower limit of frequency that they can hear. The shape of external ears of mammals is extremely variable. Subterranean mammals have poor hearing (but are sensitive to substrate vibration). Some animals (owls, bats, and toothed whales) have developed hearing well in advance of human biological or technological capabilities.[4]

Some of these findings, such as the restricted sensitivity and frequency range of bird hearing, are not well understood. Other findings make a great deal of engineering sense. Thus, the variation in mammalian *pinnae* (external ears) is the result of specialization. Subterranean mammals do not need or want large external ears. Sea mammals usually also lack pinnae, presumably for streamlining. Many prey species have found that the combination of good directional hearing and all-round eyesight provides the best warning of lurking predators. We discuss eyesight in Chapter 10, but mention it here once or twice, because there is an overlap—an interplay—between the acoustical and optical sensing of an animal. The effectiveness of ear shape as directional antennas has been established experimentally. Thus, the threshold of hearing for reindeer varies by 21 dB (a factor of 125) depending on the direction in which its ears point (toward or away from the sound source). The angle dependence of pinnae sensitivity figures prominently in our discussion of sound direction estimation in the next section. Another factor that emerges from all the findings about mammalian hearing is that the ability to sense the direction of a sound—to localize it—is linked to an animal's visual field. That is to say, a mammal can locate the source of a sound just well enough to direct its eyes to the source, so the source falls within the visual field. Animals with a narrow visual field (e.g., humans and other predators) are able to localize sound sources well enough so that they can turn to the source and see it without searching with their eyes. Other animals with wider fields of view tend to have less precise sound localization abilities. The common factor is the linking of audition and vision to detect the direction of a sound source: hear, turn, and see.[5]

There are other revealing correlations in the data. Thus, mammal head size is correlated inversely with the sound frequency that is utilized to localize source. In other words bigger animals use lower frequencies to estimate sound source direction. We will see why this makes good engineering sense in the next section. The very large variation in mammalian lower hearing limit falls into two groups: those animals who can hear below 125 Hz and those who cannot hear anything below 500 Hz. Again there is a size correlation here, with larger animals hearing lower frequencies. We will see in Chapter 12 that larger animals (not just mammals) can generate lower frequencies in their voices and so, to hear what they say, they need to be able to hear those bass notes.

How can we make sense of the data that has been gathered about bird hearing? Birds are about as capable as we are of detecting changes in frequency, though over a narrower frequency range; they are more sensitive than us to changes in sound intensity (though we're not bad: we can spot differences of only 0.5 dB), and they are much better at differentiating between notes of short duration. This latter ability clearly has something to do with bird song, which can be very complex and rapidly varying. Birds can learn songs, which is what really tells us of their hearing acumen (they can pick out individual notes of only 5 ms duration—1/200 s; we require notes that are 10 times longer). When we record a complex bird song and play it back at slow speed, we hear a lot more detail than we did first time around. Birds do not need these slow-motion replays, however. They learn the full song, with all its complexity, in real time and learn it well enough to be able to match it with their own voices. They may not be able to hear every frequency that they sing, but they can hear a lot more in their songs than we can. Some birds learn other birds' songs so well that they can match them, and the two sing a kind of duet, often associated with defending territory.[6]

The superior temporal resolution of birds appears to be the crucial difference between what birds hear and what we hear. This ability, plus the other characteristics of avian hearing summarized above, provides birds with several key survival advantages; some examples serve to show how well birds apply their hearing:

- Quails usually lay a lot of eggs. Chicks inside eggs can communicate with one another through the shells, so long as the eggs are touching. In this way the clutch of eggs all hatch pretty much at the same time—the chicks have agreed on a time for simultaneous hatching ("7 a.m. Thursday works for me"). This ability is an advantage for quails, because the newly hatched chicks are obliged to follow their parents soon after birth, and it is easier to protect them if they can remain together as a group. If the chicks hatched at different times, they could not form a cohesive, mobile group.
- Shorebird chicks vocalize in the egg so that they and their parents recognize one another's voices soon after the chick hatches, which helps with recognition in a crowded colony.
- European Blackbirds cock their heads to hear leatherjackets (Crane Fly larvae, their invertebrate prey) burrowing beneath their feet. Ringed plovers can hear marine invertebrates moving in the mud, amid the noise of a tidal estuary.
- Gannets can live in huge colonies. Gannet pairs can recognize each other's voices amid all the noise made by the great mass of squabbling birds, even though gannet squawks are quite simple. An individual bird need only hear the first tenth of a second of a call to recognize its mate.[7]

Where Is It?

First, for those readers who are neither radar nor sonar engineers, nor bats, owls, or dolphins, we need another brief interlude to get across some of the basic technical concepts required to understand sound localization.

External ears act as antennas, creating beams. We need to understand a little more about how antennas work and how they depend on the frequency of sound that reaches them. We usually think of beams as transmitters: a car headlight transmits light that is focused into a beam. However, the concept also applies in reverse. Imagine the light traveling backward, into the headlight and ending up at the bulb—in this case the headlight is a receiver and the beam it defines is a receiver beam. The same idea applies to ears, which are acoustic receivers: sound comes from outside and is focused within the ear. Beamshape varies with ear shape, and so the wide variety of ear forms that the natural world has evolved (see Figure 59) tells us that it is important to get the right beamshape. Consider a jackrabbit or hare, with tall ears held vertically. Such ears immediately tell an antenna engineer that the animal is designed for hearing sounds that originate on the ground, at the same level as the beast (assuming that our jackrabbit is on flat ground). This is because wide antennas have a narrow beamshape (in this case directed horizontally, because the ears are held vertically). Narrow antennas have a wide beamshape. So, the beamshape of each ear of a jackrabbit or hare will be narrow in elevation and wide in azimuth, like a fan laid flat. (Elevation is angular extent in a vertical plane, whereas azimuth is angular extent in a horizontal plane.) Figure 60 explains this (perhaps counterintuitive) engineering fact of life. Why do such ears benefit our loping lagomorphs? Jackrabbits and hares have many predators; those

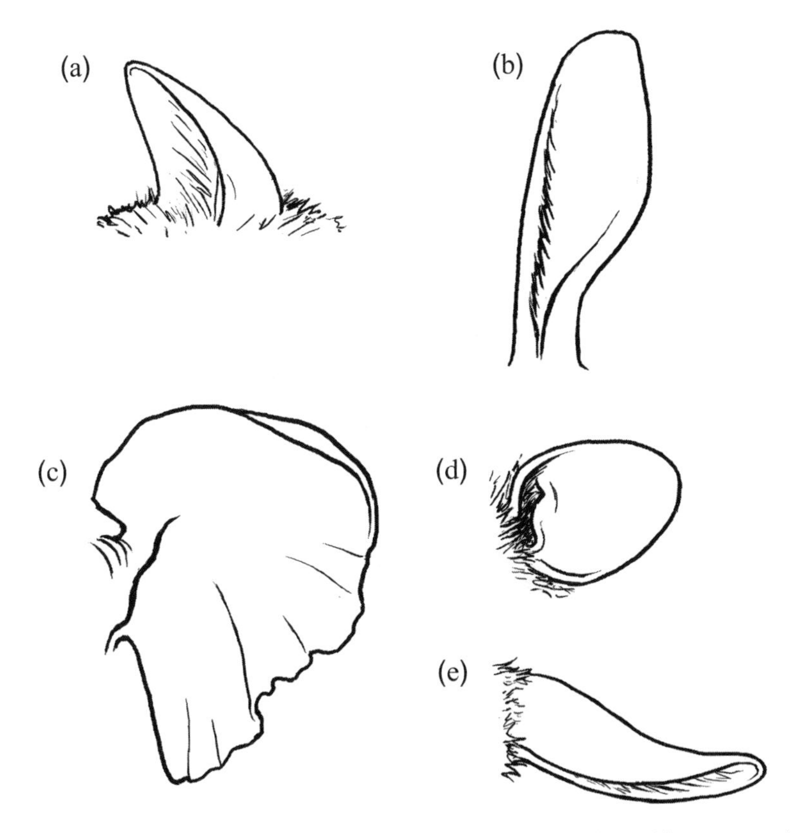

FIGURE 59 Different ear shapes: (a) cat, (b) jackrabbit, (c) elephant, (d) mouse, and (e) sheep.

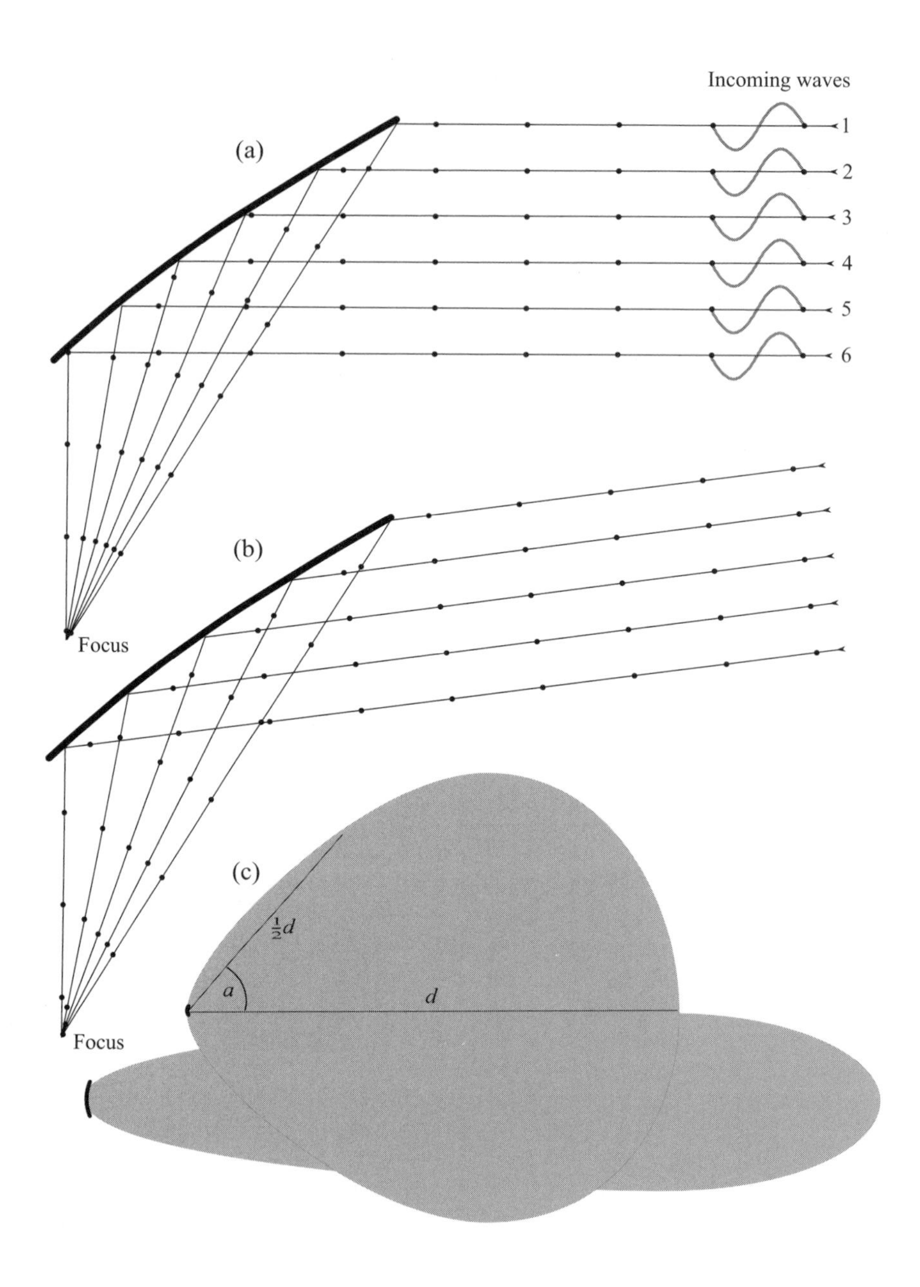
Incoming waves
1
2
3
4
5
6
(a)
Focus
(b)
Focus
(c)
$\frac{1}{2}d$
a
d

in the air can be detected using all-round vision, but those on the ground may not be seen—they may be lurking in the grass. So ears that pick up sounds in the grass are more useful than ears that are less sensitive in this direction. The cost of such directionality is reduced sensitivity in other directions, such as skyward, but a sharp-eyed hare can live with that. A mouse in the grass, however, needs to hear sounds from all directions (she has a restricted view, and her predators are as likely to approach from above as from the side), and so mice have less directional beams, formed by rounder external ears.

There are two measures of direction estimation that we must carefully distinguish. *Accuracy* refers to the ability to tell the direction of one sound source. A house mouse uses sound intensity to estimate direction within about 33°—he cannot do better than this rather broad swath, and so we say that his direction estimation accuracy is 33°.[8] The second measure is *resolution,* or *acuity.* Resolution tells us how well an animal is able to distinguish two sound sources. If two sources are separated by an angle *a,* and you can just tell that what you are hearing comes from two sources and not one, then you have an auditory acuity of *a.* If the sources were separated by a larger angle, then you could more easily tell them apart, but if they were closer together, then you would mistake them for a single source. Accuracy and resolution are sometimes the same value but not always—in fact, they can be made to be independent of each other. In the next chapter we will learn more about how a species of experts (dolphins) enhance their angular resolution capabilities by special signal pro-

---

FIGURE 60 (*opposite page*) An antenna gathers sound waves and focuses them. (a) If the antenna shape (indicated by the heavy black line) is chosen carefully, then the sound wave reflecting off different parts will arrive at the focus still in phase. We show the crests and troughs of the sound waves by circles; note how the waves 1–6 stay in step as they are funneled to the focus. (b) Now see what happens when the sound arrives from a slightly different direction. The top and bottom wave are out of step, so when all the waves are added together, they don't produce as loud a sound as before. However, two adjacent waves (say, waves 1 and 2) are still pretty much in step, which tells us that a smaller antenna—just covering waves 1 and 2—will not lose as much sensitivity with changing angle. For higher sound wave frequencies (shorter wavelengths), the angular sensitivity will be more marked. (c) So short antennas have fat beams, whereas longer antennas have thin, more directional beams. How fat or thin depends on frequency. If an antenna can hear out to distance *d* straight in front, but only to distance ½*d* at some angle *a,* then we say that the antenna beamwidth is the angle 2*a.*

MORE ABOUT BEAMSHAPE

A sound reaching you from a distance of, say, 20 m will be picked up by your external ears and channeled into your middle and inner ears, where it is sensed. The intensity of sound reaching your ears will not change if the sound source moves without changing distance—say it moves from straight in front of you to straight behind. However, even though the intensity reaching your ears is not altered, you will hear the two sources of sound differently. Sound that arrives from the front is channeled more effectively than sound that arrives from behind. This trait, like binocular vision, is an evolutionary adaptation of predator species: it concentrates the effectiveness of your hearing in the forward direction, where you are heading. Prey species, such as deer, have more mobile ears that can point backward to pick up unseen predator noises. Your ears, and those of deer and (especially) owls and bats, are directional antennas. They hear most effectively in one direction and consequently, less effectively in other directions unless maneuvered to point in a different direction.

Why bother with directionality? Why not make hearing as efficient as possible in all directions, to maximize the intensity of sounds that are made nearby? After all, any such information may be useful. There are two reasons. First, directional acoustic receiver antennas (OK, "ears" from now on) are much better at picking up sounds from the chosen direction than are directionless ears. Think of an old-fashioned hearing horn, or look at the shape of a deer's ears, and it is clear that more sound is gathered from one direction than from other directions. Second, it is often advantageous to muffle extraneous sounds. A cat that hunts a mouse in the grass may not be able to see her intended lunch, but she directs her ears forward to pick up any giveaway sounds that the mouse makes (and we will soon see how such sounds can tell her about the mouse's position). The cat does not want to hear extraneous sounds—grass rustling all around—that may smother the mouse noises, and so her ears point in the direction under investigation. We can plot a graph of the sensitivity of ears to sound direction angle, or we can sketch a picture of the receiver antenna (ear) beamshape. An example is shown in Figure 60c. It is obvious that directional ears will help an animal locate the source of a sound (by amplifying the sound from a chosen direction and muffling sounds from other directions) but in fact there are other less obvious ways in which this capability is enhanced by ear beamshape, as we will see.

cessing. For now we need only illustrate the difference between these two measures of direction estimation capability: see Figure 61.

Some animals (you included) use their ears' beamshapes to estimate the direction of a sound source by measuring relative intensity in the two ears. Another diagram will help to explain—see Figure 62. We take the example of owls, because their use of the intensity technique is unusual and instructive, though many other creatures estimate source direction via intensity differences.[9] Owls have asymmetric ears: the two ear openings in an owl's skull are at different heights. This unique adaptation (all other vertebrates have bilaterally symmetric ear openings) permits owls to estimate the elevation angle of a sound source. Most of the rest of us have ears that are separated horizontally, and so we can only use them to estimate azimuth (unless we can move our external ears independently). Owls probably use a processing technique later reinvented (in World War II) by radar engineers, called *amplitude monopulse.* By comparing the intensities of a signal in two independent beams (ear directivity patterns), it is possible to calculate direction. The idea is explained in Figure 62. If the ear directivity is chosen carefully (and owls can adjust the feathers in their ruff using special muscles, which changes the directivity), then there is what mathematicians call a one-to-one relationship between intensity ratio and elevation angle: a measured value of intensity ratio corresponds to one and only one value of elevation. An owl (Figure 63) can achieve an elevation accuracy of about 1.5° in this way.[10]

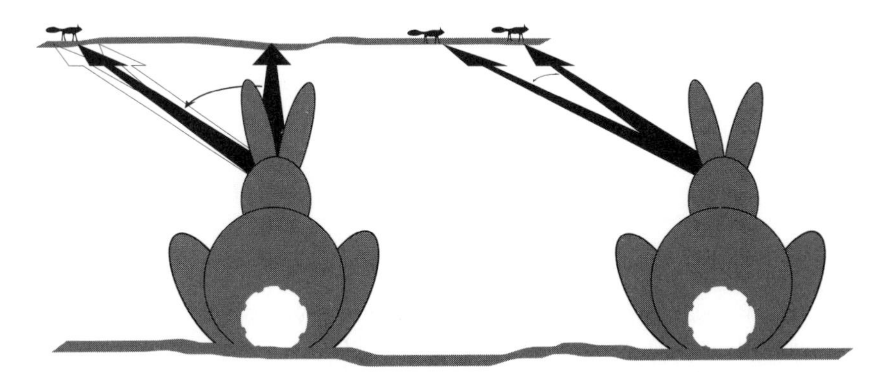

FIGURE 61 A cottontail rabbit listens to the sounds of a single fox and estimates the direction angle (relative to straight ahead, say) with an accuracy of about 27°. The same rabbit may be able to tell, by listening, that there are two foxes present, if the foxes are separated in angle by an amount that exceeds the rabbit's auditory resolution, or acuity.

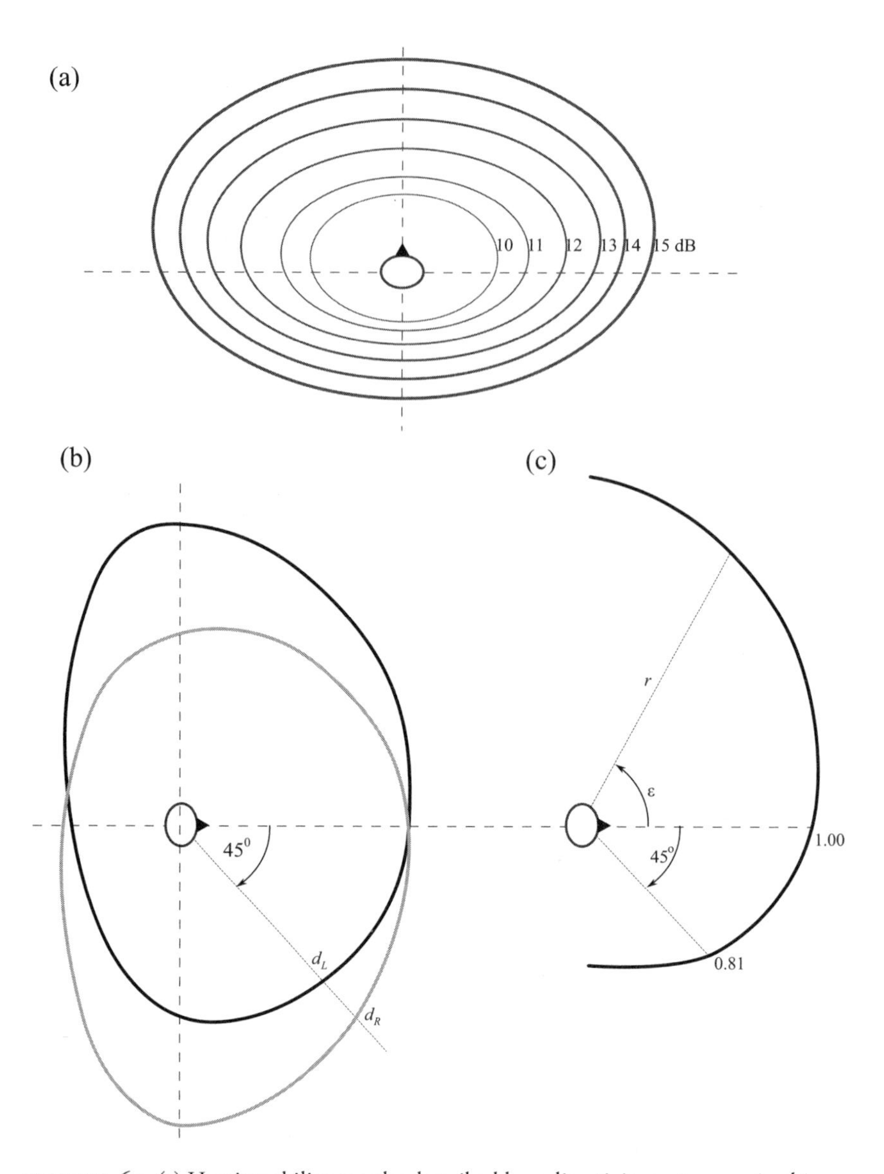

FIGURE 62 (a) Hearing ability may be described by a directivity pattern, as in this illustration. An owl is viewed from above. The gray ovals represent contours of equal intensity, and so the oval marked 15 dB represents the distance that a sound source of 15 dB intensity must be for the bird to just detect it. (b) Monopulse direction estimation by owls. Viewed from the side, one ear—say, the left one—hears better in an upward direction (black directivity pattern), whereas the other ear hears better in a downward direction (gray pattern). The ratio of intensities heard at an angle of, say, 45° is the ratio $r = d_L/d_R$. (c) The ratio r is different for each elevation angle ε, and so by measuring sound intensity in each ear, the owl can estimate elevation angle.

FIGURE 63 No, Striped Owls (*Pseudoscops clamator*) do not need to do this to accurately pinpoint a sound. Thanks to David Webster for this image.

For estimating the azimuth angle of a sound source—how far to one side it is—owls and many other creatures make use of a different binaural cue. In Figure 64, which shows an owl head from above, we see how the difference in time of arrival of a sound in each ear depends on azimuth of the source, and so timing can be used to estimate azimuth.[11] Timing works when the sound is of short duration, or when it has a sharp, well-defined edge, but not so well if the sound is a pure tone that consists of only one frequency (like the note from a tuning fork) or a narrow band of frequencies. For such a sound it is better to use the phase information of the sound wave. The waves' phase difference at the two ears is also shown in Figure 64; you can see that, as with timing, this

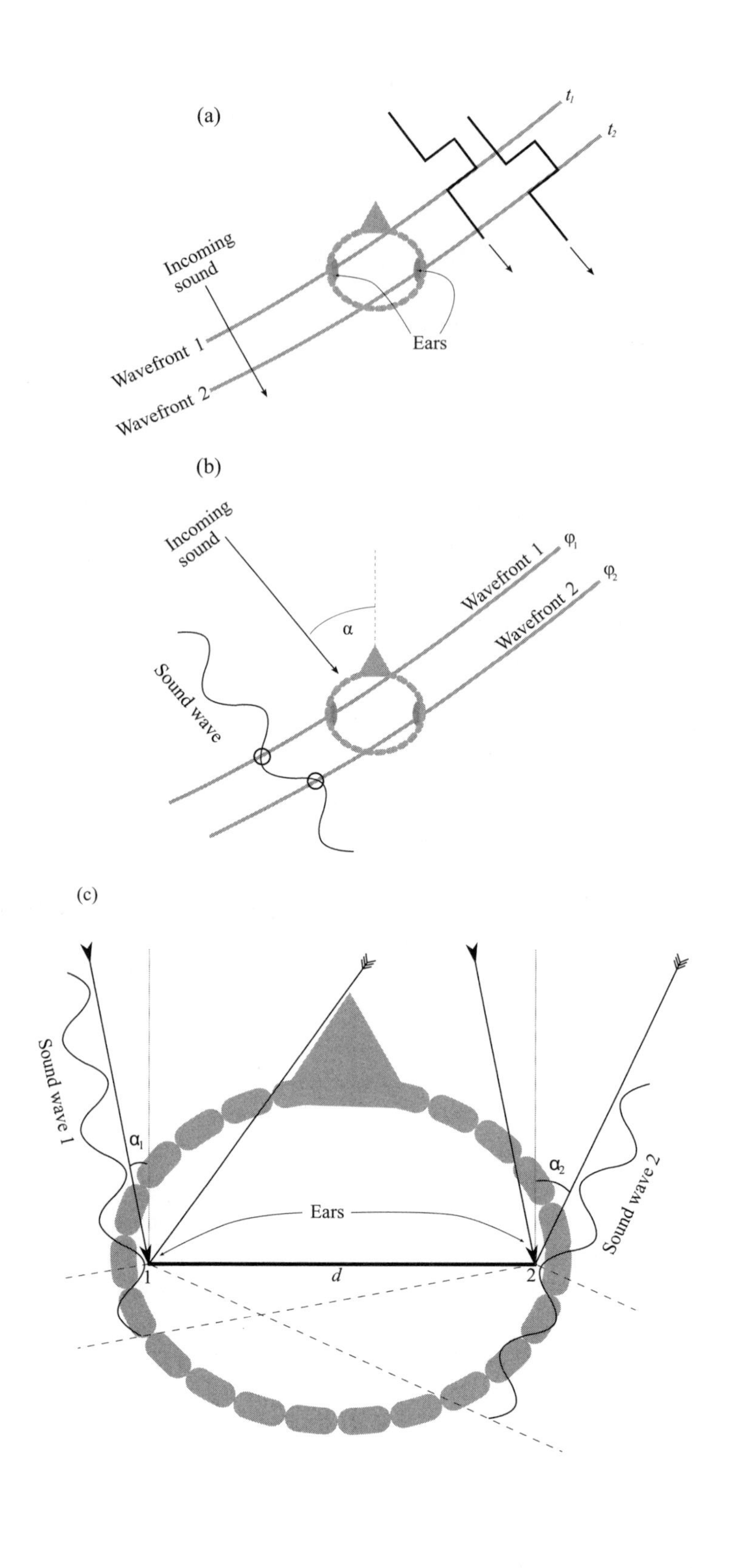
(a)
t_1
t_2
Incoming sound
Wavefront 1
Wavefront 2
Ears
(b)
Incoming sound
Wavefront 1
Wavefront 2
φ_1
φ_2
α
Sound wave
(c)
Sound wave 1
Sound wave 2
α_1
α_2
Ears
1
2
d

difference increases as azimuth increases and so is a candidate for sound localization. Some animals use phase information, others prefer timing. Some probably use both. Humans use a combination of timing and the third technique, intensity differences, and are capable of an impressive 1–2° accuracy. Owls use intensity for elevation and timing for azimuth (achieving the same accuracy in both). In general, other animals, except for echolocators, are not so accurate but still are capable of useful sound location.[12]

There are several engineering issues associated with sound localization. Interestingly, nature appears to have come up with pretty much the same solutions as radar and sonar engineers, though beating us to it by at least 30 million years. (This message will be repeated when we get to echolocation.) Issue number one: the accuracy with which an animal can estimate sound source direction depends on the direction. In particular, accuracy is best when the sound is right in front and worst when it is out to the side. This is why owls (and tracking radar, and people and other animals) turn toward a sound they want to localize. Issue number two (more serious): the time difference between sounds arriving at the ears is short. Very short. Your neurons and those of other animals work on a timescale of milliseconds, whereas the time differences that you must detect to localize a sound source to within 1–2° is about 10 μs (10 millionths of a second, or a hundredth of a millisecond). How can you do this?

It seems that you employ what signal processing engineers call "correlation processing." We do this type of calculation inside radar processors when estimating the direction of a blip seen on a radar screen. Nature got there first. We describe correlation processing in stages throughout this and the next chapter —it can do quite a lot for us—and the first stage is a description of how it

---

FIGURE 64 (*opposite page*) (a) A sharp sound, characterized by a sudden change in intensity, is represented here by a square pulse impinging on an owl's ears. This type of sound has a well-defined arrival time, so that the sounds source can be readily estimated from the time difference $t_2 - t_1$. (b) In some cases the phase difference $\varphi_2 - \varphi_1$ of the sound wave (in this case, about half a cycle) is more convenient than the time difference for estimating azimuth direction. (c) Two sound waves (two different arrows) of the same wavelength arrive from different directions, α_1 and α_2. They are different, even though the phase difference between the two ears is the same for both waves: about 3/4 of a cycle for the wave on the left and about 1 3/4 cycles for the wave on the right, as gauged from the dotted lines (but both will be detected as a 3/4 cycle phase difference). The distance between the ears (the interaural distance) is *d*.

boosts our ability to estimate time differences. If the sound that you hear is brief and has a sharp edge, meaning that it's intensity rises suddenly (like a starter's pistol), as in Figure 64a, then estimating its time of arrival in each of your ears is not too difficult. But most sounds last much longer than the report of a starter's pistol. Consider the noisy signal shown in Figure 65 (which might be the low-intensity rustling of a mouse moving in grass). Correlation processing permits even this type of signal to be timed accurately, as you can see from the figure. The signals from both ears are fed into a "correlation processor," and the output signal tells you the time difference. Your brain then converts time difference into angle estimate. How does the correlator work—how does it produce such a narrow (short duration) spike in Figure 65? By massively parallel processing, in the language of computer scientists. Imagine the noisy signal compared with a copy of itself, where the copy is shifted in time by small increments. Only when the shift is zero units will the copy align per-

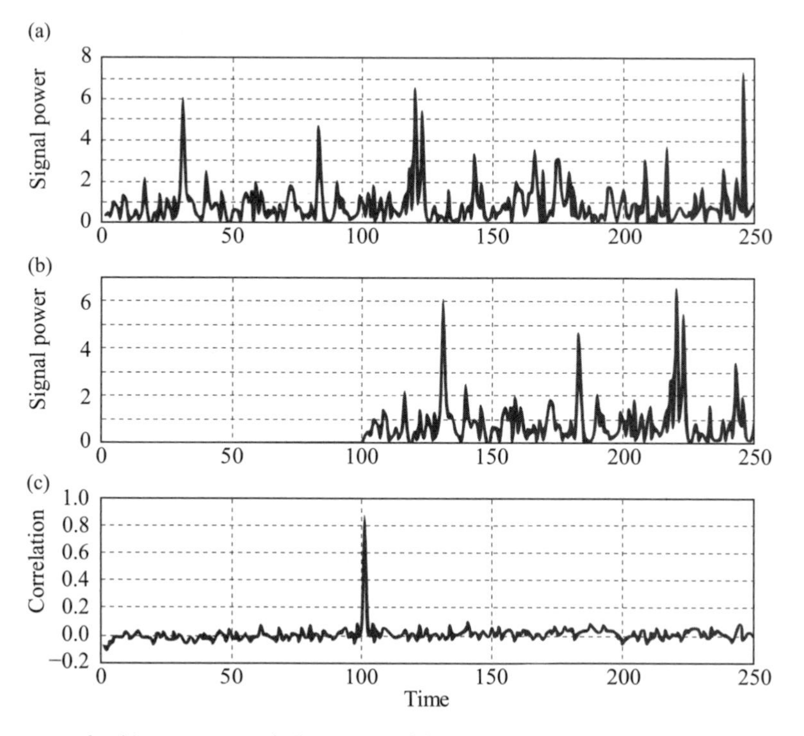

FIGURE 65 (a) A noisy signal. (b) A copy of the signal delayed by 100 time units (milliseconds, say). (c) The correlation product of these two signals spikes at the delay interval, permitting an accurate estimate of the delay.

fectly with the original; multiplying together each component of the original and copy and summing over all components produces the spike. When the original and copy are not perfectly aligned, then the multiplication and addition yields a much smaller number. In our case the original signal corresponds to a complex sound wave arriving at, say, the left ear, and the copy corresponds to the same sound wave arriving at the right ear a short time later. Now the correlation spike occurs at the delay time instead of at time zero. This sophisticated processing works just as well in your brain as it does in a radar signal processor (we used a typical radar algorithm for generating Figure 65). Correlation processing also works well in the brains of other animals, such as dolphins, cats, and wrens—yes, it is likely that even little wrens can perform this feat, as we will soon see.

Issue number three: it may already have occurred to you that smaller animals will have more difficulties than you or I when it comes to measuring time differences, because they have less between their ears. We do not mean that they have less capable brains (in this context, they may not), instead we mean that there is less of a distance between their ears. Shorter interaural distance means shorter time intervals to work with when estimating sound direction. This is, indeed, a problem. In fact, there is a lot of evidence to show that smaller animals use higher frequencies to localize sounds; that is to say, they pick out the high frequency components of a sound source to use for direction estimation. We can infer this selection from the data by plotting (after many repeated experiments on many animals types) the frequency used for sound localization against interaural distance—the result is a straight line. How does higher frequency (shorter wavelength) help smaller animals? They must be using phase difference instead of timing difference, as we saw in Figure 64b. For example, weasels use 8 cm wavelengths, whereas cows use sound of wavelength approximately 65 cm, to estimate source direction. Small wavelengths in a small skull can generate the same phase shifts as large wavelengths in a large skull, even though the time delay is shorter.[13]

If short wavelengths (higher frequencies) enhance phase differences, then why don't we all use short wavelengths to estimate sound direction? In fact, most mammals do. They can hear the high frequencies that other animals can't hear precisely for that purpose. There is a limit to how short the wavelength can be, however. In Figure 64c you see the problem of phase ambiguity: a short wavelength can arrive from two different directions with the same apparent phase difference. It can be shown (and this is a problem that radar and sonar antenna engineers have had to face) that there is no ambiguity if the interaural distance *d* of Figure 64c is less than half a wavelength.

If small animals have difficulties localizing sound via timing differences, then what about very small animals? There is a parasitic fly with a name that is much bigger than she is. The female *Ormia ochracea* lays her eggs on the surface of crickets. The hatching larvae then burrow into their cricket and feed on it. The point is that *O. ochracea* locates her cricket prey by their singing, and she can do this at night without any visual or olfactory cues. Experiments show that she can localize sounds with an accuracy of 2° or so, even though she is only a couple of millimeters long. To a biophysicist, this ability is astounding. First, the fly must work with the very narrowband (almost pure tone) continuous singing that the crickets produce; this song is typically at about 5 kHz, and so the wavelength is much longer than the fly. She cannot use timing, because the gap between her ears is only half a millimeter (and estimating by timing would require discriminating among sounds that are separated by only 50 ns—fifty billionths of a second—which is beyond even correlation processors to achieve). She cannot use phase, because the phase differences are negligible. Similarly she cannot use intensity differences between the two ears, because there isn't any difference to speak of. So how does she do it?

Researchers discovered how, in the mid 1990s, and their discoveries have led to at least one patent for new technology, as engineers seek to learn from nature. It transpires that this little fly has evolved a unique and specialized mechanism that accurately identifies the direction of a sound, so long as that sound is narrowband (which we know to be the case for the cricket song). The secret is a mechanical connector that directly links the tympana (eardrums) of the fly. Such a connection would be a problem for us, because our brains would get in the way, but it seems that these flies do not have this problem. The connecting rod is balanced on a sort of pivot, and so it can rock back and forth as the tympana vibrate due to sound waves. The mechanical arrangement is such that the connector-arm oscillations, on either side of the pivot, indicate the sound source direction. The oscillations grow in time as a kind of resonance builds up (this would likely not happen if the cricket song were wideband—the fly mechanism is exquisitely adapted to this one source of sound). One arm oscillates more than the other if the sound originates from the left side, whereas the other arm oscillates more if the sound comes from the right. Mathematical modeling of the *Ormia ochracea* mechanism has confirmed how it works and may lead to technological advances in directional microphones and hearing aid technology.[14]

Hearing and Perception

We have seen how well people can hear: we are capable of excellent sound localization (both accuracy and acuity), and we have a frequency range of hearing that exceeds that of most animals. (Don't brag about this to your pets, however: Fido has a wider acoustical receiver bandwidth—hearing range—than you do, and Felix has a *much* wider bandwidth.) We have more sensitive hearing over a wide range of frequencies than most animals. Our brains perform sophisticated signal processing that helps us with these tasks, and it does so automatically, seemingly effortlessly.

To appreciate how efficiently your brain works to let you know the source of a sound—how it does so automatically and without any prompting from your conscious mind—there is a simple test that you can perform right now. Shut your eyes and pay attention to the sounds you hear. The direction just comes to you, it is just there—you don't have to think about it. Before reading this chapter, you may not have known how you assess sound direction, but your brain knew. Now think about the nature of the sound that you have located. This sound may take myriad different forms (a brief loud click, a continuous whirr, modulated speech, etc.) and yet, whatever form it takes, you can tell where it comes from. Ordinary, everyday functioning, you may think, but really quite extraordinary when analyzed a little.

In some ways humans do have extraordinary hearing capabilities, but sound localization is not one of them. It is certainly impressive that we can automatically calculate the direction of a sound, but many other animals have the same ability, evolved over eons. Our high bandwidth is a mammalian thing, as we have seen, to help us localize sounds. Our high accuracy and acuity is a predator thing, to help us detect and track unseen prey. What about our sensitivity (low threshold of hearing)? This ability probably *is* uniquely human, and it is connected with speech. Humans are very sensitive to sounds in the 2–4 kHz band, unlike our close relatives, chimpanzees and other apes. There is paleontological evidence to show that we probably evolved our hearing to detect the very subtle features that human speech employs—features that are way more complex than the vocalizations of other animals, even birds, in part because of the mental abstractions that language, as opposed to simple communication, implies.[15]

Animals do have to perceive the meaning of a sound, so they can classify it, which involves some mental abstraction—for example, a dog knows when a sound comes from another dog. There are other subtle perception issues in the animal world of sound. Just as they cannot conceive of the complexity of our speech perception, we have difficulty imagining the owls' "sound picture," and the even more detailed acoustical imaging of our echolocating bat and whale cousins (on which there is more in the next chapter). Owls may be able to make a mental map of sound intensities in two dimensions, which would look, in some sense, like a visual image. Research over the past 30 years has established that this imaging is exactly what is going on inside the brain of barn owls; there are space-specific neurons in the owls' brains that fire when sounds are detected at specific two-dimensional locations. These neurons are laid out in such a way as to form an internal map, so the barn owls "see" what they hear. The images must be coarse compared to visual images (because visual angular acuity is much better than 1° or 2°, as we will see in Chapter 10) and must appear like a pixelated or coarse-grained photograph. But they are good enough to be able to "view" a mouse in the dark and drop down to catch it.[16]

Acoustic Attenuation

Some animals can detect sounds below our frequency range. We miss much of the songs of elephants, whales, and even of the Capercaillie (a large bird of the grouse family). Low frequency sounds travel farther than higher frequencies, and so infrasonic communication can take place over large distances. The scattering and absorption of sound waves becomes really significant when they travel through dense vegetation, such as a forest, but it shows the same characteristic: low frequency sounds travel farther. We will see in Chapter 12 that this fact has constrained the vocalization frequencies of many animals.

The physics of acoustical propagation through vegetation has two consequences of interest to us. First and most obviously, a forest animal that wants to be heard at distance needs to be louder than his friend who lives on open grassland. (Stand next to a squawking macaw and you will have no doubt that he is a forest bird who wants to be heard.) Second, because frequencies attenuate differently with distance, bird songs and calls (for example) sound different at 100 m distance than they do at 10 m distance. You can see how this works from Figure 66. Here the spectrum—the graph of sound intensity versus frequency—of a sound changes with distance from the source. Note how the higher frequencies are more strongly attenuated. This aspect of acoustics is applied by a number of forest-dwelling bird species who do not want their calls or songs to be heard by a distant predator, but only by another bird of the same type, close by. Such a bird will emit high frequency sounds, because these are attenuated more strongly. There are many examples of such birds, including the brown creepers and chickadees of the New World (and their close Old World relatives, the treecreepers and tits).[17]

It has been shown in a number of ingenious experiments that acoustic attenuation is used by woodland birds to judge the distance of a rival. As with many birds, male Carolina Wrens (*Thryothorus ludovicianus*) sing to establish and maintain territories. Because they live in wooded areas, the timbre of their songs changes with distance, as illustrated in Figure 66. (In addition to the dispersion of sound through differential frequency attenuation, songs change over distance because of *reverberation,* which is the scattering of sound off trees and other environmental objects.) By comparing his own song with one that a wren hears from a distant rival, he can judge how far away the rival is and so can decide whether to seek out and repel the rival from his territory.[18]

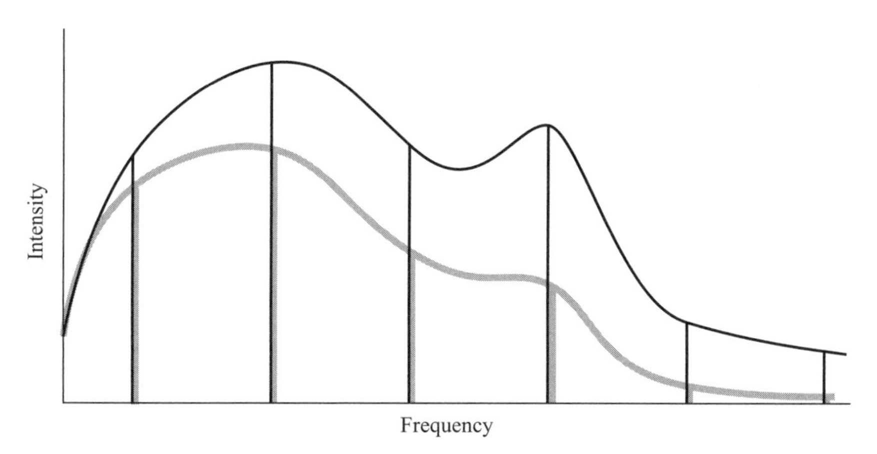

FIGURE 66 A song comprised of many frequency components (vertical lines) may exhibit a spectrum that looks like the black curve. But the same song, when heard from some distance away in a forest, will sound different (gray curve). The change occurs because attenuation increases with frequency, and so the relative contributions of different frequencies to the song change. Using the shape of a song spectrum, some birds can estimate distance to a rival singer by song matching.

The notion of adjusting a song to that of another bird, for purposes of comparison, is quite common, and is known as *song matching.* Here we have given an example in the context of distance estimation in dense vegetation, but song matching is more widespread among birds than this example suggests. Thus, pairs of European Goldfinches (*Carduelis carduelis;* not woodland birds) learn to match their songs, presumably so that each can recognize his or her mate in a noisy environment. This idea of matching a received signal with one of your own is very suggestive of correlation processing. It seems eminently reasonable to suppose that birds who indulge in song matching are performing some sort of correlation processing. Recall how gannets can acoustically identify their mates very quickly, even though a mate's voice is surrounded by thousands of others close by. Penguins in large colonies also have this ability: adults learn their chicks' voices and can later identify the voice in a crowd. It seems that even wrens and penguins, birds not renowned for their mental faculties, have some impressive signal processing capabilities.

9

Animal Sonar

Most people know that *sonar* (*so*und *n*avigation *a*nd *r*anging) is a technique warships use to sense the presence of enemy submarines amid the murky depths of the sea, where vision is impossible. They consider (rightly) that sonar is similar to radar, except instead of transmitting electromagnetic waves, sonar employs acoustic waves. In fact, sonar is in many ways exactly the same as radar; the differences between the two remote sensing techniques are almost entirely due to the different types of wave that are transmitted. Sound waves are quite different from electromagnetic waves, and these differences present different problems for the practical implementation of sonar and radar devices. The basic similarity is this: both radar and sonar transmit waves to the outside world and receive echoes. The echo signals are then subjected to a lot of clever signal processing (many radar and sonar computer algorithms are the same) to yield information about a target: range, direction, speed, and character.

Animals do not have radar,[1] but instead they make use of passive optical remote sensing (in plain language, they see), which is the subject of the next chapter. Most animals have not evolved ways to transmit electromagnetic waves of chosen wavelength and shape—hence no radar—but a few have learned how to do the same thing with sound waves. Humans talk, whales sing, birds chirp, and bats echolocate. (In fact, many whales and a few birds also echolocate, as we will see.) The word "echolocation" (or sometimes, "bio-sonar") is used to describe sonar when it is utilized by an animal. The term was coined by Donald Griffin, who first understood, in the mid-twentieth century, that bats made use of acoustic remote sensing.[2] Microchiropteran bats are the best in the animal world at echolocation, toothed whales are a close second, and two groups of birds (Oilbirds and cave swiftlets) are a distant third. (Human-engineered sonar fits in way above birds, but somewhere below whales and bats, on the scale of signal processing sophistication.) In this chapter we introduce the basics of remote sensing and then build up to our state-of-the-art understanding of the process. This approach requires that, after summarizing a few remote sensing basics, we present the animal echolocators in order: first, the crudest, entry-level processing of some birds; second, the much more impressive capabilities of dolphins and their brethren; and third, the astounding (and not fully understood) image processing achievements of a large suborder of very small bats.

Remote Sensing 101

We shall avoid nitty-gritty details: a broad brush will paint the backdrop we need. A remote sensor (be it a modern fighter plane, a dolphin, an Oilbird, or a horseshoe bat) transmits a wave (electromagnetic or acoustic) of certain power. The means of generating the wave vary and are not our main concern here.

SOUND WAVES AND VOCALIZATION

Mammals and birds make noises by using their lungs as a source of air, which is then forced through a small opening: essentially the same method by which a wind instrument generates music. In most mammals the primary mechanism for making sounds is by forcing air through the larynx. In birds the process is generally similar, though more sophisticated, with the appropriate orifice being called the "syrinx." The major difference is that the syrinx is located at the bottom end of the windpipe and has two openings—one to each lung.[3] More on this structure in Chapter 12.

Most mammals have vocal chords. These are twin protective membranes stretched across the larynx, which are opened for breathing and closed for holding the breath. By opening the membranes a little and then forcing air past them, their natural tendency to close is countered by the air pressure forcing them apart. The result is that the membranes vibrate, generating sounds. A steady flow of air through the constricted tube of the larynx is chopped up into smaller pulses by the vocal chords. This is the basis—though not the sole mechanism—of animal vocalization, including human speech. The exact details of the sound produced depend on a number of factors, including the length, size, and tension of the vocal folds. (The final sound emerging from the mouth is determined by all sorts of other factors, including the shape and manipulation of the throat, mouth, tongue, and lips.)

The judicious use of lungs, larynx (or syrinx), and vocal chords is the basis of the echolocation signals used by bats and the few examples of echolocating birds. However, things are a little different in the case of cetaceans. First, they lack vocal chords. Cetaceans almost certainly make some of their lower-frequency noises by squeezing air through the larynx, but it has long been recognized that, at least for the high-frequency echolocation signals of the toothed whales, the tubes through which wind was being pushed are the nasal passages. In these passages there are interesting structures—the phonic lips—that could be responsible for the generation of high-frequency sound, basically acting like vocal chords for the nose. However, it was only in 1997 that it was definitively shown that dolphins use the phonic lips to produce sonar clicks and whistles.[4] All toothed whales bar one have two sets of phonic lips, allowing them, like birds, to make more than one sound at a time, as when a piano plays a chord. The exception is the largest of the toothed whales—the Sperm Whale (*Physeter macrocephalus*). These animals probably use a broadly similar mechanism for generating sonar signals, although the details of the plumbing are more complex and a little more obscure—they have a single set of phonic lips in the right nasal passage, located quite far forward in the animal's head.[5]

The Fourth-Power Law

As a wave spreads out from the transmitter, it disperses, and the power contained in the wave becomes diluted. We illustrate this dilution in Figure 67a. Here the transmitter is a bat (echolocating waves are emitted from either mouth or nose, depending on species), and the target is a bug at a certain distance (*range*).[6] Sound waves emanating from the bat spread out; imagine a screen with a square hole cut in it, placed one-third of the way from the bat to the bug. The hole lets through a certain acoustic power density (intensity of sound). The power density reaching the bug is only one-ninth of this amount, however, because it is spread out over a square that is nine times as big (each side is three times as long). So, we know that power is diluted with increasing range: this is the well-known *inverse square* law.

When the bat-generated waves reach the bug, they are reflected in all directions. We sketch this process in Figure 67b. Some of the reflected waves make their way back to the bat, which picks them up in its ears. These receivers are specially adapted and very sensitive. The received power is again diluted by the inverse square law, resulting in an inverse fourth-power law. Consider the following example. Let us say that the sound intensity reflected back into the bat's ears from a small bug 3 m away is 1 unit of intensity (i.e., of power). Now the bug flies away to 3 times the range. The sound intensity heard by the bat from its echolocating signal drops to $^1/_{81}$. If the bug is 10 times farther away than its start distance, then the received signal power is reduced by a factor of 10,000. This is the brutal geometry of remote sensing: receiver power is reduced from transmitter power by a factor that falls as the fourth power of range. The same law holds for human radar and sonar, as well as for animal echolocators.

For those of you who already know something about this subject, an overpowering question may have arisen in your mind, which we had better address right away. The fourth-power law means that the received signal is much, much weaker than the transmitted signal. For human radar remote sensing, it can be 20 orders of magnitude smaller (so, the received signal is 100,000,000,000,000,000,000 times weaker than the transmitted signal). Similar comparisons can be made for human sonar or bird, whale, and bat echolocation. Question: if the receivers (ears) can pick up the reflected signal, then why aren't they totally deafened by the transmitter signal? Answer: as with human radar and sonar, echolocators protect their receivers via the technique

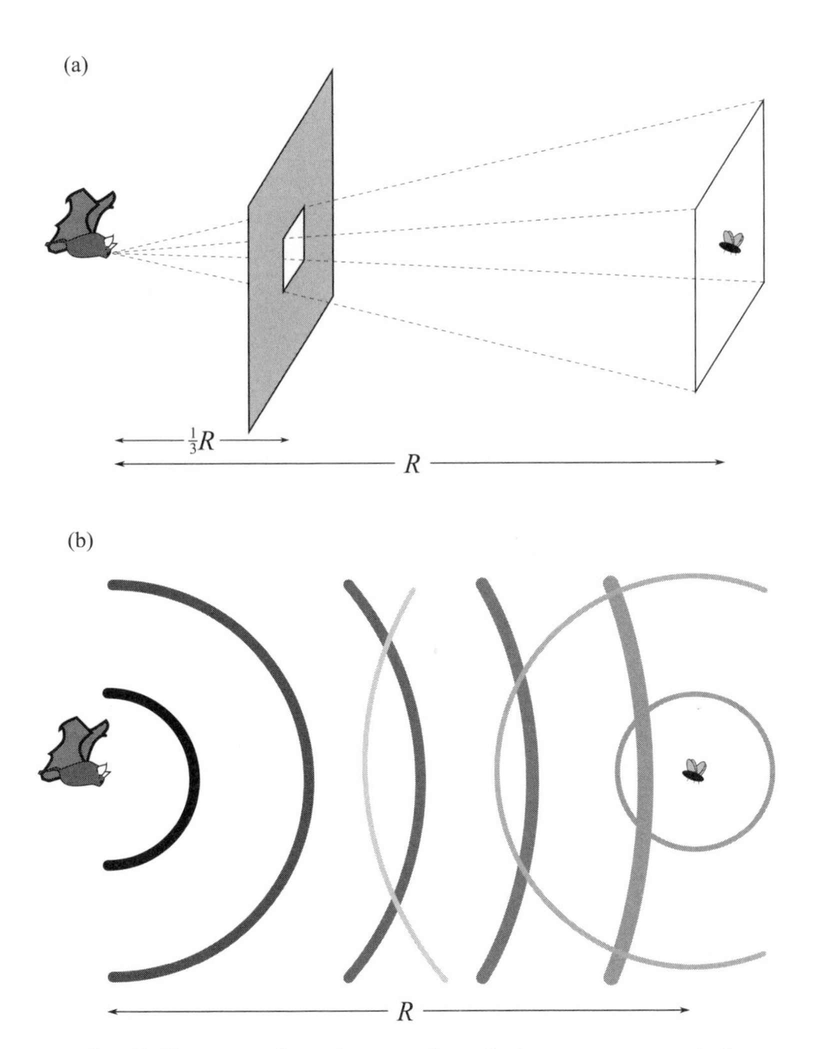

FIGURE 67 (a) The power of sound passing through the square aperture is the same as the power reaching the bug. However, at distance R this acoustic power is spread over nine times the area, so that the power density is one-ninth that at the aperture. This relation is the inverse square law of attenuation. (b) The density of bat echolocation pulse power reaching the bug is the transmitted power divided by R squared. This power is scattered in all directions by the bug, and so the power density that echoes back into the bat's ears is diluted by another factor of R squared. So, echo power falls as the fourth power of target range. For example, the echo of a target at a range of 25 m will be 625 times fainter than the echo of the same target at 5 m.

of *transmitter blanking*. Bats, whales, and the like "disconnect" their ears for a short while when they are transmitting their echolocating signals. This protection is necessary to avoid destroying their sensitive ears.[7] A bat shrieks when it transmits: it is as loud (has the same intensity) as a smoke alarm; fortunately for human sleepers, bats transmit at frequencies way above our hearing range. For whales, the transmitter power is awesome: Blue Whale echolocation transmissions are a hundred million times louder than a Boeing 747. They just *gotta* plug their ears.[8]

Attenuation

The huge variation among transmitted (shouted) and received (heard) signals is even greater than we have suggested. The fourth-power law is based on simple geometry, but physics conspires to attenuate the acoustic signal still further. Sound not only spreads as it propagates through the air (or water), but it is also absorbed and scattered as it passes through. This attenuation depends on the medium (air or water) and on the transmitter frequency. There is a third factor that can seriously influence the intensity of sound received from a transmitter. Reflection from the ground or sea surface, and refraction due to varying density of the air or seawater, can be important, especially over long distances.[9] For bats, attenuation is important and refraction is not; for whales, both attenuation and refraction effects are important, because their echolocation has a longer range. For both bats and whales, reflection effects can be important and depend on the environment in which the creature echolocates.

We consider refraction/reflection effects soon enough; here let us simply summarize the effects of attenuation. Low-frequency sounds travel through the air without being scattered or absorbed much, and so the attenuation of such waves is negligible. For low-frequency sound waves in air, therefore, the intensity is diluted mostly according to the geometrical fourth-power law that we have already discussed. At increasing frequencies, the attenuation becomes more significant. A sound wave of 30 kHz frequency is attenuated at the rate of 0.7 $dB \cdot m^{-1}$, so it loses 15% of its power for every meter traveled. A 30 kHz echolocation signal that travels 10 m (11 yards) to a target and reflects 10 m back to the source passes through a total of 20 m of air and attenuates 14 dB (i.e., loses 96% of its power) due to absorption and scattering. For very high frequencies, the attenuation losses are much higher—more than 3 $dB \cdot m^{-1}$ for frequencies exceeding 100 kHz. If we recall that attenuation losses accrue in addition to the fourth-power rule, it is easy to appreciate why the received echolocation signal power is so much less than the transmitted power.[10]

For underwater echolocators, we need to consider the attenuation rate of sound waves that travel through seawater. Again, losses are small at low sound frequencies and increase as the frequency gets higher; in fact, they are much smaller than for transmission through air, because seawater is a much denser medium and so transmits sound more efficiently. Whales and dolphins can echolocate and communicate over much longer distances than can terrestrial animals for this reason.

Two further elements of remote sensing basics will need to be introduced before we can let you loose on the echolocating creatures of the world. If this section is described as the ABC of remote sensing, then we have covered A (attenuation) and will now move on down the alphabet.

Beamshape

We examined the fundamentals of receiver beamshape in the last chapter. The concept of beamshape applies to both transmitters and receivers, as we saw in Chapter 8. Beamshape is the difference between the light emitted from an unshaded bulb and that emitted from a spotlight, for example. It is clearly advantageous for an echolocating animal to send sound toward a target rather than waste energy by sending it in all directions. Given the effects of geometrical spreading and attenuation, echolocators want to beam as much sound energy as possible onto their intended targets. So, bats have funny-shaped noses (leaf-nosed bats, horseshoe bats, etc.) that direct their echolocation sounds. Some of these amazing adaptations can be seen in Figure 68. Dolphins have a bulbous "melon" on their foreheads that probably serves the same purpose.

The *width* of a beam, whether receiving or transmitting, is important. In Figure 60 (in Chapter 8) we showed how beamwidth is specified. Narrow beams concentrate sound energy to/from a narrow range of directions. So, a narrow beam transmitter will put more sound energy on a target that is within the beam, and a narrow beam receiver will accurately estimate the target direction from the sound received. In contrast, wide beams are better at searching large volumes all around—albeit with less sensitivity. Thus, as so often happens in engineering and nature, there is a trade-off concerning optimum beamwidth that depends on circumstances.

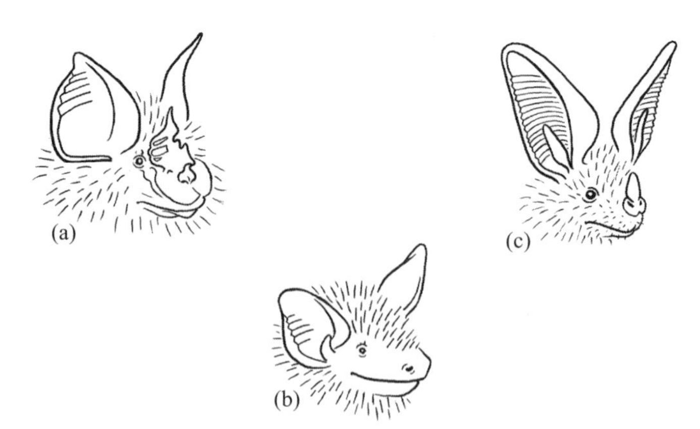

FIGURE 68 Three bat faces that show some of the variety imposed by different echolocation requirements: (a) Mediterranean Horseshoe Bat (*Rhinolophus euryale*), (b) Big Brown Bat (*Eptesicus fuscus*), and (c) California Leaf-Nosed Bat (*Macrotus californicus*).

Clutter

"Clutter" is a radar term; the exact equivalent in sonar is "reverberation." We use "clutter" because it fits in better with the central ABC theme of this section and because it is shorter and more descriptive. (Reverberation is a wonderful description of what it's like but not what it is!) Clutter is the unwanted echoes from stuff out there that reflects sound waves, but is not a target of interest. A rock, for example, reflects bat or whale echolocation sounds just as well as does a bug or a fish. Better, maybe, and therein lies the rub. Before getting you up to speed on the many and varied forms of acoustical clutter, however, we should make the distinction between clutter and noise.

"Noise" is the term given to unwanted sounds that do not originate from reflected transmitter sounds. So, the crackle on a radio is electronic noise (radar has similar gremlins, which become louder as the temperature increases). The rustle of leaves is echolocation noise to a bat that is searching for insect prey nearby. The hiss of the surf is noise to a dolphin hunting in shallow seas near a beach. Noise is characterized by its random intensity, one sample to the next, no matter how short a time interval between samples. It is also characterized by high bandwidth, meaning that many frequencies are present in a sample of noise. If we were to analyze the power spectrum of a noise, we would find that it is pretty flat—all frequencies are present with more or less the same power. Noise can originate from outside the remote sensor (as in the pounding of surf or the creaking and rustling of trees) or inside (as in electronic noise for radars, or the unavoidable random hiss of vibrating ear hairs—papilla—for bats).

Figure 69 gives an idea of the variability of clutter. You can see an airplane, say, a tank-buster, illuminating the ground with a radar beam. It may see tanks, but it will see a lot of other stuff as well—mountains, buildings, water, and land of different terrain. Two points: the area of land and water illuminated by the beam will greatly exceed the area of tanks, and some of these clutter sources are moving. The first point means that there will always be a much larger radar echo from ground clutter than from targets on the ground—and the same thing applies to sonar. A flying bat transmitting echolocation sounds downward may intend to *paint* an airborne moth with sound, but she will also paint the ground underneath.[11] There is a lot more ground than moth, and so clutter will be more strongly represented in the echo than will target. The second point is made by considering water waves and trees in a breeze, for example. Both of these move, and so their echoes will change with time. The same is true of a target, such as the flying moth, or a dolphin's fishy prey. We will see later that movement can be used to discriminate targets from clutter (e.g., rocks don't usually move), but the examples of fluttering leaves and splashing water shows that this *Doppler processing* (separating targets from clutter based on target movement) will not be a silver bullet.

Other ways of boosting the target signal—making that moth stand out from the background—will be discussed in the remainder of this chapter. Much of the cleverness of remote sensing signal processing (be it that of radar engineer, submarine sonar engineers, whales, or bats) is aimed at digging weak target signals from the echo signals, which are dominated by noise and clutter. One method may already have occurred to you from our earlier discussion about beamwidth. A narrow beam (transmitter or receiver) will scoop up less clutter in the echo signals than will a wide beam. A beam will always be wide enough to paint a target, but the amount of clutter it sees depends on beamwidth.

Figure 69 also shows, in addition to the tank-buster radar beam, a bat sonar beam illuminating a bug target against the background of tree clutter.[12] The same elemental warfare is being played out with the same high-tech weapons. Airplane and bat are the predators, tank and bug are the prey. Prey targets are detected by remote sensors and located by sophisticated algorithms that enhance the weak target echo signal at the expense of a strong unwanted clutter component in each echo. Most of the rest of this chapter is devoted to understanding how they do it: how birds, whales, and bats optimize their

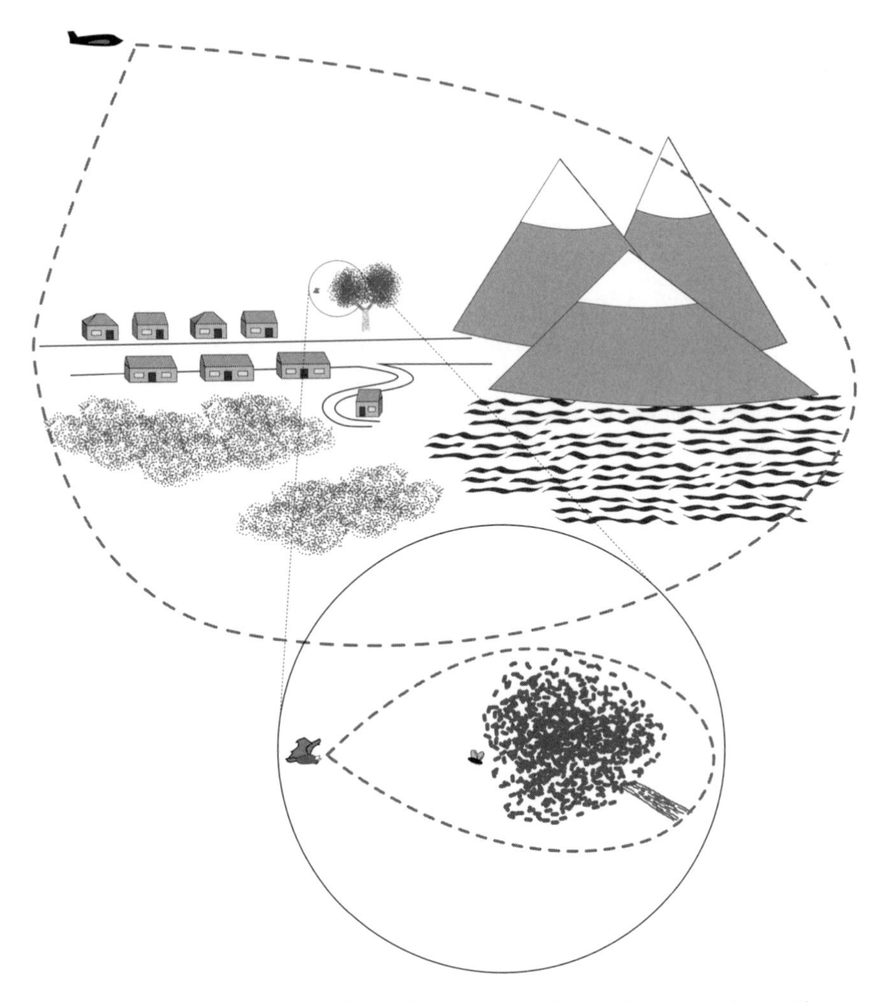

FIGURE 69 Clutter, and lots of it, makes up most radar and biosonar echoes. Clutter is the unwanted echo from everything out there that you don't want to see. The airplane's radar is looking for a tank hiding behind a mountain, but it sees whatever is in its beam (dashed curves), for example, mountains, lake, town, and vegetation. The same problem occurs on a much smaller scale for the bat, echolocating to detect a moth. In this case, the clutter consists of a tree branch and leaves. This figure also suggests something of the variability, as well as the ubiquity, of radar and biosonar clutter.

echolocation systems to help them navigate in the dark, to help them locate moving prey, and to direct them toward their prey and move in for the kill. We will move, metaphorically speaking, from echolocation basement to penthouse —from birds to bats—and while riding this elevator we will, along the way, learn more about remote sensor engineering in the animal world.

Bird Echolocation

We have seen that an echolocator emits sounds, which reflect back from the environment and are then processed; we saw earlier that this technique contrasts with, say, the owls' acoustic locating skills when silently hunting prey. In the jargon of acoustics engineers, owls and most other birds constitute *passive* sonar systems (only receivers—no transmitter), whereas echolocators are *active* sonar systems (receivers and transmitter).

Microchiropteran (small insectivorous) bats and odontocete (toothed) whales are the best animals on the planet at echolocation or sonar signal processing. Humans lag behind a little, but are catching up fast (bats have a head start of some 50 million years). Birds are also in the race, though a distant fourth, and are represented by two groups: the Oilbirds of South America and the family of cave swiftlets in South Asia and Australasia. Oilbirds (*Steatornis caripensis*) are quite big birds (they have a wingspan of 90 cm and weigh 400 g—one yard and 14 ounces); they are related to nightjars and live in cave colonies. They emerge at night to feed, locating the pungent fruit that forms the bulk of their diet by eyesight and perhaps also by smell. Thus Oilbirds never see daylight. To avoid bumping into one another and into the cave walls, they have evolved a simple type of echolocation. It seems that Oilbirds echolocate only within their caves—outside they rely on their excellent night vision. Cave swiftlets are much smaller birds (10 g—a third of an ounce) and, as their name suggests, they also live in caves, where they nest. They (and Oilbirds) use their echolocating capabilities only for navigation and not for prey detection and location. There is evidence that cave swiftlets use echolocation when approaching their nests in dark caves, much as an airplane will use radar when approaching an airport in fog. The nests are stuck to cave walls, and it is clearly in the interest of the fast-flying swiftlets to know where these walls are and whereabouts on the walls their own nests are located.[13]

Oilbirds and swiflets use echolocation to estimate the distance of an object. This estimation is, in principle, simply a matter of timing. The distance to an object is proportional to the time *t* taken for sound to travel out from the bird's mouth to the object and reflect back into the bird's ears. So, measuring the time yields an estimate of the distance. Simple processing, but the devil lies in the details, of course.

Both Oilbirds and cave swiftlets are colonial nesters, and so their echolocation signals—buzzing noises that are within the human range of hearing (unlike bat echolocation transmissions) and sound like paper tearing—have to compete with those of thousands of other birds flying in close proximity and simultaneously echolocating. Many researchers have investigated the echolocating capabilities of Oilbirds and cave swiftlets over the past half century, including Donald Griffin (and Masakazu Konishi, whom we met earlier investigating owl hearing). The result of all this research is that we now have a reasonable idea of how these birds echolocate. Their capabilities are much less than those of our radar and sonar systems, and of the whales and bats, but are nevertheless impressive. We know that, in both groups of echolocating birds, the transmitted signals take the form of a rapid series of short clicks. For Oilbirds, each click or pulse of sound lasts for about 1 ms (a thousandth of a second) and is centered on frequencies in the region 500 Hz–3 kHz. Such a short click duration means that, in theory, the Oilbird can estimate distance to an object accurately, to within about 17 cm, or 7 inches. The significance of click frequency and duration is explained in the caption to Figure 70. In fact, early tests suggested that Oilbirds could estimate distances to within about 20 cm, which is close enough to the theoretical value to show that they do, indeed, simply time the clicks and calculate distance in the same way as radar engineers. Objects smaller than about 20 cm could not be detected reliably in this way. Oilbirds are bigger than 20 cm, and so they can detect one another in their dark caves using echolocation. So, simple echolocation permits a crowd of Oilbirds to fly in total darkness without bumping into one another or the cave walls. The only complication is that each bird must be able to distinguish its own clicks from those of other Oilbirds; otherwise, it will become confused by the cacophony of clicks it hears. Presumably the birds have slightly different clicks in just the way that people have different voices and can recognize familiar voices in a crowd.

Later experiments have led to a more complex picture, suggesting that Oilbirds perform more sophisticated processing than we have indicated. Oilbirds in a dark room were observed (using infrared light, which they cannot detect) as they avoided an array of small objects of diameter 3.2 cm. The birds could

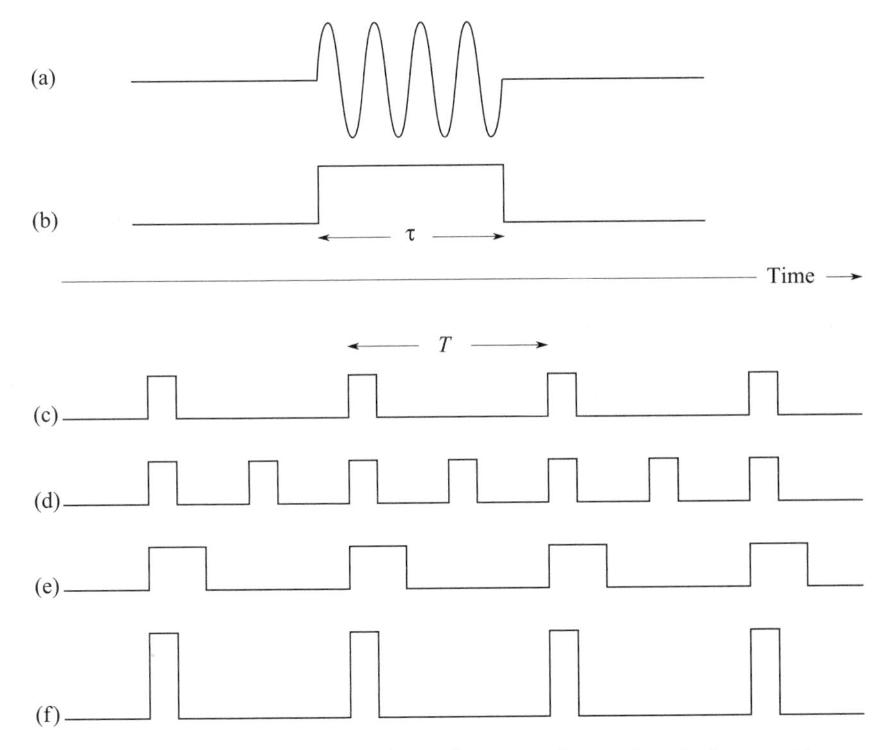

FIGURE 70 (a) An echolocation pulse or click: a simple waveform is shown, with time along the horizontal axis. (b) It is easier to represent the pulse by its envelope, shown here, representing power as a function of time. So, this pulse has a fixed power for a limited interval τ. (c) A train of four pulses, with a pulse repetition interval of T. (d)–(f) Each of these pulse trains contain twice as much power as that in (c), but they are not equivalent. (d) The maximum target detection range for this pulse train, without any signal processing, is the same as in (c). With such processing—such as adding together the echo pulses—this train will have a greater detection range than that in (c). However, it will suffer more from confusion between transmitted pulses and long-range echoes. (e) This pulse train has the best of both worlds: it can detect targets to about the same range as the one in (d) but without the pulse/echo confusion. (f) This train will detect targets to about the same range as those in (d) and (e) but without the need for complicated signal processing: it attains long detection range by virtue of transmitting loud pulses. However, it takes a more robust voice box to create such loud pulses, and in some cases—in an environment where there is a lot of clutter—loudness is no help, because it simply creates more clutter . . . unless the clutter is removed by additional signal processing. This simple example of the variety of possible pulse trains illustrates something of the complexity of remote sensing in the real world, and here we have considered only pulse duration and repetition rate—we have not allowed for variability of the transmitted waveform.

fly through the array without bumping into these obstacles. When the array was rearranged, the birds still avoided the objects. They bumped into arrays of smaller (0.5 cm diameter) objects, however, which tells us that the practical limit of Oilbird echolocation acuity is somewhere between 0.5 cm and 3.2 cm. This accuracy is much less than 20 cm, and so Oilbirds must be utilizing more sophisticated processing than we have so far suggested.

Investigations of cave swiftlet echolocation capabilities produce similar results. These smaller birds emit clicks of somewhat higher frequency and can detect objects that are smaller than we might suppose from the click duration. Despite their smaller size, swiftlets emit louder echolocation clicks than do Oilbirds. This fact is almost certainly related to their need to detect smaller objects (other swiftlets); there is less echo from a small swiftlet than from a large Oilbird, and so the swiftlets must emit louder clicks. The wavelength of each click appears to be about the same as the size of the swiftlet nest, and this suggestive fact has led researchers to conclude that the nest might resonate like an organ pipe when reflecting swiftlet clicks, making it more easily detected.

The effective echolocation range of Oilbirds and swiftlets is quite short. We can infer this range from our own experience with radar and sonar, which tells us how to calculate the maximum effective range if we know the interval between transmitted pulses (radar and sonar engineers use the term "pulse" instead of "click"). The interval between clicks for Oilbirds is about 0.1 s. An echo must return to the Oilbird before it emits another click, because otherwise, the processing becomes very complicated. How could the bird tell to which click an echo corresponded? The last one or the one before that? The bird has to know, because the echo time is different in the two cases. This ambiguity is most easily resolved by emitting a second click only after the echo from the first click has returned. Given this line of reasoning, our experience with radar and sonar tells us that Oilbird echolocation works only for distances less than about 17 m. Sometimes an echolocating Oilbird will increase the repetition rate of its clicks, from 10 Hz up to as much as 250 Hz. For the higher rate, maximum range is reduced to 0.68 m. This short range may seem to be useless, but in fact it may be valuable when approaching a nest or another Oilbird. We know that bats increase their click rate when they get very close to prey objects, such as moths, to more rapidly and accurately update their knowledge of prey location. It is likely that Oilbirds are similarly improving their acoustical information when at close range.

Before leaving bird echolocation, we share some ideas on how our practitioners might achieve better distance estimation than is expected from the length of their clicks. We approach this problem from a radar and sonar engineer's viewpoint—we do not know if these birds do what we propose, but they *could.* In Figure 71 we have simulated an Oilbird click: it has the right duration (1 ms) and the right bandwidth (i.e., range of frequencies, here 0.5–3.0 kHz). The processing simply looks at the magnitude of each click and picks the time that corresponds to the peak magnitude. An Oilbird does not reproduce a click identically each time, so we have allowed for this variation by randomly changing the relative amplitudes and phases of the different frequency components of the click. Such randomization influences the peak time, but the vari-

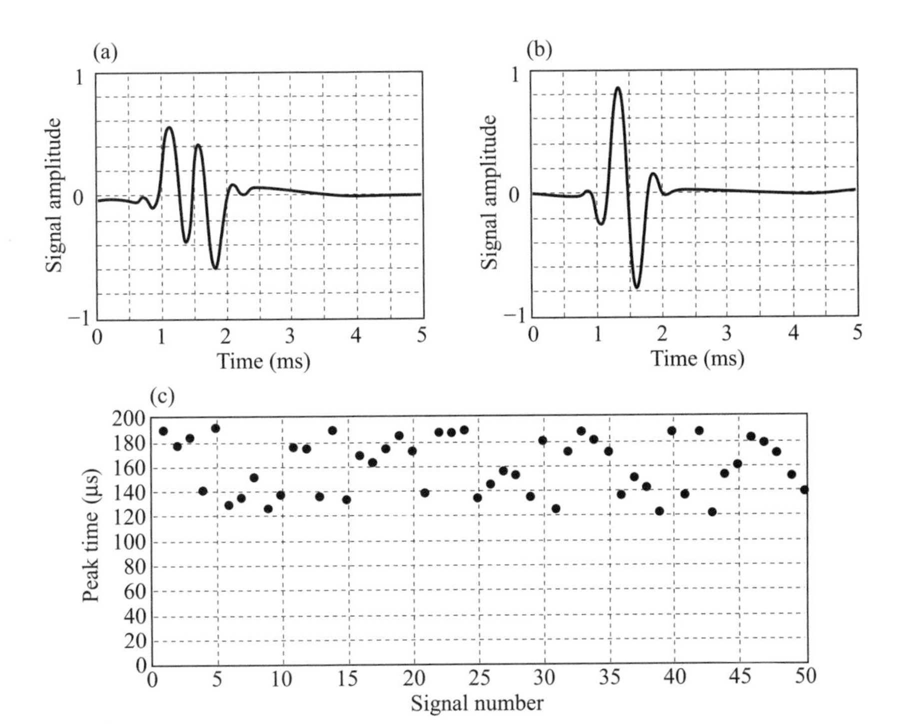

FIGURE 71 (a, b) Synthetic oilbird clicks, limited to 1 ms duration and with frequency components limited to the band 0.5–3.0 kHz. The relative amplitudes and phases of the component waves are different in the two cases. (c) The time in μs at which the peak of the signal magnitude occurs, for 50 signals. The variation of these times corresponds to a distance estimate variation of 3.8 cm—much less than the click length, which is 34 cm.

ation is much smaller than the click duration. As you can see in the figure, the distance estimation accuracy that Oilbirds could attain by this method matches the observed values quoted earlier (3–4 cm). We emphasize that there is no evidence to show that Oilbirds have adopted our proposal, but the figure demonstrates how a simple algorithm upgrade could bring significant benefits.

One basic question that may have occurred to you: if echolocating birds want precise timing, and if this timing is limited by the length of their clicks, then why not simply emit shorter clicks? That would save all this fancy signal processing. The problem is twofold. First, it may not be physiologically possible for a bird to form clicks that are much shorter than a millisecond. (It certainly *is* possible for a dolphin, as we will soon see.) Second, a short click contains less acoustic energy—it is quieter. We have seen that received echoes are much quieter than the transmitted clicks, and so, to hear echoes from afar, it is necessary to transmit as much power as you can. This is why swiftlets shriek their heads off.

Dolphin and Whale Echolocation

In moving from birds to mammals, we take a giant step forward in terms of echolocation sophistication. One authority considers, for example, that Bottlenose Dolphin (*Tursiops truncatus*) echolocation is "probably the most sophisticated target location and analysis system in existence" and compares it favorably with our most advanced military radar systems. Thus, our giant step starts from elementary echolocation, passes over the whole plethora of human radar and sonar development, and lands at the cutting edge of remote sensing—the best that life on Earth has to offer. It is hardly surprising that human remote sensing researchers are interested in whale (and bat) echolocation.[14] The U.S. military funds research into both areas; the U.S. Navy's marine mammal program has been running for decades.

The suborder Odontoceti of echolocating whales have evolved their sonar capabilities over the past 34 million years and have come up with many physical adaptations that enable them to transmit, receive, and process sound waves. The sounds that are transmitted for echolocation differ from those used for communication with other members of their species. Whales sing beautifully; the haunting and mournful-sounding songs are very appealing to many people. These songs take a very different form from the echolocation signals, which are anything but beautiful—they have been likened to a creaking door. Songs are low frequency (some are too low for humans to hear) and extended in time. Echolocation signals from the same animal are ultrasonic (way above the maximum frequency that we can hear) and very short in duration. Both can be very loud indeed.[15]

The reason for the differences between songs and echolocation signal transmissions is due to physics and engineering. We have seen how sounds of different frequency attenuate differently when traveling through seawater. Low-frequency sounds travels farther—they can be heard at a greater distance than high-frequency sounds—because they are absorbed and scattered less by seawater. When a whale sings, he wants to be heard over as wide an area as possible, and so he sings low notes. For echolocation, he requires high frequencies—we will soon see why—and so the range of echolocation is much shorter. A whale song can be heard tens or hundreds of miles away, but echolocation is restricted to a few hundred yards. A whale uses echolocation to orient itself, to find out about its surroundings (coastline, seabed, etc.). It also uses echolocation to hunt for food. Dolphins can not only detect and locate a fish by echolocation, but can also classify the type, identifying whether it is edible.

Range and Angle Accuracy

How do whales and dolphins echolocate a fish? Much of our information about cetacean echolocation comes from studying dolphins. They are intelligent, they cooperate with humans (and so are easy to train), and they are easier to keep in captivity than their larger cousins. We know that dolphins can echolocate a fish quite accurately. If the distance from a dolphin to a fish is 100 m, then the dolphin's echolocation system estimates this distance accurately to within a couple of centimeters. From Oilbirds and cave swiftlets we learned how to estimate the distance to a target by timing a pulse or click of sound. The accuracy of this estimate depends on pulse duration: shorter pulses are more precise. Amazingly, dolphins are capable of producing very short pulses indeed, only 50 μs (0.00005 s) in duration.[16] Such a short click corresponds to a distance estimation accuracy of about 1 cm—less than half an inch. Presumably this method only works for nearby targets, because, you may recall, shorter pulses contain less energy than longer ones and thus are dissipated after traveling a shorter distance through the water. (Dolphins can detect fish as far away as 600 m, though presumably with less accuracy.)

Let us return to the dolphin detecting a fish at 100 m distance. We have seen that it can estimate the distance to within a centimeter or two. What about direction? Researchers estimate that our dolphin can determine the direction of the fish quite accurately. The azimuth (horizontal) beamwidth formed by a Bottlenose Dolphin's ears is about 10°,[17] and the elevation (vertical) beamwidth is similar. So, dolphins can certainly estimate the direction of a target echo to within about 10°. However, we suspect that dolphins, like human engineers, can do better than this. It is not easy for researchers to construct an experiment that reveals the accuracy of a dolphin's target direction estimate (not least because the dolphin can update the estimate as it approaches the target, so how can we measure the original error?). We can only surmise, but it seems likely that dolphins, with more than 30 million years of echolocation evolution on their side, would have at least matched human capabilities. We humans have at least three methods of improving estimation accuracy, so that the error in estimating direction is only a fraction of a beamwidth (a tenth, or less). It seems likely that dolphins use one or more of the three methods of sound localization that we explored in the last chapter: direction estimation by measuring time delay across an array, by measuring phase delay, and by the amplitude monopulse technique of comparing intensities. Recall that many mammals adopt one of these methods for estimating azimuth and another for estimating elevation. Dolphins have all the equipment needed to exploit all of these techniques.

Angular Resolution

Spatially extended hearing is important for target resolution as well as target accuracy estimation. Dolphin echolocation angular resolution is relatively easy to measure, and researchers have obtained quite precise estimates. We saw in the last chapter that accuracy and resolution are not the same thing, although the distinction is sometimes blurred. You might be tempted to say that the angular resolution must be the same as the angular accuracy. After all, if our Bottlenose Dolphin makes an error of 10° in estimating the direction of a single fish, and if two fish have an angular separation of 3°, then surely the errors will overlap, and the distinction between the two fish will be so blurred that the echolocation system sees only one fish. Not so—accuracy and resolution are independent of each other. For example, the error in estimating direction may be systematic and not statistical—to take an extreme example, maybe the dolphin always estimates a target fish as being to the left of its true direction.

In fact, dolphin azimuth resolution has been found to depend on the bandwidth of the echolocating clicks and the distance to the target(s). For targets that are a long way off, say, 100 m, the angular resolution is in the region of 2–4° for a narrowband click. Wideband clicks lead to angular resolutions that are better: 0.7–0.9°. So, at 100 m distance, two fish will be resolved as separate targets if they are 1.5 m apart (about 5 ft). At half the distance, the fish's separation can be half as much, and they will be resolved. Here, though, is a surprise. At very close range the angular resolution improves: for echolocation targets that are only 1 m away from a dolphin, experiments reveal that azimuth (horizontal angle) resolution is between 0.03° and 0.23°, depending on dolphin species; the elevation resolution is between 0.03° and 0.8°. This is surprising, because the long-distance resolutions would have been perfectly good enough at short range. The figures suggest that some dolphins can resolve objects that are separated by only 0.5 mm at 1 m distance. Clearly, such a capability goes way beyond separating fish in a shoal; such high resolution must be for identifying features of a single fish. The dolphins are likely forming images in their brains of the echolocation target object—sound pictures. This ability would be useful for distinguishing edible from inedible fish, for example. We do not know how far out (to what target distance) this hi-res capability extends, but even if it is only 10 m, it will still be very useful in murky waters where visibility is poor.

One difference between sonar remote sensing and radar remote sensing is that acoustic energy penetrates inside the target, be it a submarine or a fish, and is reflected from internal features as well as from the surface. So, dolphins likely see a three-dimensional sound picture, with skeletal details of fish target, swim bladders, and other internal features that may differ from species to species.

Signal Enhancement

Previously we mentioned that echolocation clicks can take several different forms. Oilbirds and cave swiftlets can adjust the frequency of the clicks, and their individual durations, but dolphins and whales have much more control of the waveforms of their echolocation transmissions. These waveforms are strongly indicative (to radar and sonar engineers) of the type of sophisticated signal processing that is going on inside the cetaceans' large brains. Some pulses are of more-or-less constant frequency, but the amplitude and frequency may change from one pulse to the next. Other pulses may be formed from a chirp waveform, in which the frequency sweeps up or down over a wide bandwidth. The duration of each pulse can be altered, and the pulse repetition frequency can be changed. Dolphins and whales may send out a long pulse burst (i.e., a long series of repeated pulses) when searching for an echolocation target and then pause to listen for a response, before sending out another long burst. If they are orienting themselves in a new environment and want to learn of the surrounding seabed, then the burst of echolocation pulses that they transmit will be different from that used to detect fish. If they detect a target of interest, then they may interrogate it in more detail. They may send out a pulse burst consisting of many different types of clicks, rasps, chirps, door-creaks, groans, moans, warbles, and many other descriptive expressions. They might concentrate on certain frequencies if it seems that these will resonate within the target (swim-bladders must ring like a bell at the right frequency), or they might switch and swap around the order of the pulses—anything to get more information from the target echo.

There are two main reasons for all this variability and number of transmitted echolocation pulses (there may be hundreds of pulses in a single pulse burst). First, as we have just seen, they are trying to learn more about an already detected target. They most probably make use of correlation processing, which will help them to classify targets into known types.[18] The idea behind correlation processing, in the context of target classification, is explained schematically in Figure 72. Correlation processing can also be used, in conjunction with a chirp waveform, to provide exquisitely fine range resolution—we will investigate this subject further when we get to bat echolocation, because bats do it with particular finesse.

The second reason for the variability of pulses, and the large number in a pulse burst, is connected with the noisy and cluttered environment in which dolphins and whales perform their echolocation activities. We already mentioned the variability of clutter types in the sea, and these make the echoes complex and garbled—a small echo of interest is lost among a large echo from a nearby clutter source. Also, noise is all around the dolphin—the sea is a noisy place, with waves breaking; rain on the surface; and clicks, groans, and snapping sounds from other marine life. All these distractions make it difficult to pick out an interesting signal from the welter of acoustical garbage contained in the stream of echoes returning to an echolocating dolphin. Figure 73 illustrates the problem; note that this figure understates the smallness of a real

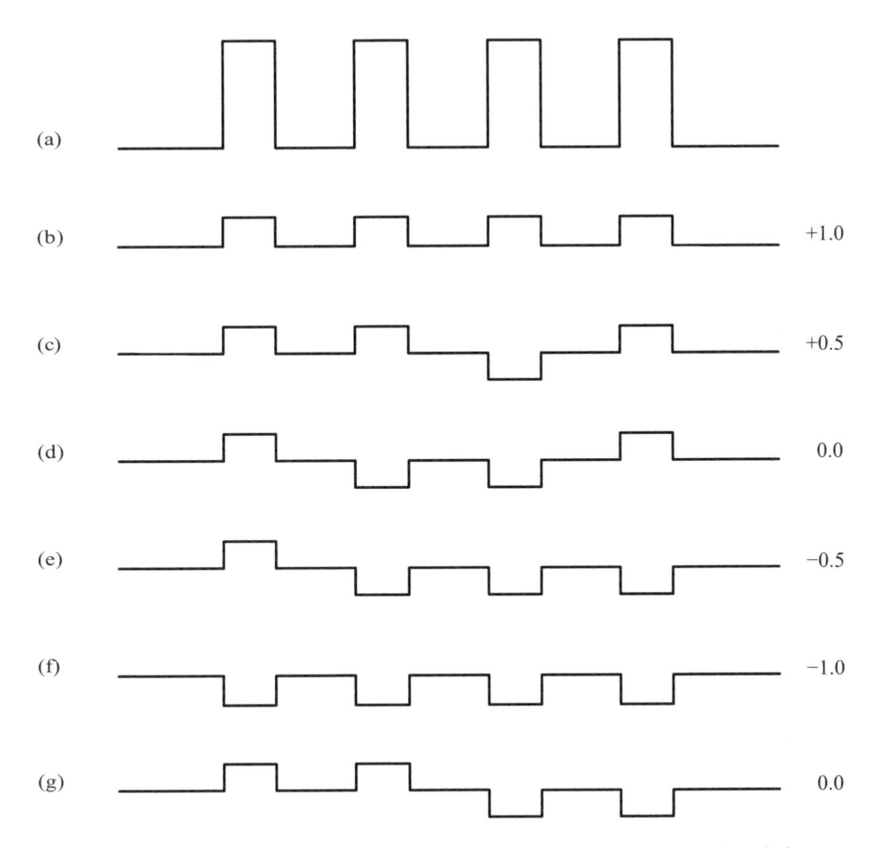

FIGURE 72 Were you a bat or dolphin, we might have provided a more detailed and realistic picture of correlation processing, but because you are merely human, we resort to this schematic. (a) A simple transmitted waveform. (b)–(f) The echoes of this waveform from different targets. Numbers on the right refer to the correlation of each echo with the original transmitted waveform. These numbers must lie between +1 (perfect correlation) and −1 (perfect anticorrelation), with 0 meaning no correlation at all—the two waveforms are independent. So, the correlation number can be utilized to determine what a target looks like when compared to the transmission. In other words, it can help to classify a target. If a herring is known to look like echo (d), then an unknown target with correlation number 0.0 might be a herring. Complication: echo (g) has the same correlation number as waveform (d), but is not a herring. Solution: transmit a different waveform, with the pulses swapped around—maybe waveform (c)—and look at the correlations again. After processing a number of such transmitted pulse bursts, you will be able to classify the target unambiguously.

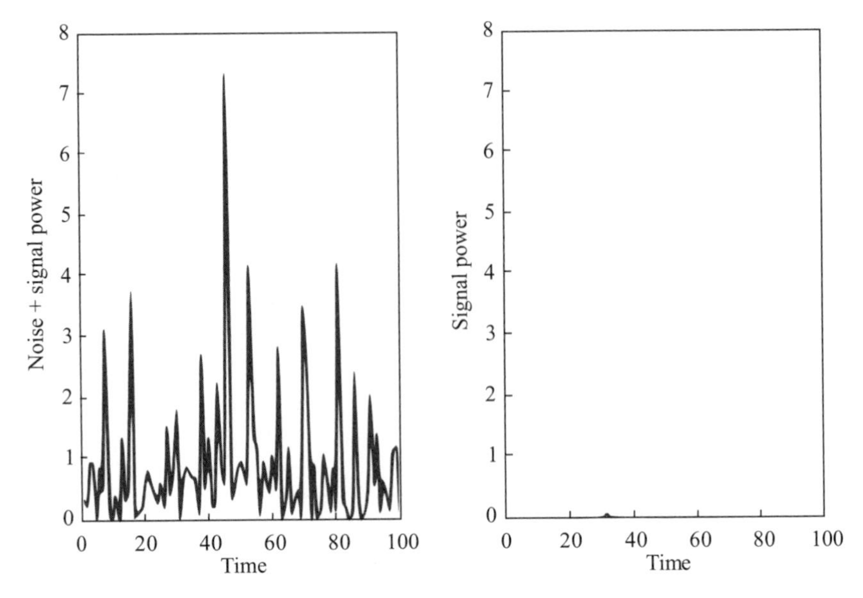

FIGURE 73 A simulated noisy echo, containing a signal with signal-to-noise power ratio SNR = –10 dB. That is to say, the signal power is 1/10 of the mean (average) noise power. (a) Noise plus signal. (b) Signal alone (see it?). The SNR for a real echolocation click can be much lower than –10 dB, and so you can see that a lot of signal processing is required if the echolocating animal is to detect the signal.

signal—many signals are hundreds or thousands of times less powerful than the surrounding noise and would not be visible on the graph shown in Figure 73. How to dig out the signal from all this noise and clutter? There are three main techniques: Doppler filtering, integration, and correlation processing. We mention Doppler processing only briefly here, because once again, bats have developed a particularly neat version of it that we examine in the next section. For now, we need only grasp the basic concept.

Dolphins are known to be able to detect the frequency of an echo and can resolve two frequencies that differ from each other by about 100 Hz.[19] One of the benefits of being able to access the frequency domain of a signal via Doppler processing is that it permits *filtering* of the data—letting through only those frequency components that are likely to contain a target and rejecting all the rest. The acoustical garbage that is contained in every echo signal, the unwanted noise and clutter, tends to be very wideband—with a much broader spectrum than most targets—and so filtering has the effect of rejecting most

PROFESSOR DOPPLER'S AMAZING SHIFT

We are all familiar with the change in pitch of a police car or ambulance siren as the vehicle passes us. When approaching, the sound is at higher frequency than when receding. Why?

What is going on is the *Doppler effect,* which is just a shift in the frequency (and thus wavelength) of a wave as perceived by an observer who is moving relative to the source of the wave. (The effect is named after Christian Doppler, the Austrian scientist who first studied it in the mid-nineteenth century. So it is the "Doppler effect," and not the "doppler effect," as many engineering publications would have you think.) If an object is moving toward you, the sound waves it emits appear to you to be compressed (i.e., shortened, meaning shorter wavelengths and higher frequencies). As it moves away from you, the opposite happens—the waves are, relative to you, stretched (longer wavelengths and lower frequencies). If the object is moving directly toward you at a constant speed, the Doppler shift is also constant and is determined by the speed. In other words, if you can measure the Doppler shift, you can work out the speed. If the object is not moving directly toward or away from you, then the Doppler shift is determined by the size of the component of its velocity in a straight line between you and it (often called the "radial component"). In other words, the Doppler shift depends only on the rate at which you and the object are separating from, or approaching, one another, and not on the sideways motion. As it approaches and then passes you, this radial component changes continuously. That's why you hear a continuously changing pitch when you are standing by the freeway.

Thus, to observe a Doppler shift, there must be a radial velocity component between observer and emitter. Either or both can be moving, it doesn't matter which, as long as there is some relative velocity between them. The Doppler effect is observed in anything that emits waves, not just sound. So, for example, light and other electromagnetic waves also show the effect. The important thing here is that, if you know what the original frequency of emission is and you can measure the Doppler shift, you can work out the radial velocity. If you were the emitter, then you know—and can control—that emission frequency. In nature the ability to detect and process Doppler shifts is, as we will see, a very useful trick that the best of our echolocators have developed to a fine art.

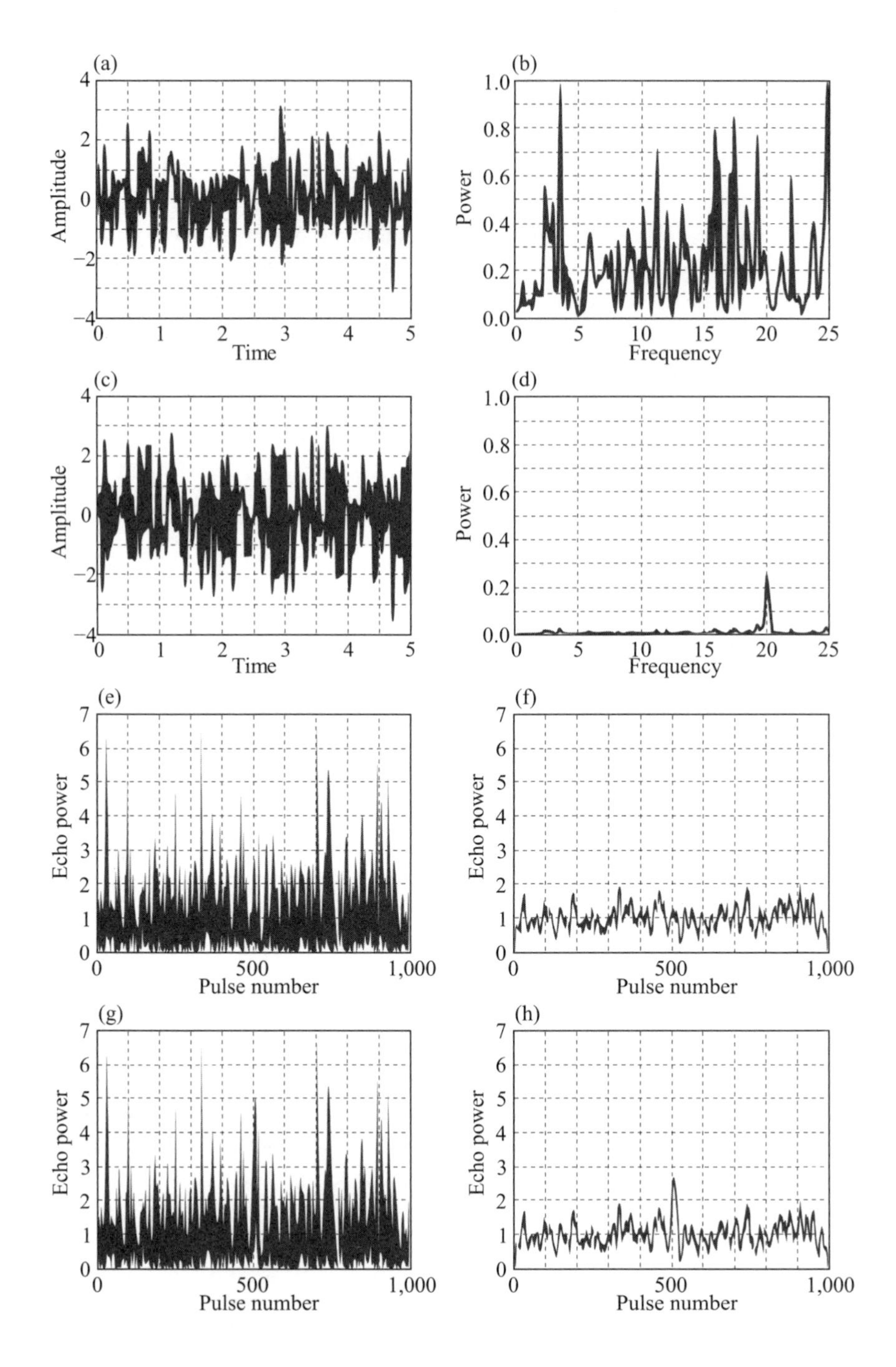
(a)
Amplitude
Time
(b)
Power
Frequency
(c)
Amplitude
Time
(d)
Power
Frequency
(e)
Echo power
Pulse number
(f)
Echo power
Pulse number
(g)
Echo power
Pulse number
(h)
Echo power
Pulse number

of the clutter while retaining the target portion of an echo signal. We can illustrate the efficacy of filtering with a computer simulation (see Figure 74).

Integration, like Doppler processing and correlation, is part of the radar and sonar system engineer's signal processing toolbox. We thought of these methods in the past 60 years; whales and dolphins have been refining them for 30 million years and bats for 50 million years.[20] Integration is a basic technique for improving the power of the target relative to the power of the noise and clutter that inevitably accompany it. The relative power of desired target signal to unwanted noise and clutter signal is often described as the signal-to-noise ratio.[21]

How do we integrate a pulse burst of echoes? The simplest method is just to add together a number of echoes. Each echo is considered to be a sample of the target (and also of noise and clutter) at a particular instant of time. Different pulses will see the same target, but the noise power changes each time we sample it, and in many cases the clutter power also varies from one sample to the next. By adding together a number of pulses, we are increasing the target power more than noise and clutter power, and thus we are increasing the signal-to-noise ratio. Integration is readily demonstrated by a computer simulation, as you can see in Figure 74.

Bat Echolocation

All the signal processing techniques and remote sensing capabilities of whale and dolphin echolocation are mirrored in the airborne world of microchiropteran bats. The suborder Microchiroptera incorporates more than 800 species—a fifth of all mammals—all of whom have well-developed echolocation adaptations and abilities. Just take it as read: bats can do all the stuff we outlined

---

FIGURE 74 (*opposite page*) Signal processing in the frequency domain and signal enhancement via integration (adding together many echo pulses). This how we humans do it in our radars and sonars; echolocating mammals do something very similar. Frequency domain: (a) An echo pulse consisting of noise. (b) The frequency spectrum of the noise pulse. (c) The noise pulse with a small 20 Hz target signal added (SNR = −3 dB). (d) Spectrum of the noise-plus-target pulse—the target signal stands out clearly. Integration: (e) A pulse train consisting of 1,000 noise echoes. Echo power is plotted versus pulse number. (f) The same pulse train but now smoothed out by integration over 10 consecutive samples. (g) The same pulse train as in (e) but with a target added for pulses 501–510. The target power is less than some of the noise echoes. (h) The pulse train of (g) smoothed. Now the target signal stands out a little more clearly. In all cases the units are arbitrary.

above for dolphins, and the associated signal processing takes place inside a brain weighing half a gram (about 1/60 of an ounce).

The performance levels are different, because bats are much smaller than dolphins and because they echolocate in air rather than water. The speed of sound in air is 340 $m \cdot s^{-1}$, which is less than a quarter of its speed in seawater. Attenuation is also different. Bats, of course, are much smaller than cetaceans, and the acoustical power they generate is much, much less. Bats can communicate (inverse square law) out to 50 m or 100 m. They can use their echolocation skills to navigate perhaps 15 m or 20 m and to hunt flying insect prey over distances of 5 m. Navigation works for longer ranges than hunting, because the environmental objects reflecting echolocation signals (e.g., trees, river banks, buildings) are much bigger than the moths and mosquitoes that bats hunt. At very close ranges (1 m or less) bats are able to form very high resolution sound pictures of the insects that they hunt (presumably to assess edibility).[22]

It is conjectured that bats evolved echolocation as a response to the evolution of diurnal raptors—birds of prey that hunt during the day. Bats were obliged to avoid these predators and adopt nocturnal hunting. Big fruit-eating bats could detect their food by smell and sight, but the small insectivorous bats could not hunt at night by sight; echolocation evolved to solve the problem. Many echolocation adaptations, both internal and external, are evident in bats. The most obvious external adaptations are seen on the face (as we saw in Figure 68), where nose and ears are often grotesquely enlarged or distorted to aid sound production, directionality, or reception. Because of their small size, the directionality of bat echolocating beams (both transmitted and received) are wider than those of cetaceans, at about 60°.[23] Despite these wide beams, bats are capable of accurately estimating the direction of an echo and of high-precision angular resolution. Thus, for example, the Big Brown Bat (a favorite of bat researchers) is capable of resolving objects separated by as little as 1.5° in azimuth, and 3° in elevation. Clearly, a lot of signal processing is going on inside their tiny skulls. They may well be adopting the amplitude monopulse approach that we discussed earlier.

However, with only two ears, the monopulse idea will only work in one direction (probably the horizontal direction, to improve azimuth resolution.). If bats had evolved more than two ears, they might be able to use monopulse in both directions, but with only two ears, this cannot be. They need an independent technique for the other direction. It has been observed that some bats wiggle their ears as they approach a target. This action is the same as scanning a beam past the target, first one way and then the other. Radar systems engineers will recognize the action: we call it "sector scanning," and we use it to improve angular estimates by moving our radar antennas. Sector scanning works by noting the change in echo signal power as the beam scans past a target. It peaks when the beam is pointing directly at the target and fades away on either side. The wiggling helps to point the receivers in the target direction, so improving both angular accuracy and resolution. Thus, it may be that bats use monopulse to improve azimuth angle estimates and sector scanning to improve elevation estimates.

The echolocation pulse waveforms vary a lot from species to species. Some species of bats have a large repertoire of echolocation calls, each adapted to a particular situation. The so-called "CF-bats" transmit loud *constant-frequency* pulses (the horseshoe bat is an example of this type). Such calls may be repeated slowly, if the bat is searching for prey a long way away (a whole 5 m) to avoid pulse-echo overlap, which would confuse the range estimation processing. The FM-bats (*frequency-modulating* bats) transmit chirps, but with a crucial and very elegant twist—to be revealed shortly. CF-FM bats transmit both types of pulse at different times. For example, to search for prey, these bats will transmit CF pulses at low pulse repetition frequency—this is the initial detection phase of their attack sequence. Then they will switch to the approach phase, in which the pulse repetition frequency is quickened to update the target distance and direction information more rapidly, and so to track the target. Then, during the terminal phase, the echolocation waveform switches to FM, which permits target classification, as we will see.

We postponed discussion of two remote-sensing techniques until this section: Doppler processing and chirp correlation processing, because bats have very elegantly refined these techniques in an interesting way. Before discussing them, it is appropriate to point out that, in general, bats have all the specialized signal processing equipment that we humans have learned and that cetaceans have evolved. Here is a partial list:

- *Transmitter blanking.* Bats would deafen themselves if they heard their echolocation pulses as they were being emitted from their mouths (or noses, in some cases). So they have evolved a mechanism that effectively disconnects their ears during transmission. Even if the bat echolocation signal pulses rapidly, the ears are synchronized exquisitely to disconnect during transmission and only during transmission.

- *Air traffic control.* Bat colonies can number in the millions, and so there exists the danger that a bat may confuse its own echolocation echoes with those from another bat. This does not happen: bats allocate frequencies, or construct unique combinations of pulse types, so that they recognize their own signals and discount those of other bats.
- *Integration.* Bats must integrate the echo signals that they receive: research has shown that the power received in a single echo is insufficient to provide bats with the target detection ranges that have been observed.
- *Multimode, multitarget processing.* We have seen how bats can switch modes during an attack sequence. There is also evidence that they can process—and track—the returns from more than one target. Such multitarget tracking is a capability that we humans have been able to build into our radars only during the past few decades.
- *Image formation.* Bats are capable of high angular resolution and range resolution, and they probably use this information to form images of close-range targets. We humans can also form high-resolution images with our airborne radars—*synthetic aperture radar* (SAR), which is responsible for many of the satellite radar images that you may have seen. Here is the rub: our SAR radar does not work in the forward direction. An airborne radar cannot form a SAR image when looking along the direction it is traveling. Bats need this capability, however, and it seems that they have evolved it.
- *Acoustical warfare.* Not strictly a bat characteristic, this eerie reflection of human electronic warfare is a property of tiger moths and other bat prey.[24] Tiger moths have antennae that listen in on bat echolocation frequencies, and if they detect a bat predator, they can then take evasive action (recall that echolocation transmissions can be heard well beyond echolocation detection range). If the bat is too close for evasive action, the moth will go to ground and so will blend in with the ground clutter. It may even send out a jamming signal at the right frequency or, in an attempt to confuse the bat, send out false echoes.[25]

Doppler Processing

CF-bats emit a fixed frequency pulse, sometimes of quite long duration—up to 100 ms or more. The echo pulse is not necessarily of the same frequency; indeed, it may have many frequency components, if the target from which the echo arose is moving. The Doppler effect discussed earlier shifts the frequency of a reflected signal if the reflector is moving. From the frequency shift, people, porpoises, and pipistrelles can calculate the reflector's speed. Suppose that a bat echolocation pulse reflects off a moth in flight. The moth's wings are moving at different speeds; at the base they move at the moth flight speed, but further out they move at a fluctuating speed due to the wing beats. Consequently, the Doppler spectrum is spread out, and different moths have different Doppler spectra, so that bats can analyze their echoes in the frequency domain to tell what type of moth the echoes have come from.

For the bats to be able to resolve these echoes in the frequency domain with high resolution, they have evolved sensors in their ears that are particularly sensitive to a narrow range of frequencies. This acoustical sweet spot has been dubbed the "acoustic fovea" by some researchers, by analogy with the optical fovea at the back of our eyes (which provides sharp central vision for demanding tasks, e.g., reading). So, echoes that fall within the sweet spot are converted into frequency spectra that the bat uses to identify, or at least classify, a flying insect.[26] Trouble is, sometimes the insect movement—or the combination of bat flight speed plus target insect movement—means that the reflected Doppler frequencies fall outside the sweet spot. Here is the clever bit: bats have evolved a way to make sure that the echoes always hit the sweet spot. They shift their transmission frequencies so that the central echo frequency hits the middle of the sweet spot. If they detect that the echoes are shifting in frequency, then they shift their echolocation transmission frequencies to compensate. This on-the-fly dynamic response is very hi-tech.

Chirp Pulse Compression

In Chapter 8 we mentioned correlation processing in the context of bird song matching. Song matching involves comparison of identical (or similar) signals, and so correlation processing is the natural signal processing tool to use. We saw in Figure 72 how cetaceans can use correlation to match a sequence of echo pulses to a template of target types as an aid to classification. There is another use for correlation processing, however, and it results in very high resolution images. In particular, the correlation processing of chirp pulses can be used to detect very small differences in echo times, and so provide very high range resolution.

Chirp pulses and correlation processing go well together, like beer and nuts. Look what happens if we correlate a transmitted chirp pulse with a chirp echo in the time domain (see Figure 75). We obtain a very short duration spike, which serves to provide excellent timing information. The width of the spike turns out to depend (inversely) on the chirp bandwidth, but it does

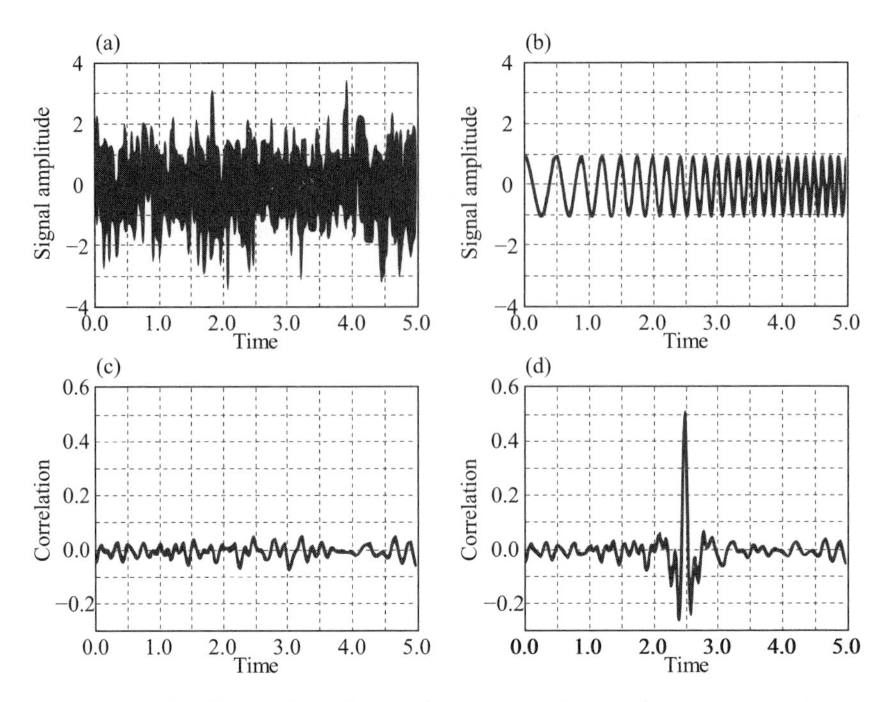

FIGURE 75 Simulation of signal strength versus time: how to obtain precise timing and hence precise target distance information. (a) This noisy echo pulse has a chirp signal buried in it. (b) Here is the chirp on its own. (c) The correlation of a chirp with pure noise just produces more noise. (d) But the correlation of a chirp with an echo consisting of chirp-plus-noise produces a sharp spike when the two chirps are aligned. So, the arrival time of the echo is measured very precisely.

not depend on the chirp duration. Recall that, for Oilbirds, range resolution depends on pulse duration. Here it does not, and by transmitting a wideband chirp, bats can achieve very precise timing and hence very precise range estimation and resolution. In fact, we can infer from the measured chirp bandwidths that FM-bats are able to resolve two objects—or two parts of the same object—that are separated in range by only 3–4 mm.

Cetaceans also transmit chirps, and so they most likely perform similar correlations to improve range estimations. However, bats have refined the technique as follows. It turns out that the physics of Doppler frequencies is not quite as we have described it so far. It is usual to talk about Doppler frequency shifts, and we have followed this practice. However, a detailed analysis of the Doppler effect shows that the change in frequency that occurs when a wave reflects off a moving object is not a simple shift but is a more complicated, nonlinear function. It is approximately linear, consistent with the idea of a frequency shift, but when the moving object has a speed that is an appreciable fraction of the wave speed, this approximation breaks down. The consequence for echo timing is that the sharp spike in Figure 75 spreads out—timing precision is degraded a little. For radar, the wave speed is the speed of light—so fast that we can always make the linear (shift) approximation. For underwater sonar, the wave speed is the speed of sound in water (1,450 $m \cdot s^{-1}$), which is usually much greater than the speed of any object moving through the water. For bats, though, the wave speed is the speed of sound in air (340 $m \cdot s^{-1}$). For most sonar engineers, the linear approximation is still good enough, but not for some bats. They have modified the linear chirp to form a *hyperbolic frequency modulation* transmitter pulse, and this slight modification corrects the problem and sharpens the timing spike again (see Figure 76).[27] Clever stuff.

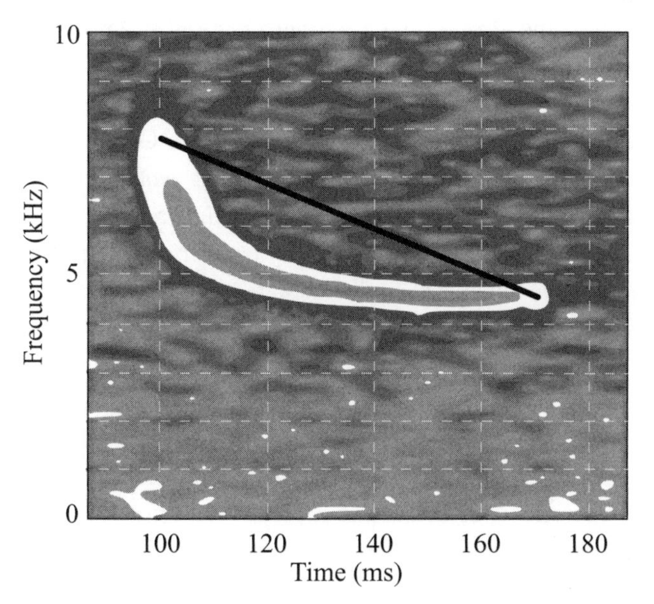

FIGURE 76 A spectrogram of a pipistrelle bat (*Pipistrellus pipistrellus*) hyperbolic FM chirp signal. Spectrograms project three-dimensional audio data (frequency, time, and sound intensity) onto a two-dimensional page—the intensity is represented by grayscale. (The simpler linear FM chirp looks like the superimposed straight line; signal frequency changes linearly with time and has constant intensity.) The complexity of this signal is evident. The boomerang shape suggests strongly that pipistrelles are performing sophisticated signal processing, as discussed in the text.

Cleverer still: some bats can achieve range resolutions as low as 0.1–0.4 mm. This extreme range precision corresponds to timing accuracy of a microsecond or so—three orders of magnitude smaller than the bats' neural response time. How do they do it? We don't know.

Other Echolocators?

It has been suggested that pinnipeds (seals and their close cousins) can echolocate in a crude fashion, though recent literature argues against this possibility. Instead, seals have evolved an effective passive acoustic detector system (in plain words: they're good listeners). In particular, they are able to pick up on the echolocation transmissions of predators, such as the Killer Whale, or Orca (*Orcinus orca*). We saw earlier that one-way sound transmission (from echolocator to target) loses intensity according to the inverse square law, whereas two-way transmission (from echolocator to target and back) obeys an inverse fourth-power law. This geometrical fact of life means that Orca targets, such as salmon or seals, can hear the Orca echolocation signals at longer range than the Orcas can hear the echoes. So a seal listens for the echolocation transmissions of its most feared predator, the Orca, and takes evasive action, hopefully—from the seal's point of view, if not the Orca's—while still out of Orca range.[28] The same reason can also be used to argue against pinnipeds evolving their own echolocation system: it can be detected at longer range than it can be used. So, a lurking Orca might listen in on a pinniped frequency band and pick up the presence of lunch (a seal), while lunch is blissfully unaware of this, because the Orca is out of seal echolocation range.[29] So, from a seal's point of view, there's little point in evolving a capability that helps you find food but gets you killed in the process.

It has also been suggested—or hinted obliquely—in the technical literature that penguins may have an echolocation capability. The evidence so far suggests otherwise; penguins use their acute vision—which is optimized for underwater operation—to detect and approach their fishy prey. We note also that, just like seals, penguins are amphibious creatures, needing to hear in both air and water and lacking any obvious specialist adaptations for sonar. They are also, again like seals, a favorite prey of Killer Whales, so sonar would not seem to offer any survival advantage.[30]

Perhaps surprisingly, a few species of shrew use a simple form of echolocation in the darkness of their underground burrows, where vision is of little use. They shriek at a very high pitch, and from the echoes can learn something about the extent of the tunnel ahead of them. We can do something similar in a cave: shouting and listening for the echo can tell us whether the cave wall is near or far. Shrews may know that their tunnel extends a long way ahead but may still send out their high-pitched signals to help navigate ("the tunnel branches up ahead") or detect a predator farther along ("rat blocking the way—let's get out of here!").

There is a species of animal that shows no signs of any structural adaptations favoring biosonar capabilities, yet can consciously learn to use active echolocation to find its way around. That species is us.[31] Blind people can very quickly become adept at the use of passive echolocation—sensing the world by listening to the noises around them and listening to the interactions of those noises with other objects in the environment. Drive a car down a quiet street with the windows open and you can hear the reflected sounds of your passage change as the style, shape, and even the nature of the buildings and street furniture around you change (e.g., a hedge and a wall sound very different). When you don't have sight as your primary sense, this ability to listen to the world is enhanced as the brain, always hungry for information, pushes its processing capabilities to new heights. Moving beyond this ability, many blind people learn to make the sounds themselves—the tapping of a cane, the click of shoes on the floor—to increase their capabilities. This process—using mechanical means to generate the signal—is the first step toward active human echolocation. Clearly, it has limitations in where, when, and how you can use it. The next step is to make the sounds organically. There are numerous people who have learned to do this by clicking their tongues, allowing them to identify size; position; and, in some cases, shape of the objects around them. In fact there are programs that teach the blind to use these techniques, and there are practitioners who can happily ride bikes, roller-skate, skate-board, and even (in a limited way) play basketball. No doubt the magnificently adaptable human brain is processing all manner of other information, including other sounds, but the clicking and listening is a major part of the process. And all this without bat ears or high-frequency signal generators to get a capability that is less impressive than that of birds but is truly astounding, because it is a consciously learned ability. Apart from enriching and simplifying the lives of its practitioners, human echolocation may also give us insights into how the masters of the craft originally evolved the capability. To get started, all you really need is a sound source, good ears, and a brain to put it all together.[32]

10

Seeing the Light

Physicists recognize four fundamental forces of nature. Two of these—gravity and the electromagnetic (EM) force—are important to biologists and engineers (the other two are locked up in atomic nuclei most of the time). The electromagnetic force manifests in three forms: as electric fields, as magnetic fields, or as a combination known as EM radiation. The animal kingdom includes members that can, among them, detect all these manifestations and use them to perceive the world around them. We concentrate here on passive EM sensors ("eyes"), because these are the most common manner in which animals utilize the EM force. Some creatures also use electric fields—they get a cursory glance at the chapter's end. Others have a magnetic sense—in Chapter 11 we will see how some migrating animals perceive Earth's magnetic field.

Optics 101

Light is EM radiation. It is that small part of the EM spectrum to which our eyes are sensitive, for reasons that will soon become clear. Before we get to eyes we must brush up on optics—the physics of light. Compared to the acoustics of Chapter 8, however, the physics is much simpler. This simplicity arises because the wavelengths of light are so short that we can, with one or two major exceptions, forget that light is a wave at all. We treat it simply as radiation that follows straight lines. This *geometrical optics* approach is entirely standard and uncontroversial, so long as we remember that the simplified physics applies only because light wavelengths are much smaller than the length scales of anything with which they interact, such as lenses or retinas.

Figure 77 introduces the nomenclature of optics and illustrates most of the optical phenomena we need to know. Transparent matter is described optically by its *refractive index.* When light passes from one transparent medium to another with a different refractive index, it changes direction in a predictable way. If the transition is sharp, as when light moves through a lens, then this refraction is sudden; one straight-line trajectory becomes another straight-line trajectory in a different direction. Increasing the difference in refractive indices of the two materials leads to a bigger change of direction. In Figure 77a you can see how a lens causes refraction that focuses an image. Biological lenses have a refractive index n close to that of water ($n = 1.33$), whereas air has a refractive index of 1.00. The lens focuses parallel light onto a focal plane, as shown by the dashed and solid lines depicting light rays in the figure. Objects that are far away emit light that reaches the eye in almost parallel rays (solid rays in Figure 77a), so that these objects focus almost on the focal plane. Closer objects are focused beyond the focal plane (dashed rays).

Some biological lenses can be made to change shape (this ability is highly developed in primates and carnivores but less so in other orders), which brings in to play some more optics. The physics of refraction tells us that the angle through which light is bent by a lens depends not only on the refractive index difference but also on the angle at which light enters and leaves the lens. Changing the angle leads to a change in refraction and so to a change in focal length, as shown in Figure 77b. This is why our vision is focused only at one distance; eye muscles automatically adjust lens shape so that light from objects at that distance is focused at the focal plane, where our photoreceptors are found. A perfect lens will focus all incoming light at a single point. A defect in the lens may result in different focal points for light in different planes. For example, in Figure 77b we see light rays in the plane of the paper focused at a certain point; if two rays in a different plane focus at a different point, then we say that the lens is "astigmatic."

There are a couple of basic reasons why focusing is imperfect, even without lens defects. Both camera and biological lenses are most easily constructed with a curved surface that is part of a sphere. In Figure 77c you can see that such lenses do not focus all light at the same place—the focal plane changes with distance of the light from the optical axis. This phenomenon, which we will meet again, is called "spherical aberration." Figure 77d shows another relevant phenomenon: *chromatic aberration* results because refractive index depends on light wavelength. Thus, a given lens has slightly different focal lengths for different colors of light. Short wavelengths refract more; blue light has a shorter focal length than red.

An Eye for Detail

We are interested in image-forming eyes, not just light-sensitive cells, and this restriction limits us to three phyla of animals: chordates, arthropods, and mollusks. We concentrate on the engineering aspects of animal eyes rather than their biological structure. To us, eyes are lens systems that provide information about an animal's surroundings, gleaned from ambient EM radiation. A *simple eye* forms an image by using a single lens. Some simple eyes (yours and those of most vertebrates and some mollusks) focus by changing the shape of the lens; others (the eyes of fish, amphibians, snakes, and cephalopods) focus by moving the lens, like cameras do. Some animals have more than one lens in each eye and can see magnified images; their eyes are like telephoto lenses. We will meet one such creature later. The majority of animals are arthropods who have *compound eyes* with thousands of lenses. In some ways compound eyes are less capable than simple eyes, but they have their advantages, as we will see.

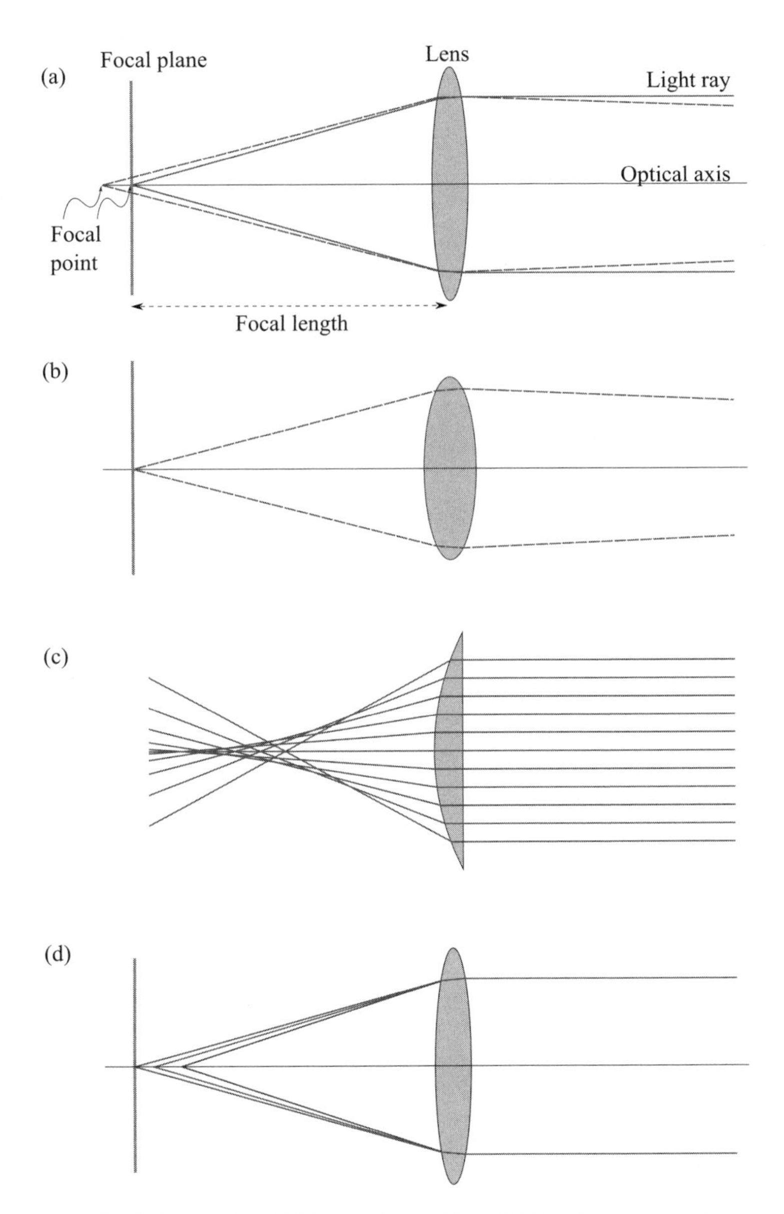

FIGURE 77 Single-lens optics. (a) Nomenclature. The solid-line light rays are from a distant object; the dashed ones are from a closer object. (b) Some animals alter focal length by changing the shape of the lens. (c) Spherical aberration: a homogeneous lens with a curved surface that is part of a sphere does not focus all rays at the same point. (d) Chromatic aberration: a lens does not focus all wavelengths of light at the same point.

Eyes are, for many animals, "the premier sensory outpost of the brain." They have evolved independently many times and are well-adapted to very different animal lifestyles and light conditions. (The resulting variability of eye shapes is made clear in Figure 78.) For example, there are at least eight independently evolved heterogeneous eyes (those with a variable refractive-index lens—reducing with distance from the optical axis—to correct spherical aberration). There exist 11 known methods of combining lenses in eyes to produce focused images; of these, 6 have been discovered only in the past 25 years. Clearly, eyesight is a very useful survival aid, and there is strong selection pressure to evolve and improve eyes. Let us look at the diversity and capabilities in and behind different eye designs.[1]

Simple Eyes

Vertebrate eyes evolved only once, almost certainly in the sea. Biologists will provide their own compelling evidence for this oceanic origin, but to an engineer it is obvious from the optical properties of seawater. In Figure 79 we plot the attenuation of EM radiation as it travels through seawater; if you recall the definition of dB from Chapter 8, then you will realize Figure 79 tells us that almost all EM radiation frequencies are stopped dead in a few centimeters; only the narrow band of visible light penetrates farther (a few tens of meters). Air also attenuates EM radiation, but not nearly so much. The fact that vertebrate eyes are sensitive to the same narrow range of frequencies that penetrate seawater points to an oceanic origin of our eyesight.[2]

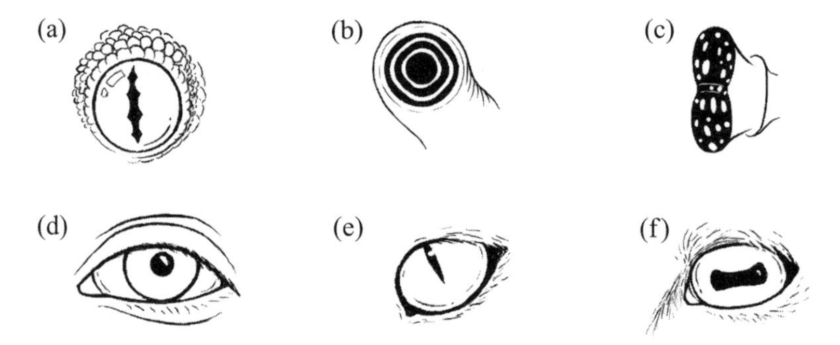

FIGURE 78 Eyes are many and varied. (a) Tockay Gecko, (b) Queen Conch, (c) Zebra Mantis Shrimp, (d) human, (e) cat, and (f) goat. The shrimp eye is compound, the others are simple.

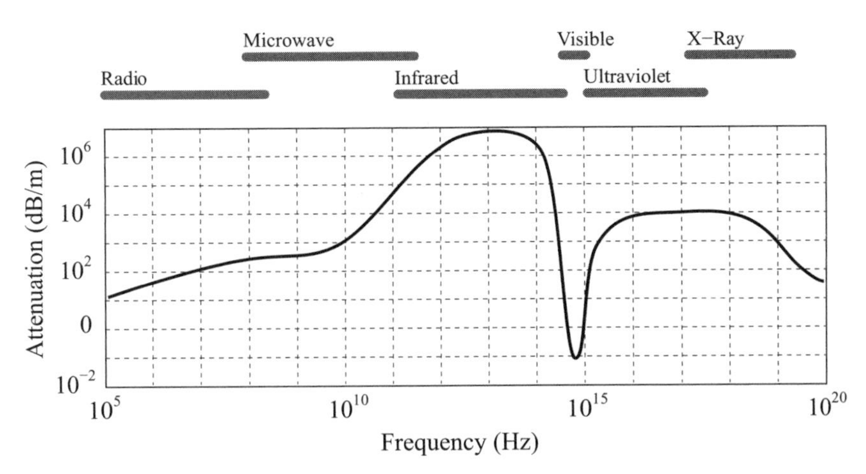

FIGURE 79 Attenuation (absorption and scattering) of light through seawater versus light frequency. Only visible light penetrates any distance, which is why our eyes utilize these frequencies.

The basic geometry of simple eyes is shown in Figure 80. Light passes into the eyeball via the transparent cornea and then through the lens to the retina at the back of the eyeball. The corneas of terrestrial animals refract light, thus helping the lens to focus the rays. In most animals the cornea is fixed in shape, and so refraction is a fixed amount, but eagles and other diurnal birds of prey can change the shape of their corneas as well as their lenses to fine-tune the focusing of light rays on the retina.[3] Also in Figure 80a,b we see how lenses change their shape (*accommodate*) to focus the incoming light at the retina. Finally, a basic fact of simple lens eyes is shown in Figure 80c: the image that forms on the retina, even if well focused, is upside-down (and also left-right inverted). This uncomfortable fact is an unavoidable consequence of single-lens eyes. Fortunately, our brains flip the image so that we see the outside world the right way up. We will have more to say about the considerable amount of post-processing that goes on behind the eyes—for now let us concentrate on the basic optics.

How well can simple eyes see? There are a number of different measures, the importance of which varies from species to species:

- Acuity (focusing ability);
- Sensitivity (ability to function in different light intensities—a sensitive eye works in low light levels);

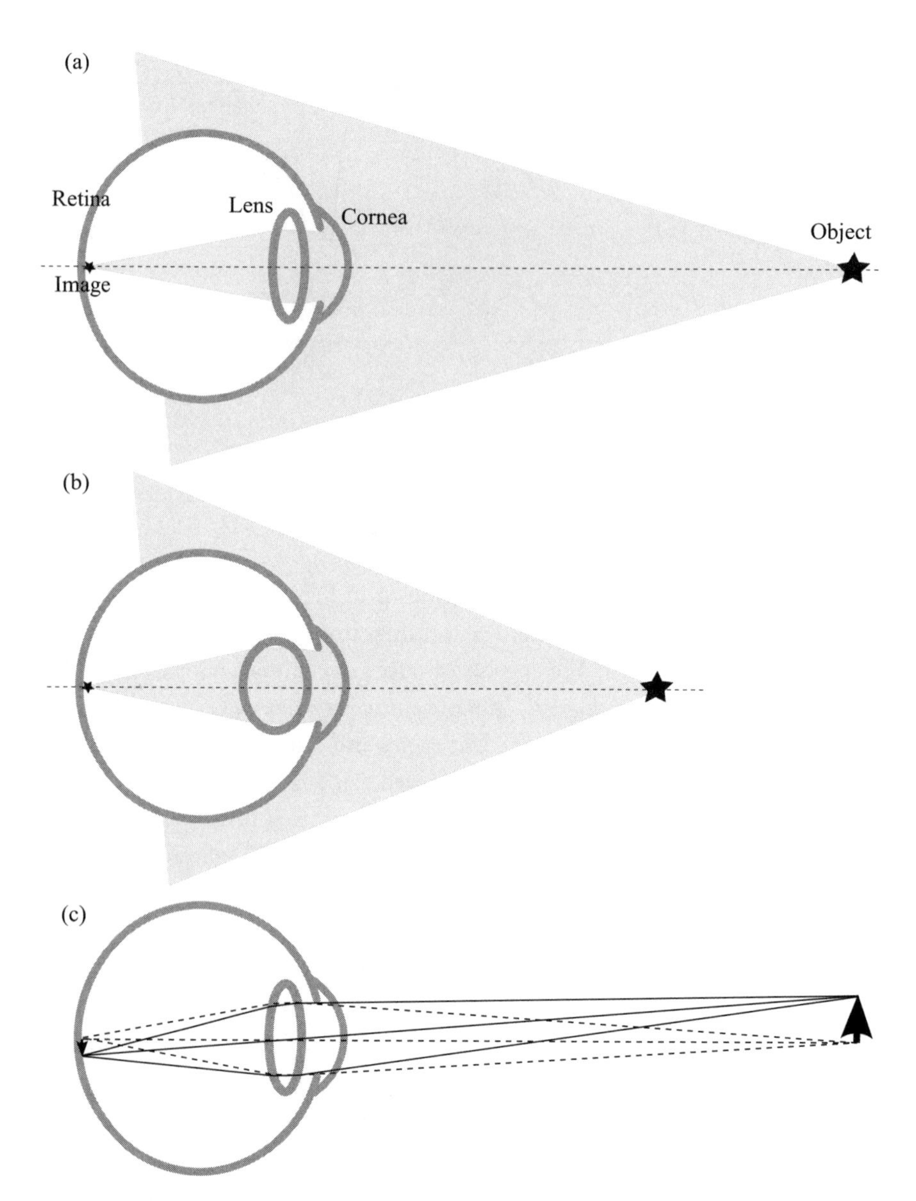

FIGURE 80 Basic optics of the simple eye. The retina is a fixed distance from the lens, and so, to focus objects at different distances from the eye, we must change the lens shape. The image appears on our retina upside down and left-right inverted.

- Color differentiation (how well the eye discriminates the spectral components of light);
- Field of view (how wide an area the eye covers, at any one time); and
- Depth perception (how well the eye informs about the distance to an object).

Acuity and Sensitivity

We take these two measures of eye capability together, because we need to mention, right at the outset, a fundamental optical constraint. How well an eye sees in dim light depends on its size: big eyes gather more light and so are more sensitive. Big eyes have big retinas—retinas contain photoreceptors that detect the light. So, the more retinal surface is bathed in light, the more sensitive the eye. Here is the catch: to form a sharp image (to have high *acuity*), the lens must focus the incoming light onto a small part of the retina, thus reducing sensitivity. There is a fundamental trade-off between eye sensitivity and acuity: these two characteristics cannot be simultaneously optimized. Nocturnal animals need high sensitivity, and so they generally have large eyes with poor acuity.[4] Animals that hunt for their food need to have sharp vision; this acuity is paid for by low sensitivity and so requires high light intensity—hunters are often obliged to be diurnal. So how do nocturnal hunters manage? The many and varied ways in which different animals get around the fundamental trade-off depends on their lifestyles, as we will see.

There is a basic equation in geometrical optics that tells us the angular resolution or acuity of a lens—let us call it the "acuity equation." This equation states that acuity gets better for shorter wavelength light and for larger lenses. For humans, the math yields a theoretical best resolution angle of 0.013° (about one seventy-fifth of a degree). This resolution occurs when the pupil is large (7 mm). In bright sunshine our pupils contract, reducing the effective lens diameter to about 3 mm and so increasing the resolution angle to 0.033°. In practice we see worse than these figures suggest; spherical and chromatic aberration reduce acuity, increasing our best (smallest) resolution angle to about 0.017°. Rats, with their smaller eyes, can achieve only 0.5° resolution.[5]

Rods and Cones

Eye sensitivity and acuity depend on photoreceptor characteristics as well as on geometrical optics. In vertebrates, the retina is made up of two types of light-sensitive cells, known from their shapes as *rods* and *cones.* We can think of these cells as consisting of photoreceptors at one end and "wiring" at the other. Light stimulates the photoreceptor and sends a signal along the wire; thousands of such wires are bundled together as a wide bandwidth cable (optic nerve) that takes data to the central processor (optic lobes of the brain). There are large numbers (from thousands to millions) of rods and cones in the retinas of vertebrate eyes. It is the density of rods that influences eye sensitivity and of cones (usually) that influences acuity.[6]

Rods are highly sensitive—experiments with rods from a monkey's eye show that they can detect a single photon—a quantum of light.[7] A dark-adapted human eye can see very low light levels (equivalent to a candle 30 miles away), and our eyes are not particularly sensitive—sensitivity (high rod density) is the hallmark of nocturnal eyes. Sensitivity results in part from the long response time; rods absorb light typically for 0.1 s before emitting a signal. Because rods are so sensitive, they saturate easily; this process is known as *bleaching,* and it leads to afterimages when rods are saturated by strong light, as when we glance at the sun. The sensitivity of rods results in many nocturnal eyes (those of cats are a familiar example) having a slit pupil instead of the round pupil of our diurnal eyes. Eye muscles can reduce a slit pupil to almost nothing in strong light (round pupils cannot be reduced in size so much) to protect the rods of those nocturnal animals that happen to be awake during the day. The eyes of a nocturnal carnivore are large, because they need to be sensitive (big lenses let in more light) and also need high acuity (from the acuity equation). If operating in sensitive mode, the pupil is dilated and the lens does not focus sharply; it bathes as large a retinal area as possible, so that the rod photoreceptors can soak up the light. To enhance sensitivity, such eyes are usually equipped with a reflective layer behind the retina. This *tapetum lucidum* reflects photons that have passed through the retina, giving the rods a second chance to absorb them. The *tapetum* is responsible for the eyeshine we see when owls or cats are caught in our car headlights. If operating in high-resolution mode, the nocturnal carnivore's lens is strongly curved to focus light to a small area of the retina. The emphasis for most nocturnal eyes, however—even those of hunters—is on sensitivity. An Oilbird's eye has high sensitivity but low acuity. A cat can see in one-sixth of the intensity of light that we require, but their resolution is 10 times worse.[8]

Cones are less sensitive—they require much more intense light to function. Each cone accepts wavelengths over a narrow range of perhaps 100 nm, and (in humans) there are three types of cones, responsive to red, green, or blue light. Thus, color vision requires cones. Cones are also responsible for acuity in many animal eyes, particularly for the high resolution of diurnal hunters. We can see from Figure 81 how cone density influences acuity.

The acuity equation tells us how lens size determines optical acuity—but cone density is also important, just as the resolution of a digital camera depends on the number of pixels per square inch (pixel density) as well as upon the camera lens. Figure 81 shows cone cells at the back of the eye. You can see how angular resolution *a* depends on how close together the cones are placed. Many animals (birds, primates, teleost fish, and some reptiles) have a small area of the retina—the fovea—that is particularly rich in photoreceptors; this high density translates into high visual acuity for the foveal region. For humans the fovea extends over an area covering about 3° of our visual field. Our acuity in this central area is high, but it is much lower in the rest of our field of view because of the lower cone density in these peripheral regions. Humans posses about 200,000 cones per square millimeter of fovea, whereas eagle cone density is 1,000,000 mm^{-2}. This increased density is reflected in increased acuity for eagles (and other raptors). The Wedge-Tailed Eagle (*Aquila audax*), for example, attains an angular resolution of 25 seconds of arc.[9]

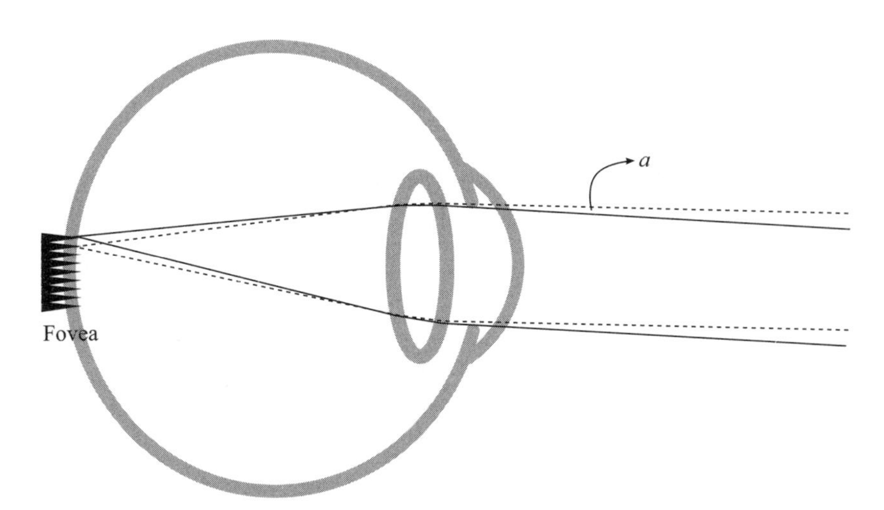

FIGURE 81 Acuity (the angular resolution, *a*) is determined by photoreceptor separation as well as by lens optics. Here the photoreceptors are cones in the fovea. The angle between two rays entering the eye, one of which ends up at the top of the fovea (black line) while the other ends up at the bottom, defines the foveal field of view. In humans this angle is about 3°.

The distribution of cones across the retina varies from species to species, reflecting the mode of living. Many animals do not need the sharp vision of hunters but instead need to have reasonably good all-round vision to detect predators. The cones of horses' eyes are more evenly distributed throughout the retina; their central field of vision is not as sharp as ours, but their peripheral vision is sharper. Humans require sharp color vision in their central field, and so we have a fovea packed with red, blue, and green cones. Our peripheral vision depends on rods (plus a few blue cones). Many rods are wired together for increased peripheral sensitivity (and consequently reduced peripheral acuity). Thus, we see sharp color pictures in the center of our field of view in daylight and fuzzy images in the side of our eyes at night. (Each of our eyes is equipped with about 6 million cones, mostly in the fovea, and 100 million rods.) Plains animals have a horizontal line of high-density photoreceptors, whereas forest animals do not; presumably plains animals need a sharp image of the horizon to see approaching ground predators, whereas forest dwellers need more all-round vision. Birds that migrate also exhibit a horizontal line in their retinas, perhaps for flight stability—seeing the horizon. Many raptors have two fovea—a central "shallow" fovea like ours and a "deep" fovea, located in the same horizontal plane but offset about 40° to one side (the outside). Thus, these birds can focus sharply in a small area out to the left side with the left eye and to the right side with the right eye, while simultaneously looking straight ahead and focusing sharply with both eyes. What function can such a capability serve? We will find out in due course, once we have learned a little more about the optical engineering of birds and other animals.

Colors—Visible and Invisible

Mammal eyes are usually dichromatic: most have cones in their retinas that can see green light and blue light, but not red. Cats do not make much use of the color capability of their eyes, because they are largely nocturnal, and rods have no color discrimination. Cats can be trained, with difficulty, to respond differently to a blue stimulus than to a green one, which shows that they have the necessary equipment to see in color, but they appear not to use it in practice—brightness levels are much more important for them. Dogs are different: they can distinguish red and blue, but not red and green. Some fish are dichromatic, whereas others are trichromatic; like us primates, they can distinguish three colors, though not necessarily the same three. Most humans can see red, green, and blue; some fish and insects can see ultraviolet. Most birds can see all four: red, green, blue, and ultraviolet—they are tetrachromatic. Pigeons, however are pentachromatic—five colors.[10]

What is the value of color perception? The physics and physiology of light reception make it clear why color differentiation is not an option at night. In bright light, however, it may convey survival benefits, and yet the degree to which different animals exploit color perception varies a lot. Pigeons must have exquisite color differentiation ability; what use is it for them to be able to separate, say, vermillion from crimson? Well, perhaps it helps to judge the plumage of a potential mate. Or choose between a Monet and a Picasso, as has been claimed. More generally, color vision is considered to assist the brain to segment the visual field into discrete objects.[11]

Color vision comes at a price, as we have seen. Chromatic aberration limits the acuity attainable by a single lens. Red and green light have similar focal lengths, but blue is significantly shorter, so eyes that focus yellow light sharply will also focus green and red light pretty well, but blue will be blurred. In cameras we add extra lenses to counter chromatic aberration and so get the best of both worlds: high acuity and color. Simple eyes adopt different adaptations to mitigate the effects of chromatic aberration. The number of blue cones (6% of the total) is much less than the number of green and red cones (31% and 63%, respectively) in humans. The narrow field of view of fovea (3° in humans) limits the light rays to a small part of the lens, and so rays that enter the eye from a wider angle are excluded from the image. This exclusion reduces chromatic aberration, because blue light scatters more through the atmosphere (hence, the sky is blue), so the narrow foveal field of view reduces further the blurry blue contribution. These and other physical adaptations, along with considerable signal processing in the brain, give us very effective edge discrimination without any conscious awareness of chromatic aberration. Adding binocular vision further enhances image clarity and depth perception. Color vision also provides more obvious benefits: for example, red-green discrimination allows us to see berries among foliage.[12]

The cones of reptilian eyes and those of birds and monotremes (egg-laying mammals, e.g., the Platypus) contain colored oil droplets that filter the incoming light and improve visibility in certain light conditions; for example, the yellow droplets in seabirds' eyes may help them see further in hazy conditions. (In some countries, car fog-lights are yellow for the same reason.) Sensitivity to ultraviolet light helps many birds to choose a mate and helps pollinating insects

OF MICE AND MEN

We can estimate the data rate or bandwidth of the signal that is sent from vertebrate eyes to the brain for processing, to form the images that are seen by the eye-brain system. We choose as our representative vertebrate the field mouse, because we happen to have all the data we need about their eyes.[13] Thus, each mouse eye contains about 75 million photoreceptors, of which 97% are rods and 3% are cones. Each photoreceptor represents a single pixel of the image viewed—a dot on the TV screen. Each rod can detect a signal that is as low as 2–5% of the maximum saturating signal, and so we say that each rod can send a signal at 50 (or perhaps only 20) different levels. Let us say, for the sake of our back-of-the-envelope estimation, that there are 32 levels. Thirty-two levels correspond to 5 *bits* of information (because $2^5 = 32$). A bit is a "binary digit" and is the basic unit of information. Cones are more discriminating: they contain about 5,000 levels, say, 12 bits. The recovery time of each mouse rod is about 0.25 s. That is, a rod can refresh or update its signal 4 times per second (12 times for cones). Putting all this together, we obtain the data rate for rods as follows: 73 million pixels per frame, multiplied by 5 bits per pixel, multiplied by 4 frames per second, yielding a data rate of about 1.5 billion bits per second ($1.5\ \text{Gbit}\cdot\text{s}^{-1}$) for each eye ($0.3\ \text{Gbit}\cdot\text{s}^{-1}$ for cones). So, two mouse eyes between them send data[14] at the rate of $3.6\ \text{Gbit}\cdot\text{s}^{-1}$. Compare this with the data rate for broadband internet, which is a quarter million bits per second, at the time of writing. So, a mouse acquires optical information 14,000 times faster than you can retrieve data from the internet. The optical data rate for vertebrates with bigger eyes (e.g., humans, eagles, and cats) is larger because of the greater number of photoreceptors.

Here is another telling figure. The United States Department of Defense will soon have available the services of the Transformational Communications Satellite (TSAT) advanced wideband orbit-to-ground laser communications system. This system is a constellation of five orbiting satellites which, taken together, will be able to communicate data to Earth at a maximum rate of between 10 and 40 $\text{Gbit}\cdot\text{s}^{-1}$. So this state-of-the-art satellite communications system will attain a data rate about the same as that found in the eyes of a family of mice.[15]

select the right flower. Kestrels use ultraviolet light to find food: the voles they hunt leave scent marks and urine trails which are visible in ultraviolet.[16]

Field of View and Depth Perception

We have seen that big eyes are better than small ones for both sensitivity and acuity. Consequently birds, with their dependence on vision, have the largest eyes relative to body size in the animal kingdom (see Figure 82). Our eyes occupy only 5% of our skulls, whereas a starling's eyes occupy 15%, and the figure increases to 50% for some birds of prey—their eyes can be bigger than their brains.[17] This leaves little room for eye muscles, and in fact birds do not move their eyes, as we do. Instead, they rotate their heads. This is why we see the whites of the eye in people but not birds or many other small animals: you can move your eyes to left or right without moving the rest of your head—thus revealing the white sclerotic coat. A bird makes up for its fixed eyeballs with a very flexible neck; owls, for example, can easily rotate their heads 180°.

The position of the eyes in the head varies with species. Predators, such as owls and chimpanzees, have eyes at the front, so that the fields of view of the two eyes overlap. Such binocular vision improves image formation—two eyes are better than one—and provides depth perception. The image formed from two eyes, viewing the same scene from two slightly different places, is no longer two-dimensional. The price paid for depth perception is reduced overall field of view. Prey animals, such as woodcock and rabbits, benefit from a wide field of view more than from binocular vision, so their eyes are placed on top or on opposite sides of the head. The size and placement of eyes tell us much about an animal: nocturnal or diurnal, predator or prey.

Raptor Runaround

A raptor, such as a Peregrine Falcon (*Falco peregrinus*), can see lunch from a mile away (the Peregrine's visual acuity is four times ours). Lunch for a Peregrine typically takes the form of a songbird, which, naturally enough, is not enthusiastic about fulfilling this role. The Peregrine thus has to chase his meal. He is superbly equipped to do so, with flight speeds that can exceed those of any other bird, especially when *stooping* (i.e., diving on prey from a height). To see a songbird from a mile away requires large eyes with high acuity. We have seen that such eyes in birds can track an object, such as a passing robin, only by moving the head—there are no eye muscles in the skull to rotate eyeballs.

FIGURE 82 The eyes of hunting birds are big. (a) Tawny Frogmouth, an Australian nocturnal hunter. Thanks to Brigitte and Norbert Holzl for this image. (b) Cooper's Hawk, a North American diurnal raptor. Thanks to Tom McDonald for this image.

In short, if a Peregrine in flight wants to focus on a bird that is off to one side, he must turn his head. Right? In fact, no, because such raptors as the Peregrine have a second fovea, you may recall. Evidence suggests that this deep fovea, offset by about 40° from the direction in which the bill points, has higher acuity than the shallow fovea of the central field. Thus, the Peregrine is more likely to first see, and can better track, a distant songbird with one of its deep foveae than with both shallow foveae. If this is the case, then our Peregrine cannot approach the songbird directly without holding its head to one side. Unfortunately, tests suggest that a fast-flying Peregrine *doubles* the aerodynamic drag it feels if it moves its head to one side during flight. So there is a big energy penalty to be paid for such movement, compared with pointing the head in the direction of motion and so presenting a streamlined, low-drag profile to the air stream.

Recently it has been shown that Peregrines solve this dilemma by adopting a spiral flight trajectory when approaching prey, instead of a straight line. Figure 83 shows that a spiral approach can maintain a constant angle between flight direction and prey direction. The Peregrine tracks the prey songbird with its (right, in this case) deep fovea while pointing its head forward to minimize drag. It flies a longer distance than would be the case for a straight line, but only about 50% longer, so the low-drag spiral flight path is more energy efficient. Another advantage is that the prey may be less alarmed if it sees the Peregrine flying at an angle rather than straight toward it, and so may delay taking evasive action. In fact, when Peregrines get close to their lunch they do adopt a straight-at-it flight path, because at such distances, the shallow foveae are good enough to see the prey. In addition, binocular vision is possible for a

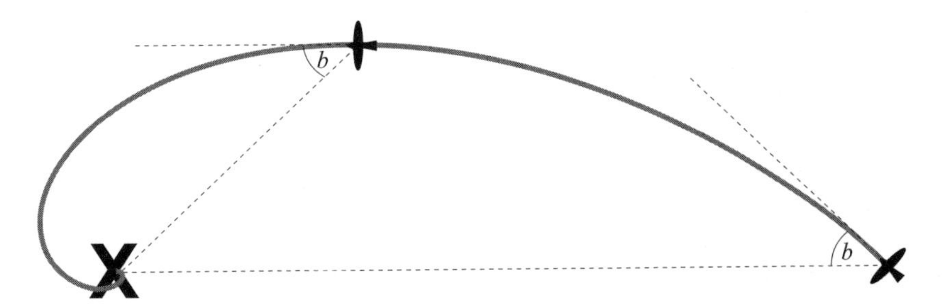

FIGURE 83 Spiral flight trajectory (solid line) of a hunting peregrine approaching prey (X). The angle b is constant, so that the prey image remains in the deep fovea of the right eye. If the prey is stationary, then the trajectory shape is a log spiral.

straight-ahead approach—both eyes can be brought to bear on the target—and so, as we have seen, distances can be judged more accurately.[18] The spiral approach of raptors to their airborne prey is an example of the constant-bearing approach discussed in Chapter 4.

The Mind's Eye

Eyes provide the data—at a great rate, as we have seen—but it is the brain that sees images. There is preprocessing in the retinal neuron connectivity, which is not usually one-to-one (one photoreceptor to one neuron), but most signal processing takes place in the brain. This processing is very computationally intensive, and yet we are barely aware of it. Animals that rely principally on vision to sense their environment devote a significant fraction of their brain volume to optical processing (typically half). This processing must be done in real time. Optical data is updated frequently, as evidenced by our flicker-fusion rate: TV pictures and movie film screens must be refreshed between 24 and 30 times each second (so the rate is 24–30 Hz), or else our eye/brain will see the flickers instead of a continuous stream of images. For birds, who live their lives at a faster pace, flicker-fusion rates exceed 100 Hz.

A lot of the processing must apply across many species. For example, we know that humans process their optical data to maintain color perception in different light conditions, and we process data to turn what is essentially a pair of two-dimensional images into three-dimensional perception. The same processing (perhaps performed using different algorithms) must be done in pigeons' brains, or in the brains of orangutans as they swing from branch to branch through a forest. Other processing is perhaps more species-specific, such as our hardwired ability to recognize faces. We are not neuroscientists, and so we will not go there (we alluded to some optical processing in Chapter 4) but will instead show the cartoon of Figure 84a as a quirky reminder of the behind-the-scenes processing that your brain carries out without telling you.[19]

One facet of human vision requires significant processing: our foveae provide sharp images over a narrow 3° angle, and yet we are barely aware of the fuzziness at wider angles. We hold sharp images in our minds of large objects that extend well beyond this 3° width because of rapid eye movements (*saccades*) and memory. Thus, you know exactly where *this* word is, and can return to it immediately at the end of this sentence, showing that focused images are tagged somehow with a spatial label in our brains and later reconstructed; your eyes will go

FIGURE 84 (a) Most people can see a difference between these two pairs of eyes, yet optically, there is none. Our brains do a lot of behind-the-scenes processing of visual information. We are grateful to Aaron Sloman for permission to reproduce this figure. (b) One method of determining whether you are a cephalopod or a vertebrate. Close your right eye, hold the page at arm's length, and look at the cross. Draw the page closer, keeping your left eye on the cross. If the dot disappears and then reappears as you continue to draw the page closer, then you have a blind spot and are a vertebrate.

straight to the underlined word, even though we have made the sentence ridiculously long and convoluted so that the word in question falls outside the area of your fovea when the sentence eventually grinds to a halt. Right?

Cephalopods and Wiring Errors

"Convergent evolution" is the term used to express the external similarity of creatures from very different evolutionary backgrounds. Thus, porpoises and sharks look very similar, because they share similar habitats. Both have adopted a streamlined form and are similarly colored to blend with their background. A closer look reveals significant differences. The same can be said of cephalopod eyes and vertebrate eyes. These very different types of animal (they belong to different phyla) independently developed simple eyes. The optics and external appearance of the eyes is similar, but the details differ. Thus an octopus or squid eye has a nearly spherical lens and a different cornea. The retina contains structures analogous to, but different from, rods and cones. Octopus eyes focus by moving the lens, not changing its shape. They can detect polarized light; some octopuses can change their color (though they see in black and white) like a chameleon, even emitting polarized light. Much of their brains must be given over to processing this complex information.[20]

There is one difference we want to emphasize between the eyes of octopuses and those of vertebrates, such as you and your dog. Our eyes contain what appears (to engineers) to be a stupid mistake, a "design flaw." This mistake must have been built in very early, because all vertebrates with retinas exhibit the same flaw, which is absent from the independently evolved cephalopod simple eye. The retinas of our eyes are packed with rods and cones, which as we explained earlier, contain photoreceptors at one end and wiring at the other, to connect the photoreceptor signal to the brain. The flaw is this: the wiring is the wrong way around! The photoreceptor end of each rod or cone is at the back of the retina (nearest the brain) instead of at the front of the retina (nearest the lens). Thus, wiring sticks out of the front of the retina, partially blocking light from the photoreceptors. The wires stream across the front surface of the retina and group together at one place, to form the optic nerve, where a hole has to be made in the retina to allow it access to the brain.

There are several consequences of this "mistake." We vertebrates have a blind spot in each eye, because the location where the optic nerve penetrates the retina contains no photoreceptors. The blind spot is located in the same horizontal plane as the foveae, displaced outward about 20°. To detect your blind spot, read the caption for Figure 84b. Different vertebrates find different fixes for this mess-up. Humans process out the blind spot, so that we are not normally aware of it. In mammals, blood vessels supply the back of the rods and cones with nutrients; but the "back" is in front of the photoreceptors, and so the blood vessels get in the way of light reception. Birds, which are more dependent than even we are on good eyesight, get around this problem by dispensing with blood vessels in their eyes. (They may supply blood in a different way, via a poorly understood structure, unique to avian eyes, called a "pecten.") Another fix is to make the retina thin, so that more light can reach the photoreceptor part of our rods and cones. This happens in the fovea (the word "fovea" comes from the Latin for "pit"). Cephalopods have their photoreceptors wired the right way around. Consequently, their vision is not obscured by blood vessels, and they have no blind spot.

Jumping-Spider Eyes

These amazing creatures have camera-like simple eyes, unlike most arthropods, and yet their two principal eyes (they have eight) are quite different from our simple eyes. Jumping spiders are the most vision-dependent of arthropods because of the way they live: they can jump 20 times their own length, and they hunt other spiders. Their six secondary eyes have a single lens and a retina with a few hundred photoreceptors. They produce a wide-angle, blurry image (with an angular resolution of about 1°); these eyes are used to sense the environment on both sides and above the spider.

The two binocular eyes at the front, the *anteromedial* (or principal) eyes, are what we concentrate on here (see Figure 85). These eyes produce a high-resolution color image (0.033–0.050°—better than a cat's) despite their small size. The front lenses, so prominent in Figure 85, are attached to the animal's carapace and so cannot be moved for focusing. Behind each lens is a long tube with a unique retina at the end. To focus, the tube moves. In front of the retina is a second lens, thus turning each principal eye into a telescope (of the refracting rather than reflecting kind): the two eyes together quite literally provide binocular vision (with a field of view of only about 2–5°). The retinas are unique in that they have four layers, one behind the other, containing a few

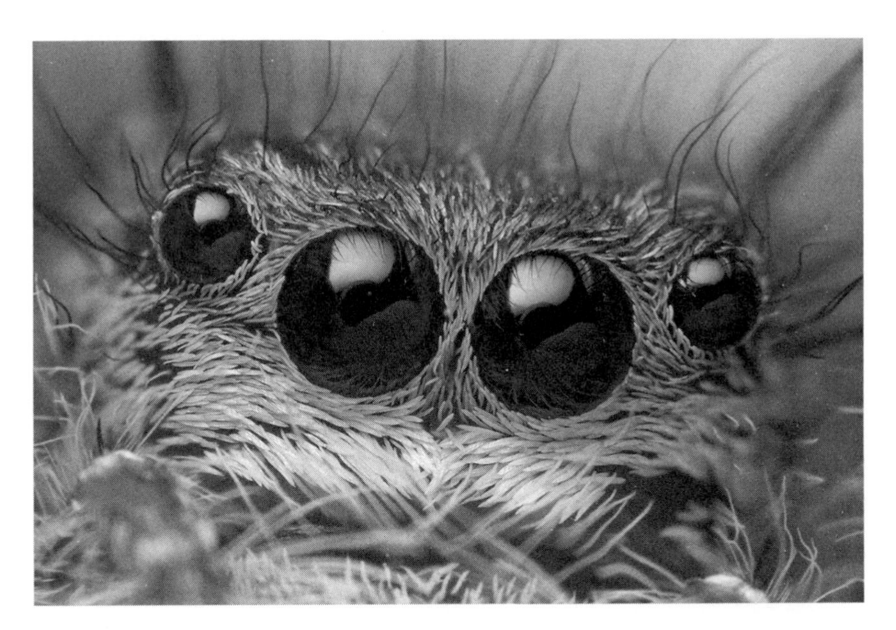

FIGURE 85 Anteriomedial and anteriolateral eyes of a *Phiddipus princeps* jumping spider. The anteriomedial eyes are for acuity and act like telephoto lenses. The anteriolateral eyes, plus four other eyes, are single-lens simple eyes for peripheral vision. Thanks to Thomas Shahan for this impressive photo.

thousand photoreceptors. The rearmost layer has the highest density of photoreceptors and so provides acuity; it is most sensitive to red light. The three layers in front are sensitive to green, blue, and ultraviolet light. It is clear what is going on here. The spider retinas are spaced so that chromatic aberration is reduced—each layer is located at the focal length of the light frequency to which it is sensitive. Thus, the jumping spider's principal eyes can see a focused, magnified image in color. Oh, and they can also see polarized light.

The cost of the extraordinary principal eye structure is a narrow field of view. When the spider is looking at a large object, it moves the tubes in complicated ways (they can be moved laterally and rotated, as well as moved forward and backward) as it brings into focus different parts of the object in front of it. Perhaps it is building up a picture—a mental image—of the object. This takes time—these spiders stare—because the field of view is so small and also perhaps because the spider's brain is small.[21]

Compound Eyes

Of all the beady eyes staring out at the world, most are not the simple eyes that serve us humans so well. You have likely been captured as an optical image by compound eyes—from the air or from a tree or from under the couch—far more often than by the simple eyes of vertebrate animals. Most arthropod eyes are compound.

The basic unit of the compound eye is an *ommatidium.* Think of a thin pyramid (with either a square or hexagonal base) with photoreceptors at the pointy end and a lens at the base. The lens inverts an object being viewed, of course, but this does not matter, because the photoreceptors do not form an image; they simply average the light level so that one ommatidium constitutes a single pixel of the compound eye. Several hundred of these ommatidia fit together to form an ant's eye; 2,000 form a housefly's eye (the compound eyes of a fly are shown in Figure 86). Each mantis shrimp's eye requires 10,000 ommatidia, whereas the biggest compound eyes, those of dragonflies, may consist of as many as 30,000 ommatidia. The simplest type of compound eye is known as an *apposition eye* because the ommatidia appose—fit together—so that the overall picture formed is a mosaic of pixels. This design works because light reaches each ommatidial photoreceptor only from a very narrow range of angles.

We can work out this angular range from basic geometry. Let us say that each compound eye of the fly in Figure 86 covers a hemisphere. The 2,000 ommatidia each cover an angular range of about 10 square degrees—say, roughly 3°

FIGURE 86 The head of a fly, showing the two large compound eyes. Thanks to Thomas Shahan for this photo.

in each direction. We expect this interommatidial angle to be comparable to the resolution capability of the compound eye. Thus, apposition eyes are of quite low acuity. In fact, whereas simple eye acuity increases linearly with lens size, that of apposition compound eyes increases as the square of size. A compound eye with the resolution of a human eye would have a diameter of about 27 m (90 ft).[22]

Compound eyes require simpler processing to form an image, and the eyes can be shaped so that they are conformal with the body. Another advantage—a consolation prize that mitigates the disadvantage of pixelated images—is that compound eyes are good at motion detection. If an object moves, its pixelated image changes markedly, thus demonstrating that good motion detection does not require high acuity.

Most insects have trichromatic vision, as we do, though at shorter wavelengths (often including ultraviolet). Some dragonflies and butterflies are tetrachromatic. Most crustaceans see two or three colors; however, the Purple Spot Mantis Shrimp (*Gonodactylus smithii*) is endowed with at least eight types of color receptor. Mantis shrimp cones are narrowband (i.e., they discriminate frequencies more finely than we do); the shrimp must utilize some form of spectral processing to improve the images they see.[23]

Diurnal insects have apposition compound eyes, whereas nocturnal insects often have *superposition* compound eyes. These eyes are the same optically, but the wiring is different. Light from several adjacent ommatidia stimulate one receptor, thus simultaneously increasing sensitivity and reducing acuity—analogous to the way that rods are wired for night vision in our simple eyes.

Thermal Imagers

The "pit" in "pit viper" refers to heat sensors that are located on both sides of the snake's head between eye and nostril. These sensors take the form of depressions in the skin surface containing cells that are sensitive to infrared radiation—heat—which is EM radiation with wavelengths longer than those of visible light, in this case 5–30 μm. It has been estimated that pit vipers (e.g., *Viperidae crotalinae*) can detect differences in temperature of as little as 0.003°C; they use this ability to detect and strike at prey; blind rattlesnakes will strike at mice up to a meter away with a similar success rate to that of sighted rattlesnakes.[24]

Clearly these snakes are using their heat sensors as alternative EM sensors. Yet their eyes are much better developed; there are no lenses or focusing apparatus in the pit organs. In fact, it is something of a mystery as to how the snakes achieve the angular resolution that they do. We have carried out a simple engineering calculation that shows that two pit organs located a few centimeters apart, with the sensitivity quoted, can estimate the direction of a heat source to within a few degrees. However, more detailed published calculations predict that resolution is much worse than this; heat flow through air in the pit chambers causes problems, and angular discrimination is seriously degraded by it. If true, then pit vipers must sharpen the blurred images they receive with considerable post-processing (to achieve observed strike rates) in ways that are not currently understood.[25]

Electric Animals

An electric sense is much less common in the natural world than an EM sense, because in practice, electric fields are much shorter range. So electric sensors have developed for specialists, in particular for aquatic animals in muddy water, where sight is of no use. If you hunt for prey in open water, then acoustics might work for you, as they do for Orcas and dolphins, but if your prey lounges around in muddy deposits or shuffles around a rocky riverbed, then sound might not work either. Platypuses have developed an electric sense that works for them, and Electric Eels have taken this sensor one step further and turned it into a weapon. Here we concentrate on the Platypus's passive electric sensor, which is more in keeping with the main theme of this chapter than the active electrocommunication and electrolocation capabilities of some animals. However, we cannot pass over this area without at least a cursory glance at the fascinating Electric Eel (Figure 87).

Electric Eel (Electrophorus electricus)

In South America and Africa there are about 250 species of electric fish (i.e., those that can generate a weak electric field, whose disturbances they use to sense their environment). The Electric Eel is different in that it can also generate a strong field, which it uses to stun or kill its prey. Not really an eel at all, this strange creature is a knifefish, a South American predator related to the catfish. An adult can grow up to 2.5 m (8 ft) long and yet all its vital organs, including its entire digestive system from mouth to anus, is in the front fifth—the remaining space is given over to three organs that act like electric batteries. The animal generates a *dipole field;* that is, it generates a positive pole at one location and a negative pole at a different location, and the electric field runs in curved lines between the two poles. Dipole fields are characterized by an inverse cube force, meaning that the field force decreases as the third power of distance from the dipole. This rate of fall-off is faster than the inverse square

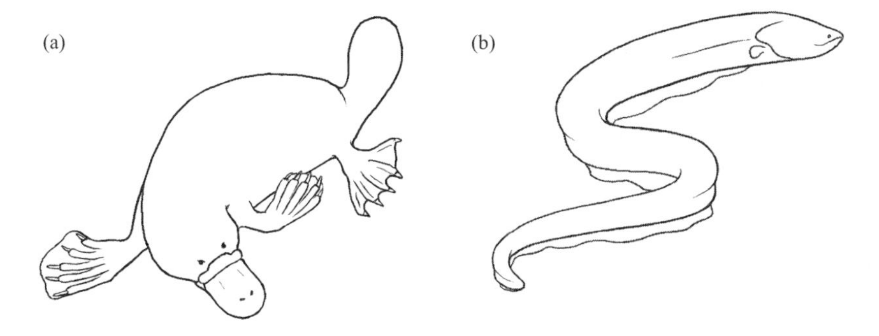

FIGURE 87 Electric animals. (a) Platypus and (b) Electric Eel.

behavior of EM radiation or of sound traveling through the air. As a result, the electric field can be used for communication out to ranges of only a few meters; for sensing the environment (e.g., locating prey), the dipole field is useful at ranges less than 1 m.

For electrolocation the field generated is typically 10 V of alternating current with a frequency of 25 Hz. Sensors along the animal's lateral line detect disturbances in the transmitted field, providing information about nearby objects. Stunning or killing prey, or defense against predators, requires a stronger field: typically 500 V at several hundred Hertz and 1 ampere current—enough to kill a human.

Platypus (Ornithorhynchus anatinus)

Monotremes are the only mammals with an electric sense; Platypuses are the most sensitive. Their electric sense was not appreciated until the 1980s, such is their reclusiveness. The Platypus eats aquatic invertebrates, which it catches in muddy waters at night with its eyes, ears, and nostrils closed. Unlike electric fish, the Platypus does not transmit a field; it passively senses the tiny electric emissions made by all animals.[26] Its 40,000 electric field receivers are located in longitudinal rows over the inner and outer surfaces of both upper and lower bills. In addition the bill contains 60,000 mechanical push-rod sensors scattered over its surface; these detect pressure differences. We will see how the electric sensor information can be used to estimate the direction of a source. Its distance, if the source is an animal that generates both pressure waves and an electric field in the water, can be determined from the time difference between reception of the mechanical signal and the electric one. This "thunder and lightning" processing is based on the difference in speed of an electric signal (which travels at the speed of light) and a pressure wave (which travels at the speed of sound in water).

A simple two-dimensional model will serve to indicate how a Platypus might estimate the direction of an electric source. Figure 88a shows two rows of sensors in the Platypus bill and a field source. Assuming uniform electrical impedance and a constant point source, the field strength at different receptors is shown in Figure 88b. The distribution of the field strength across receptors provides enough information to estimate source direction unambiguously. For example, if the source were directly in front, then a receptor on the left would register the same signal strength as the corresponding receptor

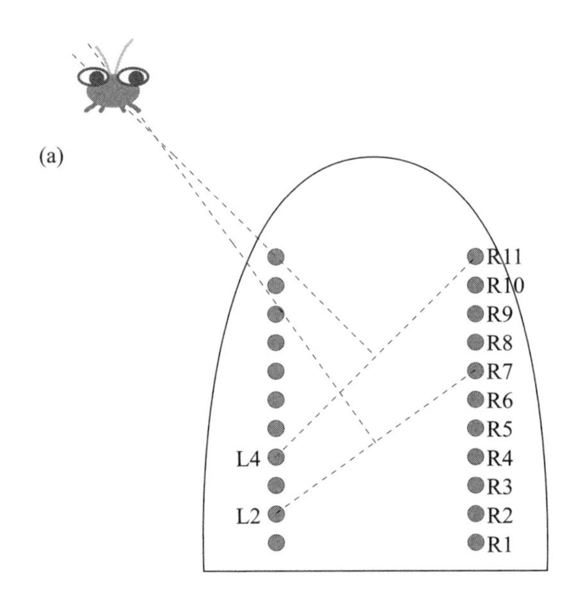

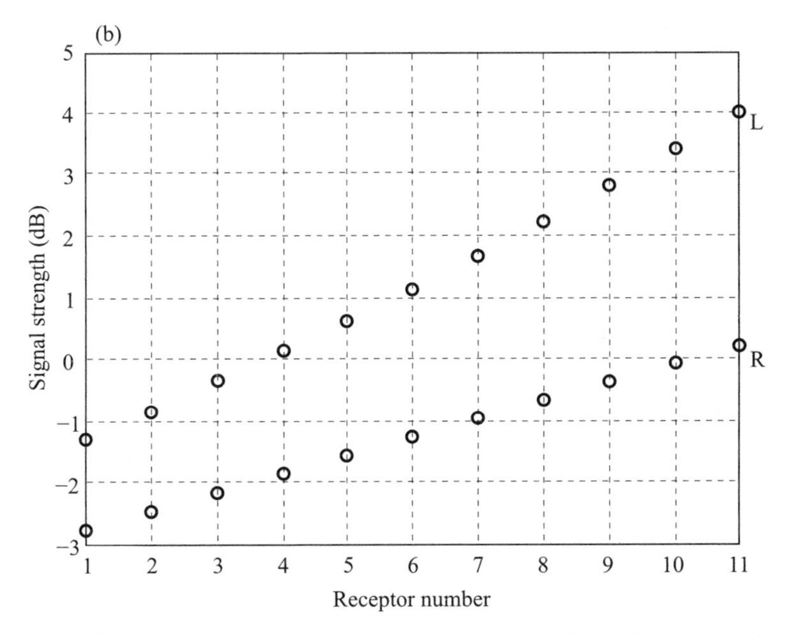

FIGURE 88 A simple model of Platypus electrolocation. (a) The bill is embedded with two lines of sensors that detect the electric field emitted by a food item. The field strength distribution over the sensors depends on the source's location. In this case, for example, the field strength in sensors L4 and R11 is the same, because they are the same distance from the source. (b) For the source direction shown, the distribution of field strengths is plotted.

on the right. For the source location shown in the figure, receptors L4 and R11 would register the same signal strength, as would receptors L2 and R7.

The oversimplification of the calculation scheme illustrated in Figure 88 serves to emphasize the considerable signal processing that a Platypus brain must carry out to interpret the data it is given. A real Platypus feeds by digging its bill into a stream bed and moving its head from side to side. Observations show that it can detect living prey and accurately estimate its direction and distance (the effective range, of course, is very short). It moves, and its prey moves; the prey electrical and mechanical signals are varying; the electric signal strengths are very weak; data received by 100,000 electrical and mechanical sensors must be analyzed in real time. All these facts lead us to the conclusion that the Platypus brain must be doing a lot of signal processing. Add to these observations one other: the Platypus's individual electrical receptors are sensitive to signals no smaller than $2\ \text{mV} \cdot \text{cm}^{-1}$, whereas the animal as a whole can detect signals as small as $20\ \mu\text{V} \cdot \text{cm}^{-1}$. This represents a signal processing gain of 20 dB (a factor of 100). So, Platypuses must process to boost the electrical signal, separate electrical signal data from electrical noise (and mechanical data from extraneous sounds and other pressure waves), process the signals to estimate source location, and do all this in a fraction of a second. One complication: the Platypus itself generates electrical signals and pressure waves as it moves. It must be able to cancel out these sources of self-generated noise.

11

There and Back Again

Animal Navigation

Animal navigation conjures images of the long-distance migrations of birds. It's an easy association to make, because bird migration is an annual event that we tend to notice. Birds are, indeed, capable of prodigious feats of navigation and so will figure prominently in this chapter. But it would be a mistake to dismiss the rest of the animal world in this context; there are four-legged (and six-legged, and no-legged) animals who are experts in many aspects of navigation science. Best known, perhaps, are Monarch Butterflies and salmon. We examine the achievements of these and many other migrants, as well as the navigational skills of many nonmigratory creatures.

The past half century has seen a revolution in our understanding of animal navigation brought about by many, many thousands of studies: radar or global positioning system (GPS) tracking observations; experiments that altered the apparent direction of the sun, or the brightness or color of light; experiments in which migrating birds were displaced several hundred or thousand kilometers; experiments in which animals were blindfolded in one or both eyes, or had their sense of smell removed, or were placed in planetariums with altered star patterns, or had magnets attached, or were banded or radio-tagged. . . . It used to be thought that, for example, a migratory bird made use of a single skill, such as navigating via the sun, but our understanding has broadened considerably and with greater understanding comes greater respect for the capabilities of migrant animals: most birds use many different cues, which they prioritize in different ways when these cues conflict. We explore all the ingenious strategies that different animals adopt to find out where they are and in what direction they should travel, but (and here is the disclaimer to give us some wriggle room) we will not cite every single primary source that we consulted on the subject. You may wish to dig deeper into this fascinating subject, and to that end, a few references are provided. But the literature is far too large for us to be comprehensive, and so, for the sake of readability as well as brevity, in general we do not discuss how the secrets of animal navigation were discovered or who discovered them.[1]

Far-Flung Examples

The Arctic Tern (*Sterna paradisaea*)[2] winters in the windy southern oceans surrounding Antarctica and breeds in the short arctic summer of Alaska, northern Canada, Greenland, Scandinavia, or northern Russia—an annual round trip of 35,000 km. This bird therefore spends more of its time in daylight than any other living creature. Hatchling Loggerhead Turtles (*Caretta caretta*) scurry from the east Florida beaches where they were born and swim far out into the Atlantic Ocean, in a sweeping 13,000 km arc that takes them past the Canary Islands, the Cape Verde Islands, and the west coast of Africa. They return as adults to the very beach where they were born to lay eggs. Monarch Butterflies (*Danaus plexippus*) from southern Canada and the northeastern United States migrate 4,000 km each fall to very specific mountain fir forest sites in central Mexico; in spring they begin the return journey, which is completed by their offspring. The Bar-Tailed Godwits (*Limosa lapponica*) of New Zealand migrate to the Yellow Sea nonstop. Repeat: nonstop. This 11,000 km journey is the longest single-hop migration that we know about (see Figure 89). They then fly on to their breeding grounds in Alaska. Atlantic Salmon (*Salmo salar*) from western Europe and eastern North America leave their

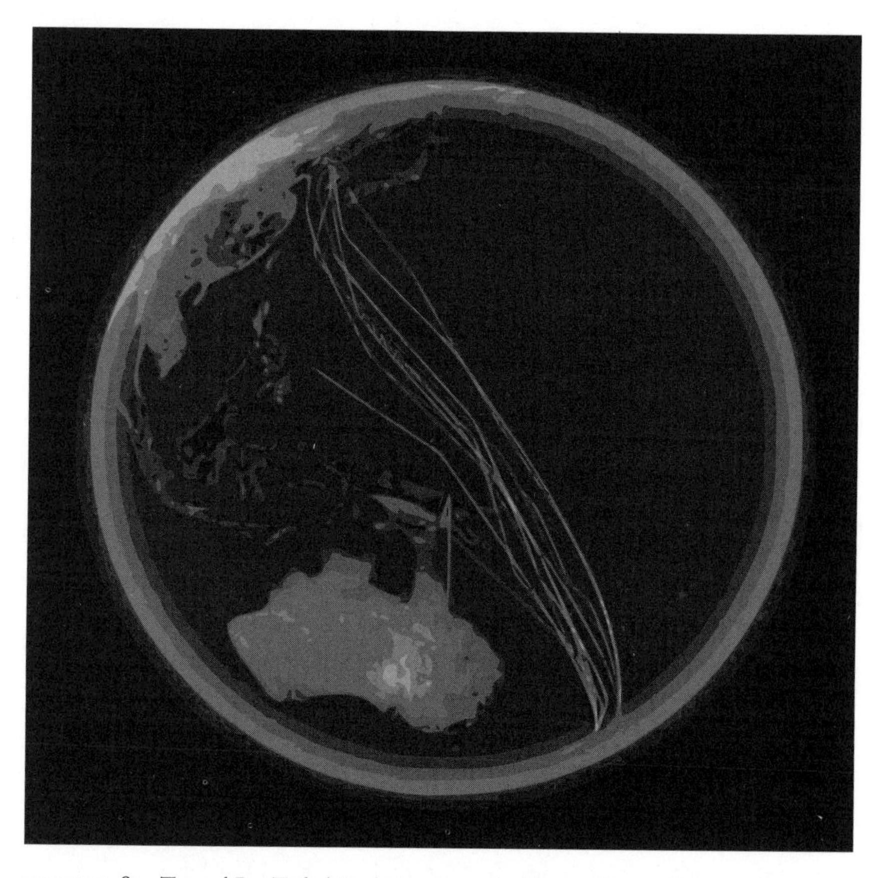

FIGURE 89 Tagged Bar-Tailed Godwits are tracked by satellite on their nonstop migration from New Zealand to the Yellow Sea, which takes them 9 days. After pausing by the Yellow Sea, they fly on to Alaska. Adapted from a United States Geological Survey image.

spawning streams and swim into the Atlantic Ocean, finding their way to the waters off southern Greenland before returning, years later, to the very streams where they were born, to spawn and die. A recent publication suggests that several species of dragonfly cross 3,500 km of open ocean in an annual migration between India and the Maldives.[3]

These feats—of endurance as well as navigation—almost defy belief. We could go on, because there are many more examples. Here are a few: the Ruffs of Senegal in western Africa fly almost halfway around the world to their breeding grounds along the coasts of northeastern Asia. Humpback Whales breed in winter in equatorial waters before heading to polar regions in the summer to feed. Barn Swallows that nest in Britain during the summer migrate to South Africa for the winter before returning, often to the very same nest site of the previous summer. North Atlantic Bluefin Tuna breed in the Gulf of Mexico and migrate to the coast off Scandinavia to feed. The Lesser Golden Plover (*Pluvialis dominica*) is another long-distance migrant; some fly from Pacific islands, such as Hawaii, to the Arctic. Those that winter in Hawaii must travel over thousands of kilometers of featureless ocean to find a pinpoint on the map—the Hawaiian Islands are among the most isolated in the world.[4]

Not all feats of animal navigation are associated with long-distance migration. Indeed, much of what we know of bird migration comes from experiments and observations of the Homing Pigeon (*Columba livia domestica*)—a nonmigrating species. The difference between migrating and homing is this: migration is a seasonally driven urge to move to different territory, whereas homing (for pigeons) is returning to a loft to roost. Migration is driven by species survival logistics to be discussed in the next section, whereas homing is an individual response that occurs all year around. The pigeons' "home-on-demand" behavior is very useful to biologists who want to investigate navigation in birds. Homing and migrating may be driven by different instincts, but they employ the same navigational skills. Pigeons can be displaced several hundred kilometers from their lofts, but they usually return home very quickly, even if they have never previously been to the release site. Similarly, albatrosses forage over millions of square kilometers of featureless ocean, yet they return unerringly to a speck of island to feed their young. One Laysan Albatross (*Diomedea immutabilis*) was displaced from its home island of Midway across the Pacific Ocean to Whidbey Island, near Puget Sound in Washington, and released: it found its way home across 8,200 km of ocean in 10.1 days. Several

Manx Shearwaters (*Puffinus puffinus*) were displaced across the Atlantic Ocean to Massachusetts from their home in Wales and released: most were back in their nest burrows 12.5 days later after a journey of 4,900 km. On a smaller scale, bees forage far from their home and yet, when fully loaded, they are able to make a beeline straight for the hive. Sahara Desert Ants (*Cataglyphis bicolor*) likewise forage in all directions away from their nest site; they, too, can turn and head straight for home when they want to.[5]

What is going on here? The Desert Ant does not simply leave a scent trail and retrace its steps—it turns toward home and marches straight there, even though home may be out of sight. The pigeons, albatross, and shearwaters were released in unfamiliar terrain in parts of the world well off their beaten tracks, and yet they made their way home so fast that they must have known the way.

Maps and Compasses

There are two aspects to navigation. Before heading for home our intrepid wanderer needs to know its position and it needs to know the direction of its goal. We'll assemble the skill sets required for goal location by starting with the simplest.

Piloting consists of using landmarks—fixed features—to orient and guide. These fixed cues work only if the animal can perform sufficient mental abstraction to build a map in its head. In a well-known experiment, it was demonstrated that female digger wasps (*Sphex* spp.) used landmarks near their nest sites that guided them to its entrance. If a nearby pebble was moved to a different location, the wasps would be confused and have difficulty locating the nest entrance. Similarly, squirrels use landmarks to locate food caches. Landmarks may be used by long-distance migrants, but obviously in such cases they are not the whole story—far from it. A migrating bird may use topographical cues (e.g., mountains, rivers, and coastlines), ecological cues (e.g., different vegetation zones), meteorological cues (e.g., reliable trade winds and changes in humidity) to gain information about its location. Some whales use coastline cues, and fish can use currents and salinity gradients: any relatively fixed feature can serve to indicate location. Piloting skills require a mental map and memory but not a compass. Piloting works on its own for short distances or small areas, but not for large distances over unfamiliar or featureless territory.

Now we move up a level. Many animals have an internal compass and an internal clock, which are combined to give information about position and homing direction. There are many different natural compasses, as suggested in Figure 90, which we will unpack in stages throughout this chapter. We postpone a discussion of how such compasses work and concentrate on how, given compass and clock, it is possible to migrate long distances or return to home after foraging. The simplest method (the most inflexible and consequently unreliable) is used by many novice migrants. A young European Cuckoo (*Cuculus canorus*) must find her way to her African wintering ground, but, in the nature of cuckoos, she must make the journey alone, unaccompanied by parents. Many Old World warblers, such as the Willow Warbler (*Phylloscopus sibilatrix*), also make the journey alone; their parents went on ahead, leaving the youngsters to fend for themselves. These birds have a simple vector navigation package hardwired into their brains. Instinct tells them to, for example, "head southwest for 9 days, then turn due south until you reach a desert, then head east for four days and then south until you run out of gas." Such genetic instructions may work—Willow Warblers return to northern Europe year after year, so they must have survived their first migration south—but the number

FIGURE 90 Dead-reckoning requires the skills suggested in this illustration—and more.

that falls by the wayside is huge. A side wind may blow the novice off course; a head wind may slow her down, wrecking her schedule. In a few millennia the desert may become grassland, so the genetic instructions must also evolve to handle the change.

A famous experiment involved displacing a group of migrating European Starlings (*Sturnus vulgaris*). These were heading southwest across Europe to their usual wintering grounds in northern France. The group was displaced several hundred kilometers south and then released. The young, novice starlings who had not migrated before resumed a southwesterly flight, wintering in northern Spain. More experienced starlings knew they were too far south (how?—read on) and adjusted their course to northwest, taking them to their traditional wintering area. Clearly, the novices were just reading the instructions that their genes handed out to them, whereas the more experienced birds were overriding these with learned cues, enabling a more flexible response to changed circumstances. The novices would eventually have learned these new cues, but in later years they returned to northern Spain, not northern France, because their learned experience then taught them only better ways to get there, not where they should go.[6]

Another built-in routine is displayed by Blackcaps (*Sylvia atricapilla*)—Old World warblers that migrate from Europe to Africa. Austrian Blackcaps have genetic instructions to "head southeast and then south," whereas German Blackcaps are born with the instructions "head southwest and then south." The difference means that Austrian Blackcaps skirt around the east side of the impassible Alps, whereas German Blackcaps skirt around the west side. Cross breeding these birds—Austrian with German—produced young Blackcaps with mixed instructions. Many of these birds tried to migrate due south, but could not because of the mountains. This experiment shows that different genetic migration routes lock birds into particular geographical areas and inhibit successful interbreeding. Eventually, Austrian and German Blackcaps may become distinct species.[7]

A more sophisticated example of clock-and-compass directional movement is *dead-reckoning.* For the centuries that preceded navigational understanding and instrumentation, dead-reckoning was used by sailors to find their way around the world. If you have a compass—and here we adopt the sun compass, favored by many animals, including human sailors—and a clock, and if you can estimate how fast you are moving, then you can keep track of where you are.[8] This is vector navigation. The log book of an Age-of-Sail ship in the open ocean might contain such entries as: "Proceeded due east at 5 knots for 3 days. Wind southwest. Changed course to northeast for 8 days." Knowing your speed, you can, with a clock and compass, estimate how far you move in a particular direction. Then, if you change direction, you calculate your new movement and add it to the previous movement. Keep adding subsequent movements over each day (or hour or minute) and you will know your position and direction relative to your starting location. The process of adding all the contributions of your movement, carefully accounting for changes in direction, is known as *path integration.* As we shall see, it can work very well, but there are pitfalls.

First pitfall: the sun moves across the sky at the angular rate of $15° \cdot hr^{-1}$ and so, to make use of the sun as a directional reference, an animal must have a clock and must compensate for the sun's movement. The only exception is when the sun reaches its highest point at noon, in which case it is due south (viewed from the Northern Hemisphere); in this case absolute direction is provided without the need for a clock.[9] Apart from this case—and many animals will need to change direction more than once per day—the sun's movement must be clock-compensated. Second pitfall: the sun is not always to be seen (at night or under clouds). Third pitfall: measuring speed may be subject to systematic errors. For example, ship speed in the Age of Sail was determined by throwing overboard a log attached to a cable and measuring how many knots on the cable slipped through the navigator's hands during a timed interval. (Hence "log book" and the nautical unit of speed, "knot.") This method provided a rough estimate of speed relative to the water. The problem was that ocean currents caused the ship speed relative to land to be different. Birds will have similar problems estimating their flying speed (relative to the ground) in a wind. Fourth pitfall: the path integration process is subject to accumulating errors. Biological clocks are not perfectly accurate, and so estimation of sun location may be wrong; even if it is right, then measurement of heading direction relative to the sun may be in error; estimation of speed or distance may be in error (try walking a short distance—say, 100 paces—estimate the ground covered and then measure it). These errors are for just one straight-line segment of a route, and they accumulate as the segments are added together.

Let's illustrate the difficulties of dead-reckoning position estimation by considering desert ants. These scavengers forage far from their nests and then, when they find a food source, head straight back home without retracing their steps —they path integrate using the sun as a directional reference—see Figure 91 (some of the skills required for dead-reckoning were suggested in Figure 90). They are not faced with the problem of an unseen sun, or of systematic errors due to shifting water or air, but they do face the problem of accumulating measurement errors. How accurate do an ant's angle and distance estimations need to be for it to find its way home? We can simulate this problem on a computer by assigning random errors to each angle and distance measurement made by the ant. Then we see where it ends up after a certain number of paths have been added together.

Results of such a *Monte Carlo* simulation are shown in Figure 92. We find that for the case of scavenging ants or any other animals who return home after a day toiling in the fields, the estimates of angle and length of each path segment do not have to be very accurate at all. The accumulating errors partially cancel in this case, and the ant gets home guided by her dead-reckoning. (In our simulation results she gets to within 5 or 10 path lengths of her nest, and we assume that, from there, she is guided to the front door by landmarks,

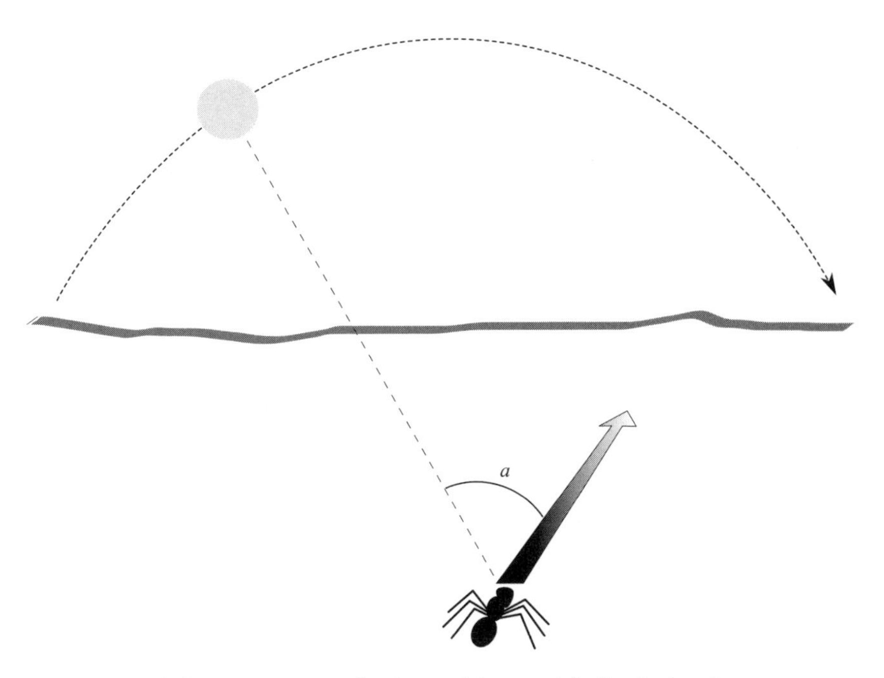

FIGURE 91 A desert ant ensures that it travels in a straight line by keeping a constant angle *a* between its path and the sun. It must compensate for the sun's movement across the sky.

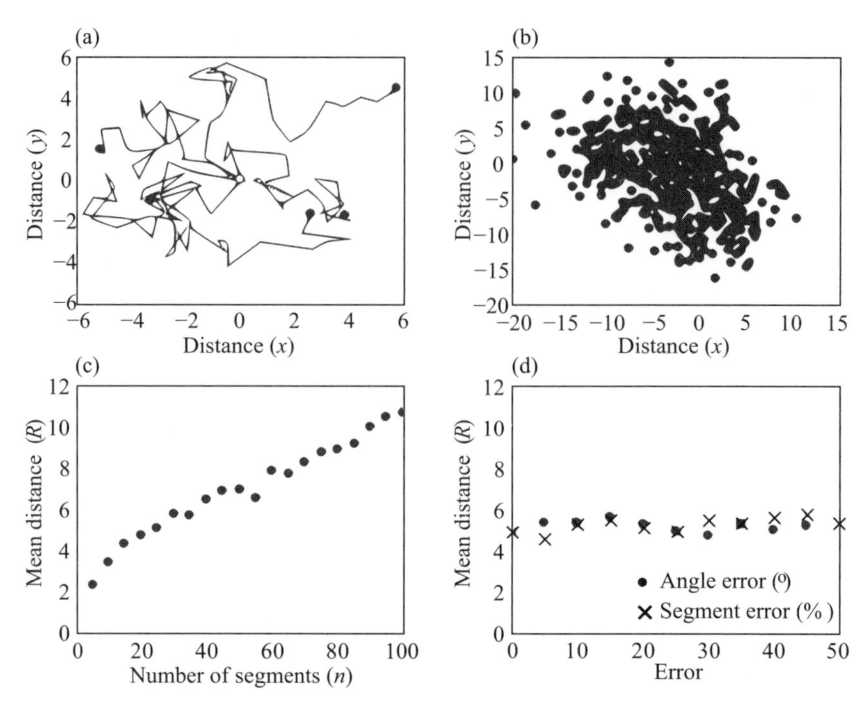

FIGURE 92 Monte Carlo simulation of ant homing behavior. (a) Five ant trails, each starting at the nest (open circle). The positions (closed circles) after 25 path segments are shown. Each straight-line path segment is of mean (average) length 1 unit (e.g., 1 m, or 100 m), and the direction of each segment is random. (b) The ant estimates each path segment length and direction angle, and then heads home. Because of the estimation errors, it does not end up exactly at the nest; the terminating positions of 500 ants are shown. (c) Mean distance from home increases with number of path segments, due to accumulating estimation errors. Here we plot the mean distance *R* from the nest after returning from a path consisting of *n* segments—*R* increases proportionally to the square root of *n*. (d) However, dead-reckoning error is not sensitive to estimation accuracy if the ant is trying to return to its nest; here *R* is plotted against the root mean square of the angle estimation error and of the segment length estimation error. (Root mean square is a standard statistical measure of error magnitudes.)

visual cues, or scent.) What matters is that the average path length (the distance between measurements of direction angle) is not too long and the number of paths (i.e., number of direction changes) is not very large. In other words, dead-reckoning works quite well over short distances, so long as you get home before sunset.[10]

What if an animal uses dead-reckoning as a guide to get from A to B, where A and B are widely separated—say, A is a bird's northern breeding ground and B is his southern wintering ground? Such long journeys cannot be made solely by referring to the sun for directional information. Also, as we will see, the estimation errors for any dead-reckoning scheme for long journeys (not just one that is based on the sun) accumulate relentlessly, in contrast to the case of our foraging ants. Long-distance migrants require true navigation—piloting and dead-reckoning on their own won't cut the mustard—and so it is to migration and navigation that we now turn.

Why Migrate?

Migration may have evolved from shorter foraging trips in species with greater spatial memory and better navigational skills than the norm. Indeed, we now know that migratory birds have better memories than nonmigratory birds. Clearly, memory is required to build up cognitive maps—internal representations of spatial relationships among objects in the migrants' surroundings. (Think of Barn Swallows returning to the same nest site year after year, or sea turtles returning to their natal beach after years in the open ocean.) But why migrate at all? Not all species do so, and in some instances the case for migration is on a knife-edge, with finely balanced pros and cons. Generally speaking, the purpose of migration is to escape from deteriorating conditions and to exploit seasonally favorable habitats elsewhere.[11]

Consider birds that breed in the far north. There is much more land in the northern hemisphere than in the south, and so there is an abundant insect food supply during the short breeding season—a food supply that is not depleted during the winter months by residents. The northern summer days are longer, which gives these birds more time to gather food. However, there is a year-round food supply in the tropics. So migrant insect eaters get the best of both worlds: bountiful food in spring and summer when they need it most—rearing a brood of youngsters—and sufficient food in winter. They cannot stay in the far north all year, because the temperature drops both increase the cost of living and reduce—or eliminate—the source of food. Hence the birds' spring and fall migrations. The main disadvantage of migration is, of course, the hazards involved with such arduous journeys: adverse weather, adverse habitats en route, exhaustion, and predation all take their toll.[12]

Sometimes the cost/benefit analysis is finely balanced, and we find similar species with very different migration patterns; indeed there are several degrees of migration, shading continuously from sedentary species to regular annual long-term migrants. Some migrant birds, such as Golden Plovers, fly over large expanses of open ocean, and so they fly nonstop; others, such as American Robins, migrate over land—such birds may pause during their migration, or even reverse direction, if conditions are unfavorable. Robins are exceptional in that their spring migration northward is a leisurely affair—they follow the 3°C (37°F) isotherm, which advances about 60 km per day. (Generally birds migrate north in spring faster than they migrate south in the fall—presumably to claim breeding territory.) There is differential migration within a species, irregular irruptive migration, and migration patterns of birds that change with age.[13]

Navigation

Given that an animal needs to migrate, how does it find its way? We constructed a Monte Carlo simulation showing that the clock-and-sun dead-reckoning method won't work: a more flexible system of navigation is required. We took as a simple example of dead-reckoning navigation a bird that flies south 1,000 mi to a specific wintering ground. She attempts to find her way in the same manner as the desert ant: clock-sun direction estimation and path integration. Let us say that she flies south in one nonstop journey, and she checks her direction every 10 mi by reference to the sun or some other celestial object. The dead-reckoning calculation in this case is particularly simple: our bird need only check that she is flying south and count the number of path segments (here 100 segments are required). For clarity we assume that her estimate of distance is perfect—without error—and that she estimates direction with an error that always lies within a restricted wedge, as shown in Figure 93. Every 10 mi she checks which way is south, and adjusts course as necessary. If she passes within 10 mi of her wintering site, we assume that she can see it and so find her way without further path integration. How does she do?

Not very well. In Figure 93 our bird and 99 others fly south as described above and look around for the wintering site after 100 path segments. If the estimate of southerly direction had been perfect, then 100 path segments each of 10 mi should take the flock due south 1,000 mi, as required. However, due to direction estimation error they end up short of the goal and spread out over a wide area. They carry on for a few more path segments: some of them find

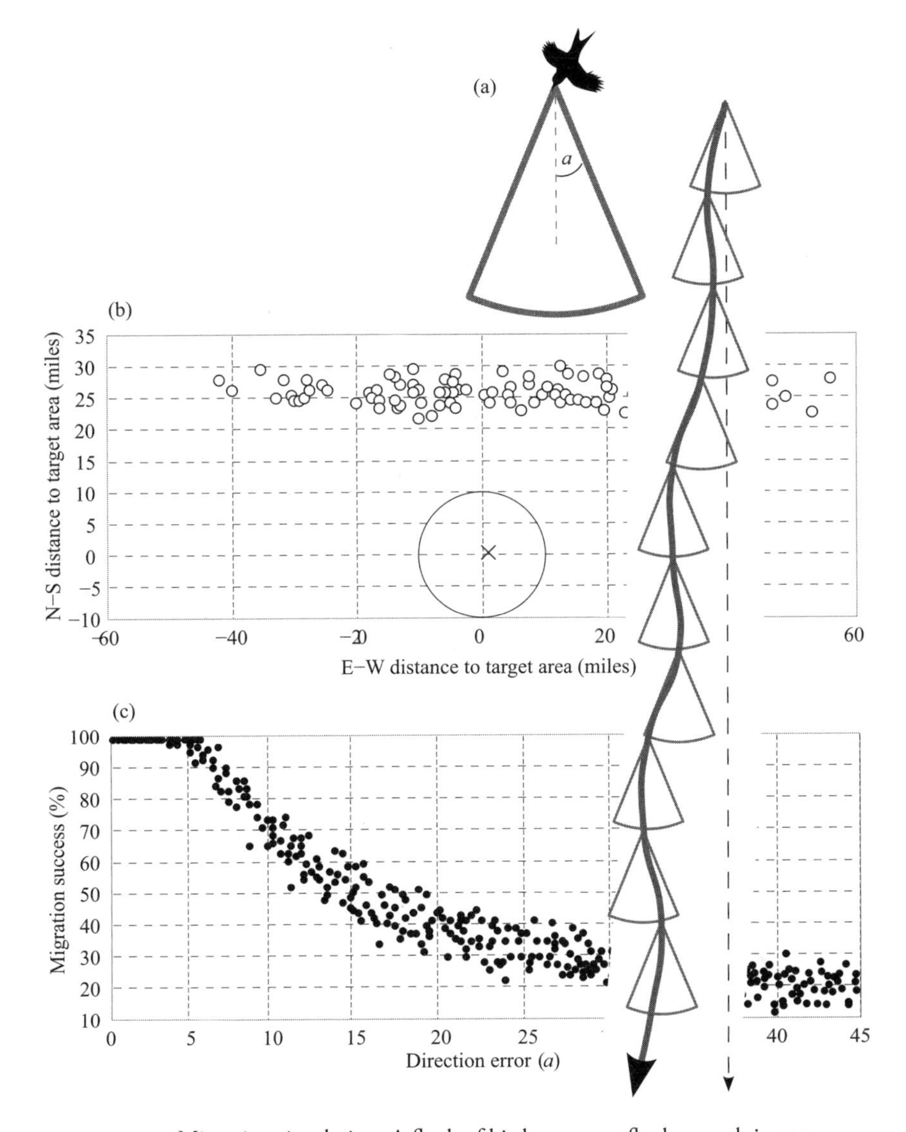

FIGURE 93 Migration simulation. A flock of birds wants to fly due south in 100 stages. (a) A bird estimates direction with an angular error that is random, unbiased, and falls within an angular wedge defined by the half-angle *a*. After estimating direction, the bird flies south 10 miles and then checks direction. It progresses by stringing together 100 such flight stages (as suggested schematically by the wavy line). (b) After flying 100 stages, each of 10 miles, the simulated flock of 100 birds (small open circles) is spread out east-west over nearly 100 miles. On average they are still 25 miles short of the target area (large circle with "x" in it). If the birds average their individual estimates, then they hit the target area. (c) Simulation results showing the percentage of birds that successfully find the target area as a function of direction estimation error *a*.

home—get within 10 mi of the goal—but many fly too far east or west; they overshoot and miss their wintering ground. Figure 93c shows how the percentage of birds that navigate successfully falls with increasing angular estimation error. The larger the error is, the fewer birds make it to the wintering ground. For this linear migration route there is no partial cancellation of estimation errors as for the scavenging ants; here errors accumulate remorselessly.

Note that if the flock members talk to one another during their flight south and compare notes, they do much better. If the 100 birds make independent direction estimates every 10 mi and average these estimates to obtain a "flock direction estimate," then they will very likely hit their target. It is just about possible that real flocks reach a consensus about which direction to go, but this method only works if the individual estimation errors are unbiased—just as many to the east as to the west—and if there are no systematic errors affecting them all, such as would be induced by side winds. And what about solo migrants? Many birds migrate at night on their own. Either they have near-flawless direction estimation capabilities or they are doing more than clock-and-compass orienteering. In practice many migration routes are much longer than 1,000 mi, and the routes are more complicated. Birds are chased by predators that wait for them at migratory bottlenecks (e.g., Central America for New World migrants, or Gibraltar for western European migrants), and many migrants pause en route to find food. It is asking too much to expect them, under these circumstances, to keep track of all their movements and path integrate so that they can set off in the right direction when they resume migration.

So, piloting via landmarks and dead-reckoning orienteering will not work over long distances, and there must be more in the route-finding locker of long-distance migrants than these two techniques. Those Manx Shearwaters who returned home after being displaced across the Atlantic must have navigated home. The release site was unfamiliar to these birds—there were no familiar landmarks for piloting. They did not make the outward journey under their own steam, and so could not have made any dead-reckoning calculations.[14] Piloting and orienteering require knowledge of the start position as well as the goal position. Navigation consists of using external cues alone to find an unseen goal—only the goal location needs to be known—and so it is a step up in sophistication compared to piloting or orienteering. Let us now consider the arsenal of cues that animals use when navigating.

Clues about Cues

Biologists now know that animals generally use a multitude of cues to help them navigate. We have already discussed piloting via landmarks; another type of fixed feature, chemical gradient (waterborne and airborne odors, or salinity gradient), is a powerful aid that many different animals utilize to estimate their position. A type of acoustic landmark is provided by infrasound, which, as we will see, may provide positional cues over hundreds of kilometers. We have seen how the sun can be used as a compass for direction estimates: so can stars and the geomagnetic field. This last cue—Earth's magnetic field—is so important that it merits a separate discussion; in this section we show you some of the intriguing clues about navigational cues—see Figure 94—that experimental biologists have coaxed from many different types of animal over the past half century.

Polarized Light

The use of polarized light for navigation is a special case of the sun-compass technique. It permits those animals capable of detecting the polarization of sun-

FIGURE 94 Avian navigation: dangerous, marvelous, and extremely skillful. Here we see some of the cues that long-distance migrant birds use to navigate: coastlines, star patterns, and the sun (in particular, the setting sun).

light to record the sun's location up to an hour before sunrise and after sunset, and in murky meteorological conditions where the solar disk is not directly visible in normal light.

We have seen that light is a type of EM radiation, which means that it consists of two coupled waves: an electric wave that oscillates in one direction (say, left-right) and a magnetic wave that oscillates in a perpendicular direction (say, up-down). The combined EM wave moves at the speed of light—of course—in the third direction (say, straight ahead). Oscillation of the two coupled waves is not always up-down and left-right—the combination may be, say, shifted 45° either way. Sunlight reaches Earth in an unpolarized state, meaning that the electric component is randomly oriented—all angles are represented equally in any sample of light rays that you choose to test. Polarized light shows a preference for a particular direction, for example, 90% of the light has the electric component oscillating left-right and only 10% oscillating up-down. So if sunlight is unpolarized, then what is the use of being able to see polarized light? The answer lies in what happens to the polarization when light is scattered. If you have polarized sunglasses, then you will see light that is scattered mostly in one plane; scattering cuts out the light that is polarized in a perpendicular direction. Sunlight reflecting off a smooth surface, such as a body of water, becomes polarized. By orienting the polarizer in your Ray-Bans to be perpendicular to the polarization of the reflected light, you cut down on the glare from reflections but can still see much of the ordinary, unpolarized light.

Sunlight becomes partially polarized by scattering off the atmosphere and (here is the crucial point) the intensity of polarization is directional. Light that travels from the sun to Earth's atmosphere and then scatters 90° before hitting the retina at the back of a bird's eye will be most strongly polarized. Light that scatters only 20°, or that backscatter 160°, is very weakly polarized. So a bird that can sense the polarization of light will see different light intensities in different directions, depending upon the sun's location. So long as there is light in the sky, the bird will be able to tell where the sun is, even if the solar disk itself is below the horizon (say, after sunset) or is obscured by clouds.[15]

Several nocturnal migrants, such as the Yellow-Rumped Warbler (*Dendroica coronara*) and the Savannah Sparrow (*Passerculus sandwichensis*), are known to use polarized light from the setting sun to orient. In the latter case it is thought that the polarized-light orientation system is calibrated by the bird's magnetic sense (of which more later) but that at sunset the dominant cue is polarized light.[16] Birds are not the only creatures with this "Ray-Ban" sense: certain ants and bees that are active at dusk and dawn have crossed-polarization analyzers in their eyes that provide polarization information.

Stars in Their Eyes

Most songbirds migrate at night. A journey of 3,000 km will take them only three or four weeks and, on nights when they travel, they cover 300–600 km. This is faster than daytime commuters, who interrupt their flights for feeding. Also, flying at night means less chance of being seen by predators. Given the advantages of nocturnal migration, it is not surprising to learn that birds, like ancient sailors, evolved a navigational cue based on celestial observations.

On a clear night the moon and stars alone provide sufficient information for a navigator to estimate position and direction; celestial navigation requires no clock. The main drawback, naturally, is that stars are not always visible. Even on clear nights, a bright moon can obscure nearby stars, rendering stellar navigation less accurate. So navigating via the stars is, for most birds, an opportunistic cue that is just one of many tools in the toolbox.[17]

How do they do it? Experiments with birds in planetariums show that nocturnal birds during their first summer learn the position of the Pole Star (Polaris, the North Star) by observing the rotation of the night sky—the Pole Star remains fixed. The stars in a planetarium can be changed; experiments with the celestial rotation centered on a different star (Betelgeuse) led to birds trying to navigate using Betelgeuse as a reference. To show that the center of rotation is what matters in this context, not the stellar distribution across the sky, a random distribution of stars was substituted; the birds learned to navigate by the new "pole star"—the one closest to the center of rotation. A planetarium experiment also showed that some migrating birds do rely on the stellar distribution across the night sky. Warblers' migratory restlessness showed that they wanted to fly south, and so each night the planetarium sky was "moved" 300 km south. After two weeks the migratory restlessness ceased, because the birds thought that they were under a tropical sky, 4,000 km to the south.[18]

It's quite a striking image: a nestling bird just a few weeks old staring up at the sky, night after night, learning his trade. If only our kids were that dedicated when doing their homework.

Infrasound

We saw in Chapter 9 that low frequency sounds are attenuated very little in air, and so infrasound travels a long way through the atmosphere—hundreds or thousands of kilometers. The same holds true in water. Whales communicate over very long distances via low frequency (long wavelength) sounds, as discussed in the next chapter. So, whales must be able to hear these frequencies. Salmon can, too—they are very sensitive to infrasound in the band 0.1–10 Hz (well below our hearing threshold). Salmon do not vocalize, so their infrasonic hearing has nothing to do with interspecies communication; rather it helps them navigate by detecting velocity differences between two layers of water sliding past one another.[19]

Pigeons are also acutely sensitive to very low frequency sounds: they can detect acoustic vibrations at frequencies as low as 0.05 Hz—one cycle every 20 seconds. Below 10 Hz the pigeons' sensitivity is 50 dB (100,000 times) better than ours. Strong hints that these birds use infrasonic information as a navigational aid are provided by those very rare disruptions of pigeon races in which most of the birds fail to find their way home to their lofts. Usually a pigeon can make it home from 1,000 km away, and fast (averaging 50 $km \cdot hr^{-1}$), so it is not normal for most of a flock to go AWOL. For one particular race, the time at which the pigeons appeared to lose their bearings was correlated with the infrasonic shock waves from a Concorde supersonic jet passing by. So, how can infrasound aid navigation?[20]

We know from radar tracking experiments that migrating birds (and Homing Pigeons) follow major topographical features, such as coastlines and mountain range fronts, even at night or when these features are visually obscured by clouds. We also know that these features emit infrasound (why?—see next paragraph). The implication drawn is that migrating birds are following an infrasonic path. The idea of an acoustic avian map of infrasonic cues may help explain one of the enduring mysteries of bird navigation: how do displaced birds determine their location to know which direction leads home?

To understand why prominent topographical features emit infrasound, and how this fact can give rise to a large-scale acoustic map, we begin with a brief summary of the acoustic landscape in which we all live—blithely unaware of its existence. *Microseisms* are low frequency seismic waves continuously generated by ocean waves that form an ever-present background in all seismic recordings. The frequency corresponding to the largest amplitude (intensity) of these acoustic waves is only 0.16 Hz, corresponding to one cycle every six seconds. The wave amplitudes are small: about one-thousandth of a centimeter (10 μm) in coastal regions and only 1 μm in continental interiors. (Recall that very low frequency sound travels a long way, and so these microseisms, which originate in the oceans, can be detected hundreds of miles inland.)

The idea of an infrasonic map becomes plausible when we realize that large topographic features, such as coastal cliffs and mountains, act like giant sounding boards, amplifying and directing the infrasound that impinges on them. Indeed, steep-sided and relatively flat-faced rock structures will be effective acoustic antennas for infrasound: the key feature required for a directional antenna, from the antenna engineer's point of view, is that the radiating elements be arranged on a grid that is large and flat, at least on the length scale of the radiation emitted by the antenna. "Large" means long compared with the wavelength emitted. Flatness is required so that the radiating elements (in the case of mountains and cliffs, these are the individual facets of rock) emit the sound in phase, so that all the sound emitted from all elements adds up constructively instead of canceling out (recall the description of antenna beamforming in the caption to Figure 60). Infrasound with a frequency of 0.16 Hz has a wavelength of 2 km, and so any long cliff, ridge, or prominent coastline with no indentations larger than, say, 400–500 m will be an effective directional antenna. If a cliff face is rough at this length scale—if it curves, is cut by deep indentations, or exhibits promontories that are bigger than half a kilometer—then it will radiate 0.16 Hz microseisms *isotropically* (evenly in all directions), but may radiate lower frequency (longer wavelength) infrasound in a preferred direction.

So, prominent geographical features radiate the infrasound energy that impinges on them from ocean waves, and the radiation may or may not be directional, depending on the frequency of infrasound and on the size and shape of the geographical feature. Figure 95 shows a coastal mountain as an acoustic antenna with two different beamshapes (corresponding to two different infrasonic frequencies). In reality such a mountain would be radiating many infrasound frequencies and would act as a differently shaped antenna for each one of them, but for clarity we show only two such frequencies in Figure 95. An infrasound-sensitive bird flying over this mountain would pick up the acoustic radiation with different intensities at different frequencies, and this perceived spectrum of infrasound would change with bird position, as suggested in the figure. The bird could use this information to establish which mountain is down there and to find its position relative to the mountain.

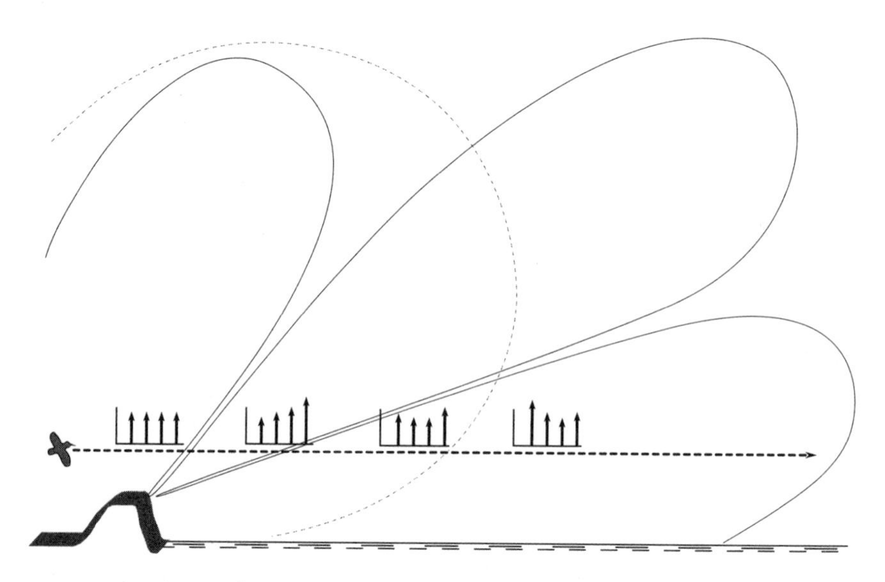

FIGURE 95 A coastal mountain (wide black band in lower left) acts as an acoustic beacon. Two beamshapes are shown, for two frequencies. These can be of simple shape (dashed gray line) or more complicated (solid gray line), depending on mountain shape and infrasound frequency. Along the flight path of a bird (dashed black line) the acoustic spectrum of the topography changes (bar plots, looking like cell-phone reception indicators, appropriately enough).

There are lots of prominent coastal (and inland) topographical structures that act as infrasound antennas. Infrasound travels long distances through the air. Put these two facts together and you see that large areas of landmass throughout the world are covered with acoustic landmarks—beacons that beam out navigational information to those who can interpret it. Perhaps a pigeon that is displaced to territory it has never seen before still knows where it is, because it can pick up the sounds of several familiar beacons, even though these may be hundreds of miles away. Imagine a Parisienne who is kidnapped and transported blindfold to an unfamiliar part of her great city. She escapes, but does not recognize the street she finds herself in. Then she sees the Eiffel Tower in the distance. She also sees the sun to the south. With this information, she finds her way home. She doesn't have to be smart: given the acoustical equivalent of such prominent visual beacons as the Eiffel Tower, even a bird brain could work out its position.

Odors

Animals can use their sense of smell to home in on a goal—such as home. Salmon sniff their way to their natal streams, pigeons to their lofts, and tube-nose seabirds to their nest burrows. All use other techniques to navigate when far from home, but when home is so close that they can smell it, they do.

The home stream odor imprints itself in a young salmon's brain; the adult recalls this odor 2–5 years later when returning from the ocean to spawn. Experiments have shown that, when an artificial odor is introduced to a stream containing many young salmon (known as *parr* or *smolt*), they remember it; years later they pick out the stream with the same smell. The point of the experiment is that the crafty biologists changed streams while the salmon were away in open ocean: they discontinued the artificial odor in the spawn stream once the smolt had departed and introduced it to a different stream years later.[21] In birds it may be the case that olfactory cues are important a considerable distance away from the odor source; some researchers claim that pigeons can make use of odors to find their way home from as far away as 700 km, whereas starlings can do so from 240 km. These are statistical results (groups of birds with an impaired sense of smell orient themselves less well than do groups of birds with an intact sense at the stated distances), and the numbers probably reflect the experimental measurement setup rather than some intrinsic maximum range—the point here is that olfactory navigation may work over hundreds of kilometers.

Some researchers think that the olfactory sense and the magnetic sense are the two most important for animal navigation. Magnetic? It has become increasingly clear from a deluge of data over the past 40 years that many different species of animal can sense the geomagnetic field lines that envelop our rotating planet. We now turn to a discussion of the natural world's magnetic compass.

Animal Magnetism

That birds might navigate using the geomagnetic field was suggested as early as 1855 but was not taken seriously until the mid-twentieth century. Now we know that many species of birds, fish, mammals, amphibians, insects, and reptiles have a magnetic sense, and we are beginning to unravel the complex ways in which they use it.

First we need to describe the phenomenon of geomagnetism. The rotating metallic core of our planet acts like a giant dynamo and generates a dipole (north-south) magnetic field, as sketched in Figure 96. The field lines make an angle with Earth's surface—the *inclination angle*—that varies with latitude, being steep near the poles and falling to zero at the magnetic equator. The intensity of the geomagnetic field is measured by the density of field lines; near the poles the field lines are close together, and the magnetic field strength is about twice that at the equator. The direction of magnetic north is not quite the same as that of celestial north (along the axis of planetary rotation), as can be seen in Figure 96. Geomagnetism is a dynamic effect, and the field strength

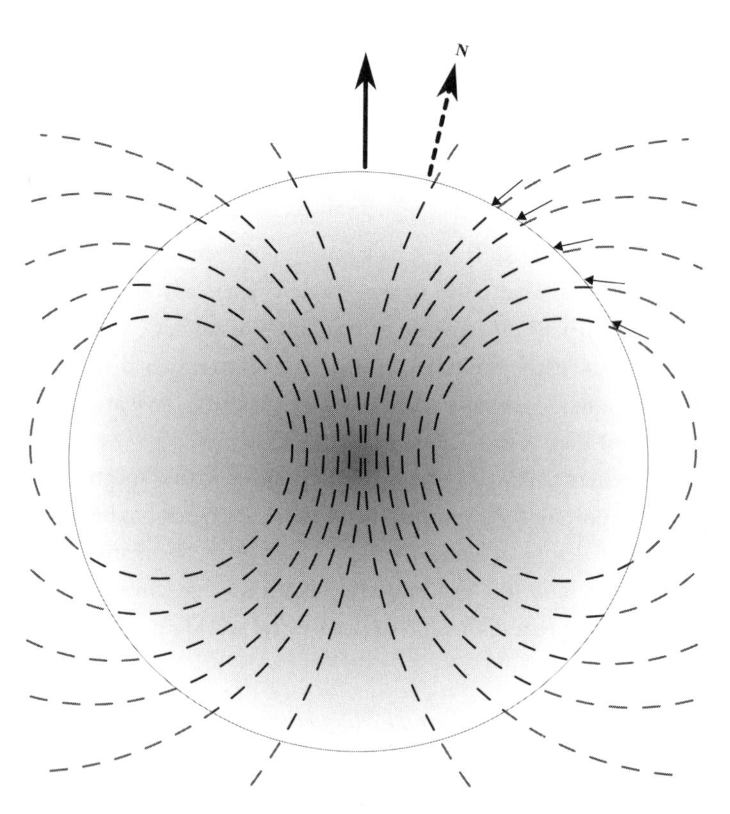

FIGURE 96 Schematic of geomagnetic field lines (dashed). These define a magnetic north (large solid arrow), which differs from true north (dashed arrow) by a declination angle of about 11°. The intersection of field lines with Earth's surface (small arrows) defines the inclination angle, which increases toward the poles.

and direction are never constant. Currently magnetic north and celestial north differ by a *declination angle* of about 11°, though magnetic north wanders several tenths of a degree annually. The field strength also changes: it has fallen about 10% over the past century.

The dynamic, ever-shifting geomagnetic field varies over short and long timescales. Over intervals of milliseconds to hours the field varies due to the effects of electric currents in the ionosphere, which are induced by solar activity. There are regular diurnal variations caused by the solar wind; there are irregular magnetic storms associated with solar flares. Terrestrial thunderstorms can also distort the local magnetic field. Over longer timescales (100,000–1 million years) there are reversals of the dipole field—the north and south poles change position. (There have been 10 such polarity flips in the past 3 million years.) These flips take a few hundred or thousand years to complete, during which intervals the field lines are very confused. Generally speaking field lines run more or less north-south, except during flips. The magnetic equator (defined by zero inclination angle) is normally a wavy line lying close to the celestial equator, except during flips. Local anomalies (due, e.g., to a mountain of iron ore) bend the field lines. Field distortions also characterize the seafloor near plate boundaries where the seafloor is spreading. These terrestrial and submarine anomalies create a magnetic map, and the field lines create a magnetic compass, which animals have evolved to exploit.[22]

For half a century biologists have been heavily influenced by the map-and-compass hypothesis, which states that an animal commencing a long journey must first determine its position relative to the destination (map step) and then set a course toward it (compass step).[23] In this view it is easy to understand why migrating species have developed a navigational sense that taps into the geomagnetic field. The magnetic information available is enough to provide a rough estimate of location (magnetic map—arising from anomalies in the field) and a pretty good estimate of direction (magnetic compass—arising from the field lines). There are two distinct types of magnetic compass: a *polarity compass* senses the direction of the magnetic field (e.g., the compass arrow points north), whereas an *inclination compass* senses the inclination angle of the magnetic field (it points equatorward or poleward—though it cannot tell which pole). Nature, by and large, seems to have opted for inclination compasses. Among birds, for example, there is evidence for inclination compasses in European Robins (*Erithacus rubra*), four *Sylvia* spp. of warbler, Pied Flycatchers,

Dunnocks, Indigo Buntings, Savannah Sparrows, and Bobolinks. Swainson's Thrush may use a polarity compass. Among other animals, Loggerhead Turtles and *Tenebrio* beetles certainly have an inclination compass; Eastern Red-Spotted Newts may use both inclination and polarity, whereas bats appear to be sensitive to magnetic polarity. Other creatures with a magnetic sense (not yet fully identified) include *Pieridae* butterflies, honeybees, sand-hoppers, such amphibians as the Natterjack Toad, the American Alligator, whales, Sockeye Salmon, tuna, Rainbow Trout, and Homing Pigeons.[24] This list will probably grow as more and more long-distance migrants are investigated.

There is a complication with inclination compasses for animals, such as birds, that migrate across the equator. Let us say that a Bobolink sets off south from Canada to its wintering ground in northern Argentina. It initially heads in the direction its onboard inclination compass says is toward the equator. When our Bobolink reaches the magnetic equator, there is no vertical component of geomagnetic field, and so the inclination compass is useless. Passing south of the equator, it must switch to the direction its inclination compass says is away from the equator. (If it continued equatorward, it would be heading back north.) Things would be much simpler if the Bobolink utilized a polarity compass ("head south"), and yet many birds have opted for the inclination compass and the more complicated instructions "head equatorward until you cross the equator, then head poleward." Another problem with magnetic compasses (which applies to both polarity and inclination compasses) is that of declination. Savannah Sparrows migrate to the far north to breed in summer. In these regions the magnetic compass is not true: it is a poor guide to the direction of true north, because true north and magnetic north are separated by 1,000 km. Savannah Sparrows have learned to adjust for this magnetic declination by using the directional information provided by stars to recalibrate their magnetic compass.

Although our understanding of geomagnetic sensing is incomplete, it seems that a consensus has been reached after "years of controversy and very mixed experimental results."[25] Many data have been gathered about bird magnetic sense, and this confusing mass of observations has been largely untangled. The story it tells is complicated (hence the time taken to untangle it) but fascinating. Many birds have, in effect, two independent magnetic senses: a "magnetic direction sense" provided by an inclination compass connected to the right eye, and a "magnetic map sense" arising from little magnets within the brain. Let's deal with the map sense first.

Magnetite is a naturally occurring iron oxide consisting of small single-domain crystals that are tiny permanent magnets; these occur naturally in the diets of many animals.[26] Clusters of magnetite crystals have been found in the brains of birds and other animals with a magnetic sense. The tiny magnets twist to align with the geomagnetic field, and the twisting force is proportional to geomagnetic field strength. It is thought that the field strength—the intensity—provides magnetic map information. The presence of such a map sense would certainly explain the displacement experiment results noted earlier for shearwaters, starlings, and albatrosses. Sea turtles provide another example: after years in the ocean, they make a beeline for their breeding ground, independent of current or location, as satellite tracking experiments show—a clear indication that sea turtles have a map (and they are known to have a magnetic sense).[27] Another example of magnetic map information (actually, misinformation in this case) comes from whale strandings. Whales occasionally beach themselves, seemingly on purpose, often with fatal consequences unless they are assisted by well-meaning humans. Here is an intriguing statistical correlation: whales beach in regions with low geomagnetic field intensities. We know that whales move to the tropics in fall and winter to breed, and that these equatorial regions have lower geomagnetic field strengths than the polar regions where they feed in summer, so it seems reasonable to suppose that whales navigate by somehow sensing magnetic field intensity. We speculate that they beach because magnetic anomalies fool the whales into thinking that a beach is in fact a route to their breeding areas.

In birds magnetic direction is generally provided by an inclination compass linked to visual information. They can see geomagnetic inclination compass information, perhaps as a turquoise background that is stronger toward polar directions and weaker nearer the equator. In European Robins (and probably many other birds) the magnetic field is sensed through the right eye only—so that the information is processed in the left hemisphere of the brain. We used to think that only humans exhibited such lateralization (differences in function between the left and right sides of the brain), but we now know it is widespread among vertebrates. The connection between optical and geomagnetic senses suggests that robins can use their magnetic sense for navigation only during the day. Experiments show that an artificial strong magnetic pulse does not affect the inclination compass of birds but does mess up their magnetic map. Presumably, the pulse twists the magnetite but does not influence the birds' visual capability.[28]

Sensor Fusion—Putting It All Together

The pigeons' sun compass is calibrated to its innate geomagnetic compass early in life. A similar pattern of orientation and calibration is observed in nocturnal migrants that use the stars.[29] Here we have the main reason, in our opinion, animal navigational senses are so good: redundancy. For many animals there is more than one navigational cue, and each individual knows to check one cue's predictions against another. Such versatility permits successful navigation in changing circumstances—for example, through unusual weather during migration, or during a geomagnetic flip. The hierarchy of cues may differ among species; that experienced animals navigate better than novices may, in part, result from their learning how to weight the navigational information gleaned from different senses.

Pigeons navigate using landmarks, olfactory gradients, sun compass, and geomagnetic cues. Salmon use olfactory cues near home and a magnetic sense in open ocean. Butterflies use a time-compensated sun compass and also landmarks to compensate for wind drift. Amphibians also have a redundant multisensory orientation system based on acoustic, olfactory, magnetic, and visual directional information. They can boast two independent compass systems based on time-compensated celestial cues and a light-dependent magnetic inclination compass. They can beacon along acoustic and olfactory gradients and pilot along visual landmarks, and at least one species (the Eastern Red-Spotted Newt mentioned earlier) can truly navigate with the aid of a magnetic map. Some bird species learn to navigate on their own, with only their genes to tell them what to do the first time. Others, such as geese, migrate in large flocks led by experienced birds. In both cases the successful migrants are learning from their experience; they build up landmark maps for use next year; they learn the night sky; they develop a hierarchy of senses ("Ah, so polarized light is more reliable at these latitudes than my inclination compass . . .") and calibrate one sense against another ("North-northwest? No way the stars are right. My magnetic map and compass say otherwise. Those pesky ornithologists have been messing up my brain in a planetarium . . .").[30]

Of course, birds and other animals know nothing of geomagnetism, time-compensated compasses, or celestial mechanics. They know what their genes and their senses tell them, and they learn to weight this information based on experience. We know that a bird's eye can see polarized light and magnetic field intensity, but the bird doesn't know this: it sees patterns of light that arise from these sources and its brain interprets them as directional cues.

12

Talk to the Animals

In going about their daily business, all animals cannot help but emit signals of one sort or another—chemical, auditory, visual, or electrical. As we have seen, other animals have evolved senses to detect those signals. However, the emission of a signal detectable by another creature is not necessarily communication. Entire libraries have been filled with discussions of what is, or is not, communication, but we keep things simple.[1] For our purposes communication is about passing messages between animals. It requires two things—

deliberate intent and encoded information. By this we mean both that the signal is emitted intentionally and that it contains information that can be processed and utilized by the recipient. The intent is not necessarily conscious intent—*you* may not know you are signaling, but your body means to do it.

Two complications arise. The first is signals intelligence: an animal intercepts, interprets, and uses a communication intended for someone else. The second is information warfare: the message deliberately deceives the recipient in a way that benefits only the transmitter. Animals use these techniques all the time, both within and among species, and we point out examples as we make our way through this chapter.

Talk the Talk

Animal communication takes the form of displays and badges. Displays are things the animal does to send a message. They may be conscious vocalization or ritualized movements or they may be the unconscious emission of chemicals. Badges are passive messages encoded in physical structures—such as the tail of a peacock or the mane of a lion. The structure itself is the message. It may provide information on the health, fertility, strength, or status of the owner, but the animal doesn't have to do anything to put its message across.

What displays and badges share is that information is encoded, transmitted, received, and interpreted. The recipient may then act on that information. This observation gives us our working definition of communication: the intentional transfer of apparently useful information. Originally this must have been a mutually beneficial exchange, but evolutionary pressures have allowed a modicum of eavesdropping and deception to creep in, as we will see. So that's communication. But before we look at how animals do it, we should first consider why they do it.

It's Good to Talk

Why do animals communicate? Fundamentally, we're back to sustenance, security, and sex. Consider this: animals need to reproduce. In most cases that means finding a mate.[2] To do that you have to be able to identify another creature as being of your species, which involves an exchange of information, and as being ready and willing to mate with you, which involves the exchange of more, and different, information. As animals become more complex, their needs become greater, leading to more elaborate interactions and the requirement for more sophisticated communications capabilities.

A primary purpose of any communication system is to allow animals to find one another. An individual may not only advertise its presence but may also include information on how it is likely to react to others. Recipients of the message may then choose, for example, whether they wish to respond or run away.

Building on these needs, communication also allows animals to identify one another, both in terms of their species and as individuals. Knowing who, and what, transmitted the communication can affect both how you interpret it and how you react to it. Lions roaring at dusk will tell wildebeest that the pride is about to start its hunt, but another pride of nearby lions may extract more information from the same message.

When animals do meet, communication helps them to do so without trying to kill one another. This communication can vary from aggressive displays (which persuade one participant to give way before serious fighting occurs), to reassuring displays (in which an animal signals its willingness to avoid conflict), to acquiescent displays (which preemptively signal acceptance of another's dominance).

Other forms of communication are used to sort out who is (or is not) ready to breed and who they are willing to breed with. This may include forms of communication intended to persuade potential mates that you really are their best option. In some cases the recipient's response may even include an automatic change in physiological state, making it more receptive to breeding. Mating-related communication may also act within groups or colonies of individuals, in some circumstances to suppress breeding behavior and in others to ensure everyone breeds at the same time, or even that young are actually born at the same time.[3]

The ability to communicate information allows animals to use one another as extensions of their individual sensory systems, to detect, identify, and locate things that they themselves are not in a suitable position to observe. This ability is particularly useful for the detection and avoidance of predators, where one animal's instinctive response may indicate that it is about to run for the hills, thus warning its neighbors to do likewise. Those with more sophisticated social groupings may go even further and transmit unique messages that convey threat-specific alerts and warnings. In a similar sort of way, albeit not always in their own best interests, animals may also communicate that they have found, or know where to find, a particularly interesting source of food.

In addition to all the above, communication is vital in establishing and maintaining relationships among animals. These relationships may be as simple as maintaining a suitable separation from other creatures to actively engaging in cooperative ventures. The more sophisticated the relationship, the greater the role communication plays in its maintenance by letting the various members know, not just what each member knows and wants but what each might be about to do next. This information can vary from wolf cubs whining to tell their parents they are still hungry to a goose giving a special call to let the rest of the flock know it is about to take flight. You will, undoubtedly, recognize examples of all these aspects of animal communication in the people with whom you interact every day.

Of course, animals not only communicate within their own species but, just as importantly, they intentionally communicate among species. Much of their communication is about avoiding conflict. Thus, distinctive markings may warn a potential predator that the badge holder is not worth tackling, because it is poisonous, dangerous, or just tastes horrible. An interesting spin-off is the number of entirely harmless creatures that mimic more dangerous ones' badges in the hope that they, too, will be left alone—information warfare used to deceive an opponent. Behavioral displays can also be used to avoid conflict. When a hare spots a fox, it may run a short distance, then stop and sit in plain view. This isn't because it thinks it is invisible. A healthy adult hare will almost always outrun a healthy adult fox, so this instinctive behavior has evolved to tell the fox it has been spotted and shouldn't waste everyone's time and energy on a doomed pursuit. In a similar vein, as we saw in Chapter 1, African antelope jump in place to let predators know they are full of energy, primed for a run, and not worth chasing.

Some interspecies signaling may be more positive in nature, although it is difficult to be sure if the communication is truly intentional. Many species respond to another's intraspecies signals indicating the approach of predators or the existence of a food resource, but this behavior may just represent an eavesdropper exploiting signals intelligence. Can we see an unambiguous message being communicated from one creature to an entirely different creature, resulting in palpable benefits for both? Indeed so: many large predatory fish, such as the grouper in Figure 97, allow themselves to be cleaned of parasites and dead skin by a variety of "cleaner fish," like the wrasse. These cleaners are comparable in size to the groupers' normal prey, but congregate at mutually recognized cleaning stations, where they offer their services. When a grouper

FIGURE 97 Mutually beneficial interspecies communication. Tiny wrasse clean a massive grouper, secure in the knowledge that, here and now, it won't eat them.

approaches the cleaning station, it adopts a specific posture, indicating it is not a threat and wishes to be cleaned. The grouper gets a wash and brush-up, and the wrasse gets lunch—a clear case of mutual communication producing mutual benefit.

The Medium Is Not the Message

To communicate, animals must be able to generate and transmit messages as well as receive them.[4] Reception was achieved using the external sensor systems, but they also had to develop mechanisms for generating and transmitting signals. At some point, every known sensory mode has been adapted to support a corresponding communication function.

For communications purposes, each mode has its limitations, largely determined by physics, not biology. Thus, different modes may have different energy requirements; their range and clarity may be affected to a greater or lesser degree by the medium (air, water, or ground) through which they must be transmitted; they may or may not require a direct line of sight between transmitter and receiver; they may produce a signal whose source is difficult to locate in space or time; they may produce a transitory signal or one that lingers; they may be capable of encoding a large quantity of information or just that the signal itself exists; and so on.

Not all animals use all possible modes, and they certainly don't all use them in the same way. Equally, many communications require simultaneous multimode signals to convey their full meaning. Thus, aggression may be signaled by a combination of visual posturing, loud sounds, and chemical emissions. Fear may be signaled by the same sounds and postures, but different chemicals. Miss one part and you don't get the full story.

Chemical Signaling

Not surprisingly the earliest, simplest, and most widespread communication system is based on chemistry. Chemicals used to encode messages are called *semiochemicals* and the best known are pheromones, complex organic molecules used for communication among members of the same species.[5] They are used to provide identification, both directly (when detected on the individual emitter) and indirectly (through scent marking of objects or places) to indicate or manipulate sexual behavior, mark ownership of territory, regulate social behavior as diverse as feeding young and fighting rivals, lay trails, signal alarms; call other species members to a specific place for a specific purpose, or just to let them know you are there.

Animals are chemical machines, and so pheromones probably have their origin in molecules that leaked out—hormones; body wastes; chemicals released on injury; or even those absorbed from food species, whether plant or animal. Animals that could respond to one another's leakages had an evolutionary advantage and their response thus passed to their offspring.

As we saw in Chapter 6, eusocial insects are enthusiastic pheromone users. An ant experiencing a crisis will emit an appropriate alarm pheromone for as long as the crisis persists. The small, volatile molecules rapidly diffuse and so are detectable several centimeters away (a long distance in ant terms). Detecting a low concentration draws other ants to the scene, where the higher concentration triggers their response, which includes emitting more alarm pheromones to attract reinforcements. As the crisis abates, the rate of emission lessens and the molecules quickly dissipate so that, once the crisis is over, the ants return to their normal tasks. In contrast, the pheromones used for identifying one another or to encourage a general gathering together are less volatile and so more persistent. This leads us to a general principle found across all species: for short-lived messages, use a small, volatile molecule; for long-lasting messages (e.g., territorial markers), use a larger, less volatile one.

Other semiochemicals are used to send signals to members of a different species. Collectively known as *allelochemicals,* they can be divided into three broad types. *Allomones* benefit only the emitter; they include chemical repellents or defenses as well as those used for purposes of propaganda or deceit, such as luring prey within range. *Kairomones* benefit the receiver. These are often emitted as pheromones but intercepted by another creature, especially a parasite or predator with an ulterior motive. *Synomones* benefit both emitter and receiver, such as the chemicals sea anemones use to attract their symbiotic partner, the anemone fish.

Tactile Signaling

Touch may be the next simplest mode: the hardware was already available, and animals just had to find a way to encode and interpret messages. Here we specifically mean tactile communication in which the source animal physically touches the recipient. If it involves vibrations transmitted through an intermediary medium—gas, liquid, or solid—we consider that a form of acoustic/vibratory communication, discussed later. Tactile signaling has the advantage of providing instantaneous feedback: the toucher knows what it is touching, can feel the response, and modify its signal accordingly. It also works in the dark and is difficult for others to eavesdrop on. On the downside, you do have to stay in contact, and it is difficult to communicate with a large number of recipients.

Many insects have poor vision and hearing, so touch, usually with the antennae, becomes their second-most important mode of communication after chemical. Meaning is encoded in the manner and location of the touching and, unsurprisingly, it appears to be especially important to the highly organized eusocial insects. However, it is difficult to establish how much of the transmitted message is purely tactile and how much is due to associated chemicals, especially as insects also detect smells with their antennae.

Physical touch is important in many vertebrates, especially mammals, to encourage mating, provide reassurance, and establish dominance. Many mammals engage in mutual grooming, which serves to establish and reinforce social bonds within a group. Apart from the immediate effect on the participants, grooming may also serve a crossover purpose in that the act of grooming one another is visible to other group members, who can use it to infer the mutual status of the participants.

Visual Signaling

Once you've got vision, using it for communication is remarkably simple. You send a signal just by doing something or even just by being there. The very act of signaling tells the recipient where you are, and, unless you're being deceitful, what you are. Unfortunately, visual signaling has its drawbacks. For one thing, you need a direct line of sight between the participants. Not only can that be obscured by environmental factors (weather, terrain, vegetation, or even intervening creatures), but it also means that the recipient has to be looking at the source, which limits their freedom to do other things and may even endanger them. Furthermore, to establish communication in the first place, the source has to attract the recipient's attention. That can attract unwanted attention from predators or other opportunists, who can intercept the signal for their own ends. A signal that involves prolonged posturing or display only increases the risks.

On the plus side, unless it involves elaborate and exhausting displays, visual signals are relatively easy to maintain, so long as you have nothing better to do, and can communicate with several individuals at once. If you've invested in a badge, you can even communicate without actually doing anything, leaving you free to take care of business. Or, like a peacock displaying, you can use the badge to accentuate the signal. In the most extreme cases the badge might not even be a part of the animal. Remember the bower birds from Chapter 5? The bower's only purpose is to signal an individual's quality as a mate.

The persistence of visual signals can be exploited in other ways to create communications that don't require the presence of the signaler. Thus, when bears rub themselves against trees to mark their territory, they also leave visible scratch and bite marks. Mammals as diverse as the rabbit and the hippopotamus place piles of dung as territorial statements. Initially the associated chemical signals may be more important, but the visual impact lingers after the scent has faded.

For sighted creatures, visual signaling is so elementary that you don't actually need any specialized equipment to generate a message. Unless you want to do it in the dark, which, inevitably, some creatures do. And they've found ways to do it: generate your own light source and signal with that. Humans have flashlights, but animals have to get by with bioluminescence.

Bioluminescence is the emission of light due to a chemical reaction within a living organism. Many creatures have this ability, and some have developed ways to signal with it. Although the greatest numbers of bioluminescent animals are found in the dark ocean depths, perhaps the most familiar example is the firefly, which is really a generic term for several different, but related, species of beetle, not all of which emit light. For those that do, light emission is mostly concerned with finding a mate, with the pertinent signals being encoded in the duration, intensity, and sequence of pulsed light emissions. A human observer can't see it, but each flash is a sequence of rapid pulses, whose exact structure depends on the species. The males flash and watch for an appropriate response from a suitable female. Although glowing in the dark might seem like a good way to get eaten, doing it with a bunch of friends greatly improves your chances of surviving the night.[6]

Acoustic Signaling

Acoustic signaling offers many advantages. You don't need to see, smell, or touch one another to do it and, in many cases, you can do something else at the same time. The signal can carry information over long distances, around corners, and past environmental obstacles to reach multiple recipients, but it doesn't linger to interfere with subsequent transmissions. Frequency spectrum, duration, volume, and repetition rate can all be varied. More variables allow more information to be encoded but also allow the signal to be adapted for different environmental conditions. There is inevitably a downside. Making noises gives away your existence, and possibly location, to opportunists. Eavesdroppers may intercept and use information intended for others. Making noises also burns energy. Attenuation can be a problem—as we saw in Chapter 9, intensity falls off rapidly with distance from the source, and noisy environments reduce signaling range even further. Making louder noises to compensate burns off even more energy. There are other physical and environmental constraints: in general, smaller creatures produce higher sound frequencies, and high frequencies don't carry as far (see Chapter 8). Figure 98 shows a master practitioner.

Animals generate the noise themselves, use external instruments, or combine these two approaches. The frequency at which a creature can make noise depends on the physical properties of its sound-generating mechanism. For example, basic physical principles suggest that bigger creatures will have deeper voices, with the frequency varying inversely with body weight to the power –0.4.[7] Experimental data show this scaling relationship holds up well for air-breathing vertebrates over a range of 0.01–10,000 kg in weight and at least 20–5,000 Hz in frequency. Because of the different ways in which real animals generate and manipulate sounds, specific animal types do show a lot of variation around the theoretical frequency—elephants, for example, will

FIGURE 98 This male Black Howler Monkey's mane may advertise his fitness as a mate, but he is named for his impressive vocal displays: an enlarged hyoid bone in the throat acts as a resonator, making him one of the loudest-known land animals. We thank Anita McFadzean for this image.

readily vocalize from 15 to 35 Hz. And there are exceptions—notably, horses, whose voices seem too high-pitched for their size. Allowing for the effects of animal size on the ability to both generate and detect sounds, more detailed analysis shows that, among members of the same species, the effective communications range varies directly with body weight to the power 0.6.

Insects do things differently, but similar scaling rules still apply. Analysis shows that the sound frequency varies inversely with body weight to the power –0.36, surprisingly close to the vertebrates' result. The net effect is the same: the smaller the creature is, the higher the frequency of sound it makes and the shorter the range over which it can communicate.

Remember *C. ridiculus* from Chapter 1? For these hypothetical varmints, the adult male was three times the weight of the female and twenty-seven times the weight of the young. Assuming they are air breathers and the adult male is roughly rabbit sized—say, 1 kg—then he should vocalize around the 1,200 Hz mark, with his smaller mate being closer to 2,000 Hz and his offspring at about 4,700 Hz. The deeper tone of the male lets him signal other, distant males, whereas the higher squeaks of the young don't travel far enough to alert predators to the nest site. If our critter was more like a large mouse—say, 0.1 kg—the relevant figures would be nearer 3,000 Hz, 5,000 Hz, and 12,000 Hz, respectively. Rabbit-sized males' communication ranges would be about four times that of mouse-sized males: the bigger the creatures are, the farther apart they can wander while still staying in touch.

Sound is just a vibration: to make it you have to move things.[8] Most insects and other noisy arthropods achieve this by stridulation—rubbing one body part against another. Generally this action involves a body part that acts as a file and another that acts as a pick: ideally the pick is connected to some sort of resonant structure, such as wing panels. The file is dragged across the pick (or vice-versa), and a noise is generated, as shown in Figure 99.

The detailed frequency structure of the produced sound depends on the frequency at which the file is dragged, the frequency with which the pick impacts the file (determined by the number of individual picks on the body part), and the oscillation frequency of the attached vibrating structure. As a sound-generating mechanism, this structure is not terribly efficient—radiated acoustic power is a lot less than the input muscle power—except at high frequencies, where the sound wavelength is comparable to the dimensions of the body parts involved, which leads us back to small creatures making high frequency noises that don't carry very far. Some insects, like cicadas, get around

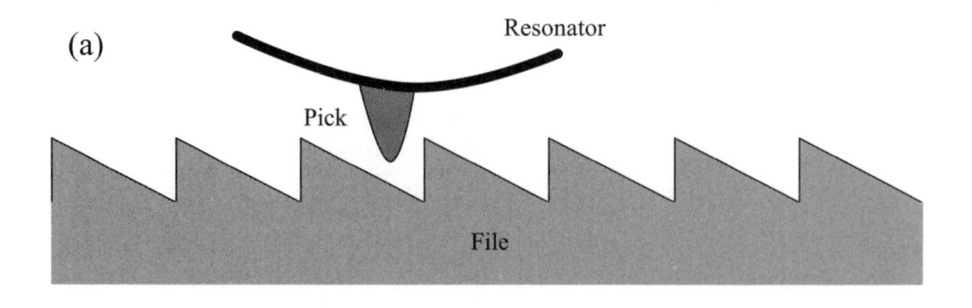

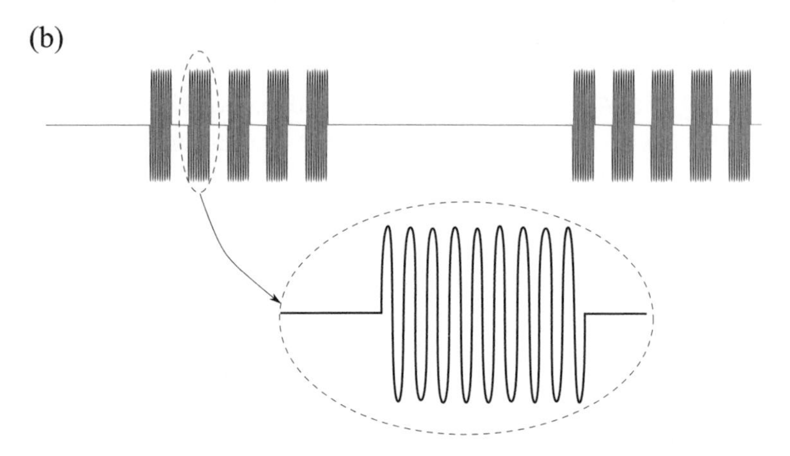

FIGURE 99 How to make a noise if you don't have lungs: (a) The file-pick structure used by many arthropods. (b) The results of three strokes of the file, showing the individual oscillations produced by impacts of the pick on the file teeth. Below this, individual pulses are expanded to show the resonant oscillations produced by successive pick impacts on the teeth. Across species, variations in numbers and spacing of teeth, as well as the speed of the pick, allow for very different signals.

this problem by having large, air-filled abdominal cavities lined with flexible, ribbed cartilage they can vibrate under muscle control. This mechanism is more efficient and exploits the resonant properties of the cavity to produce louder sounds at those useful lower frequencies.[9]

Lung-equipped air breathers have the advantage of an internal air reservoir whose volume and pressure can be readily controlled by muscles. To recap on Chapter 9, forcing that air through a valve with mechanically resonant flaps causes the valve to vibrate, producing an oscillating air flow and thus making a sound. The details of the sound are determined mainly by the geometry, mass, and tension of the valve—air pressure is less important. In mammals, for example, this valve is the vocal chords, whereas in birds it is the syrinx.[10]

The basic difference here is that vocal chords constrict one air passage (the trachea), whereas the syrinx consists of twin structures constricting the bronchus from each lung, as shown in Figure 100. In both cases varying the tension in the valve causes variations in the fundamental frequency, and resonances in the air column of the vocal tract produce further variations in the output sound spectrum. In most cases the sound is then radiated through the mouth, where other structures (tongues, lips, or beaks) can modify it even more. Although birds' beaks and tongues are more limiting than the equivalent structures in mammals, birds do have an advantage in that the two parts of the syrinx can often operate independently to produce a complex, pulsating airflow rich in harmonics, which we call "birdsong."

Reptiles and amphibians are more varied: some have vocal chords (or equivalent structures) in the larynx, and some, such as snakes, don't. Those without can

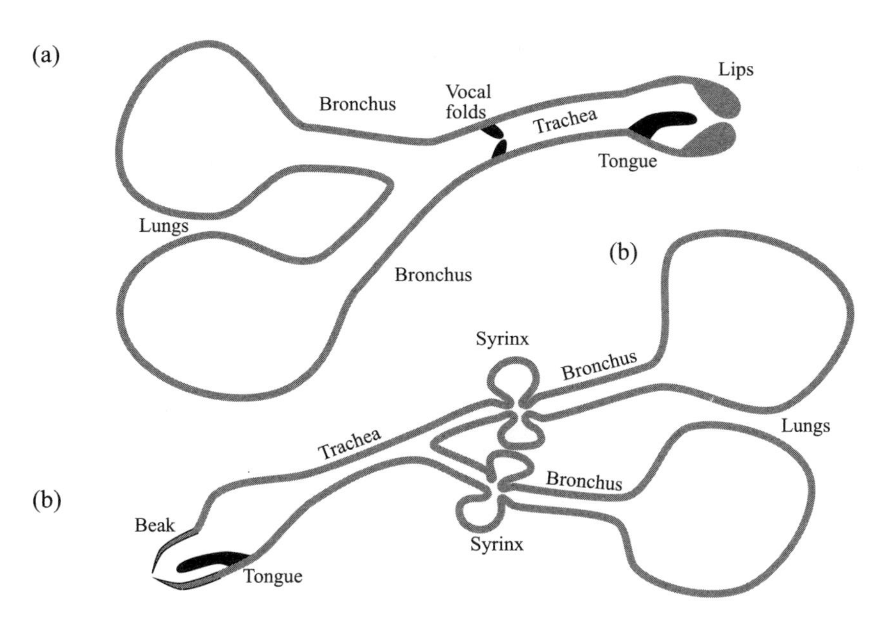

FIGURE 100 How to make a noise if you do have lungs: the vocal tracts of mammals and birds, showing the components that affect the output sound. The bird's twin syrinx allows greater vocal variety than the vocal folds of mammals can.

still hiss by forcing air out of the lungs. With a few notable exceptions—such as frogs—modern reptiles and amphibians tend not to be very vocal.

Some animals (especially among the frogs and birds) employ inflatable sacs in the upper vocal tract that resonate at the vocal frequency, thus acting as an amplifier. The final sound is sometimes radiated through the thin membrane of the sac itself rather than the mouth, which favors transmission of the lower frequency components. Opening the mouth can let out a higher frequency sound. Aquatic air breathers, like Chapter 9's cetaceans, who want to communicate under water, generally avoid drowning by using a closed system that moves air between internal reservoirs, but the basic principles are the same.

Water breathers have fewer options. Not surprisingly, most crustaceans go for some form of stridulation. So, too, do fish that lack an air-filled swim bladder. Fish with a swim bladder work a bit more like cicadas—as we saw, this is more efficient than simple stridulation, which means they tend to be louder. Some can even make noises by expelling air from the swim bladder, although it's not always vocalization as we know it: at night, in the darkness of the ocean, herring communicate by farting.

The use of instruments generally comes down to hitting something to make a better noise. Thus kangaroos, hares, and rabbits all thump their hind legs on the ground to signal danger. Beavers do the same by slapping the water with their tails. Among birds, woodpeckers and sapsuckers go further and drum on trees with their beaks. They know that dead, hollow wood, or even tin roofs, make a better noise. Many arthropods follow suite: Deathwatch Beetles (*Xestobium rufovillosum*) seeking a mate bang their heads against their wooden tunnels in the beams of old houses. Most of these animals are, in principle, capable of self-generating sounds. So why use environmental infrastructure to generate messages? Inevitably, it's all about making a suitable noise that carries far enough. Remember—high frequency, bad; low frequency, good. A woodpecker's vocalizations wouldn't carry far in a dense forest, but its lower frequency drumming will.[11]

Some small creatures take long-range acoustic signaling a step further: they make the noise themselves, then use an external instrument to amplify it. Crickets are especially good at this, with many species modifying leaves to use as baffles. Their small size means they should produce sound at high frequencies, but instead they use an energy-inefficient mechanism to generate a much lower frequency sound, which the baffle then amplifies so that it carries farther. As we saw in Chapter 5, others, such as mole crickets, like *Scapteriscus acletus,* go beyond even that and dig a special horn-shaped burrow (shown in Figure 101), which is carefully tuned to the resonant frequency of their low efficiency but low frequency sound production.[12]

Larger animals may have also found different ways of extending their communication range. In the open oceans the effects of temperature decreasing and salinity and pressure increasing with depth combine to produce a layer of water, known as the SOFAR channel, typically about 1 km down, where the speed of sound is lower than in the water above and below the layer. Acoustic

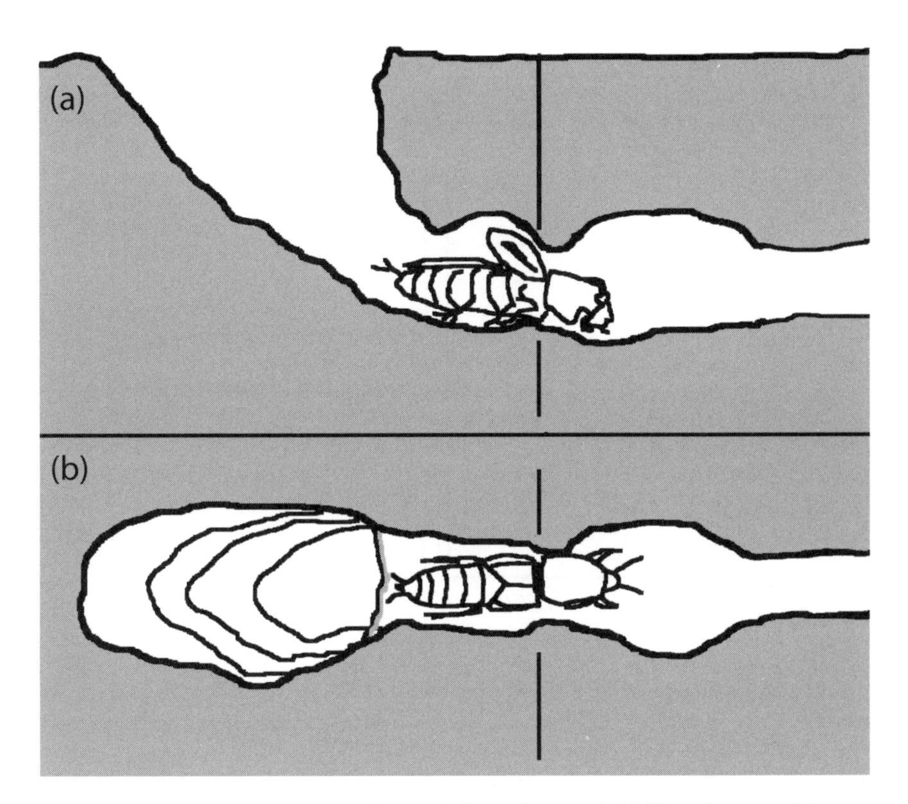

FIGURE 101 Getting more range by amplifying the signal. (a) The side view of the burrow of the mole cricket *Scapteriscus acletus* shows the contour lines in the horn. The optimum position for sound generation is in the constriction between the bulb (on the right) and the horn. (b) The burrow in a top cutaway view. Other species favor a twin horn configuration with two external openings.

waves propagating through this layer are reflected from the upper and lower boundaries, as shown in Figure 102. The effect is that the acoustic signal spreads —more or less—in two dimensions, not three, and the signal therefore travels much farther. Humpback whales seem to use this effect to broadcast their songs across the vast expanses of the oceans, although we cannot be sure if they truly make intentional use of the layer. A similar effect can occur over land when cold air is trapped close to the ground, with warmer air above it. Trapped in such a temperature inversion layer, sound can travel much farther than normal, allowing elephants on the African plains, for example, to be heard

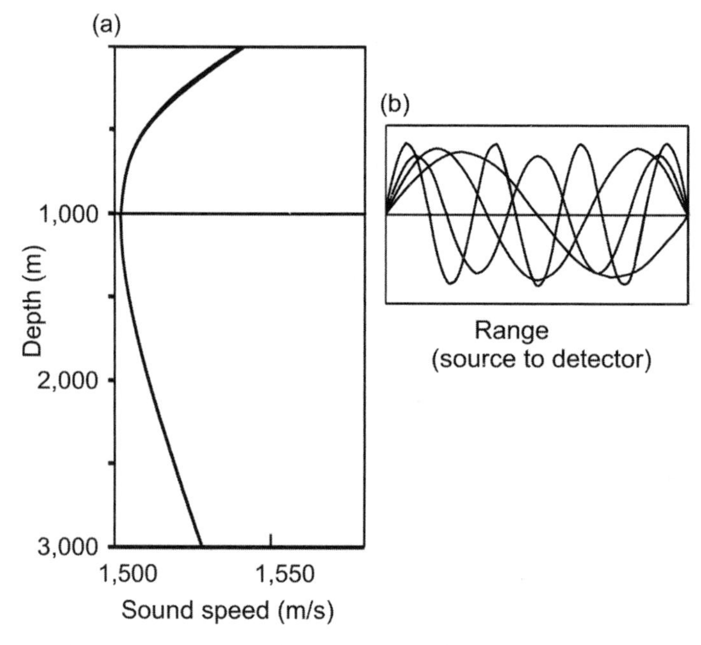

FIGURE 102 Getting more range by focusing the signal. (a) The SOFAR channel traps acoustic signals in a layer of lower sound speed, allowing it to propagate over longer distances. (b) Signals from the same sources follow different paths to the detector. As a sound wave moves away from the SOFAR channel axis, the part farthest from the axis moves faster, so the wave turns back, producing a path that oscillates about the axis. Different paths take different times, so a single transmitted pulse of sound is received as a series of pulses. The path closest to the central axis is shortest but corresponds to the slowest speed, so that signal arrives last and is usually the loudest. At higher latitudes the SOFAR channel axis is closer to the surface.

over enormous distances. As with the whales, we don't know for certain if this use is deliberate.

Electrical Signaling

Animals capable of generating and detecting electrical fields should be capable of using them for communication.[13] Some species of electric fish do signal their species, gender, social status, breeding condition, and current level of aggression by varying their electrical discharge. As with bioluminescence, the relevant information is probably encoded in the type, frequency, duration, and pulse structure of the emitted signal.[14]

Getting Your Point Across

Animals don't like wasting energy. If it seems like they are, look closer: there will always be a good reason for it.[15] Communication is no different. Communication mechanisms only evolved because they allow useful information to be transferred. Ideally, that should occur with the minimum expenditure of energy while exposing the participants to the minimum external risk. There is, as ever, a cost/benefit trade-off. For many species a signal that results in a successful mating is going to be worth it, even if it kills you. For others one high-cost signal may produce long-term benefits or reduce the effort needed for subsequent, related, signals. Establishing dominance might entail a vigorous work-out to demonstrate strength and stamina, but maintaining it could be a more leisurely affair. The initial high set-up cost is offset against the low maintenance cost.

More important messages justify spending more energy on transmission. Those sent over long distances or through noisy or cluttered environments will also need more energy input. Some messages are more time-critical than others: they must be sent, interpreted, and acted on quickly. That might justify a greater energy budget, but might also require a much shorter signal. If a signal has to be brief but is also very important, like a warning of the rapid approach of a predator, it must also be unambiguous. There is no time to ponder subtle nuances of meaning.

Most animals have quite limited processing power, which restricts the number of distinct displays they can support. So does the need for each extra display to be distinctive, which probably means more elaborate, more time and energy consuming, and more risky. As a result they favor more general messages, reserving the specific ones for those special life-or-death occasions. Combining different modes of communication, such as sound and vision, helps expand the repertoire and increase the chances of getting a message through a troublesome environment.

Other than species—or even individual—identification, two of the more important items of information that a message often contains are where you are and what you're doing or thinking of doing. Signaling location is an interesting problem for something that doesn't want to be eaten. If your species habitually shouts out where it is, you can bet a predator has evolved the ability to exploit that information—signals intelligence at work again. Consequently, many creatures are coy about revealing their whereabouts. Something small and edible that sees a predator may run away while giving a signal that warns others of the presence of the threat but gives little information on the signaler's location. His friends don't need to know that; they just need to know he's running. They might not even need to know why. The predator can't home in on the signal, and so the signaler has alerted his colleagues at little cost to himself. However, if signaling for a mate, a creature may choose a format that does make him easier to find. We see this in the mating calls of some frog species. A solitary male uses a call that provides only vague information on his location. While a suitably interested female will be prepared to search for him, a predator can't directly target him. When several males are in the area, he wants the female to find *him* quickly, but because he benefits from safety in numbers, predators are less of a personal threat. Hearing the nearby voices of rival males, he switches to a call that makes him easier to locate.

When signals are concerned with current or potential behavior, the context can often be important—a similar display means different things under different circumstances. If you are waiting to pull out at an intersection and an oncoming driver slows, flashes her headlights, and gesticulates, then her meaning may be very different than if she performs the same display *after* you have just pulled out right in front of her. In particular, displays indicating aggression and fear can be very similar, perhaps because aggression and fear are very closely related. Knowing which is being communicated can be very important.

Among both birds and mammals there are many communication displays that are used to indicate the signaler's desire to perform some act but inability to currently achieve it. The creature wants to evade a domineering individual, but cannot; it wants to attack something, but can find no opponent; it wants to mate but cannot find a partner. Animals communicate frustration, too.

The precise form of a rather generic display may indicate the probability of a creature's general response to another's signal, without specifying the precise response that will be chosen. Thus, a creature can indicate that it is too busy with something else to respond to a specific communication, or it can hint that it might be amenable to a more direct approach. Why waste time or energy on a detailed response when a more general "Not now. Busy." will suffice?

Would I Lie to You?

Animals lie. Like humans, they are deceitful, disingenuous, and downright dishonest. And like humans, they lie because, at least in the short term, they believe that the benefits outweigh the costs.[16]

Males of some frog species croak to tell each other how big they are. The bigger the frog, the deeper the croak, and the sound alone can scare off a potential sexual rival. However, some small males will produce a croak much deeper than their body size suggests, effectively making themselves sound bigger. Frogs that could easily dominate them don't even try, because they are fooled by little guys talking tough. Talk is cheap and the mating benefits well worthwhile, but the smaller frog has to accept the risk that his bluff might be called and he'll get beaten up.

Animals as diverse as shrikes and chimps are known to lie to keep others from a tasty food source. The shrike emits fake alarm calls to scare competitors away, but the chimp is more subtle and pretends to have found nothing of interest, faking apparent disappointment and boredom until the rest of the troop move away.

So why don't they all lie all the time? Basically, because it is to no-one's advantage to do so. If every shrike with food issued false alarms to scare rivals away, then the alarm signal would become meaningless, putting everyone at greater risk. The behavior that is passed to offspring tends to improve their chances of surviving to breed. The total cost of making the signal must be balanced against the benefits it confers. That cost is more than just energy and resources required but must also account for the risk associated with making it. If you're lying, you have to factor in the risk of retribution if you're found out. If those around you sometimes lie, your reaction to their signals has to factor in the risk that you lose out by treating an honest signal as a lie.

In contrast, deliberately deceiving other kinds of animals might really pay off—if you can get away with it. Being ultimately derived from the emitter's physical state, pheromones generally transmit "honest" signals. Well, within a species they do, but information warfare encourages some species to mimic another's pheromones for their own nefarious purposes (technically, the chemical is therefore an allomone). The female bolas spider mimics the female sex pheromone of the moth species on which she preys. Male moths seeking a mate will, instead, meet a sticky end.

Showing an aptitude for deception, many bolas spider species protect themselves from predators by taking on the appearance of uninteresting snail shells or bird droppings. This passive defensive mimicry is more camouflage than communication,[17] unlike the more up-front mimicry of pretending to be something much more dangerous, poisonous, or horrible tasting in the hope that you will be left alone. This is known as "Batesian" mimicry and can involve mimicking the actual appearance, sound, or behavior of another creature or just reproducing the scariest part of it. Thus, some moths have patterns on their wings which make them look, vaguely, like the heads of predatory birds. The moth reacts to threats by flashing its wings and hopefully scares attackers away. For the same reason other creatures' body markings mimic the heads of dangerous snakes. A similar effect, named for a different nineteenth-century naturalist, is Mullerian mimicry, in which two different, but potentially harmful (or at least foul-tasting), species mimic each other's warning signals to deter their common predators.

It's not only spiders that are able to fool their prey. In Central and South America some species of forest falcon will lie in wait and emit sounds like those made by birds mobbing a predator. As small birds are drawn to join in the fun, the falcon strikes. Interestingly, the technique seems to work best against visiting North American migrants, who are presumably unaware of local customs.

As ever, what is at work here is natural selection: if you are born with the ability to make a signal that fools prey into being more easily caught, you will be more likely to prosper and pass that ability on to your offspring. The cost of making the signal is worthwhile; the consequences of being caught out are negligible. Well, unless your act of deception makes *you* more susceptible to predation. But natural selection would take care of that, too.

Dancing with Bees

Ants lay pheromone trails to guide their colleagues to things of interest, but for their honeybee cousins, pheromone trails through the air just aren't reliable enough. Instead, they communicate through dance. Well, strictly speaking, it's a combination of tactile, vibrational, and chemical signals, but to humans, it looks like dancing in the dark.[18]

Honeybee foragers mostly communicate information about food sources, although they also deal with water supplies and potential sites for new hives. They do this by providing samples of what they have found and dancing in total darkness on the vertical surface of the honeycomb to show the distance and direction at which they found it. To get the full message, other bees must follow the display by maintaining careful contact with their antennae. Dancing bees also emit pheromones, which draw others to witness the dance and possibly provide information on the quality of the source.[19]

For a source close to the hive (say, within 30–50 m) the bee performs a round dance, turning in circles, alternately right and left. For an intermediate range (50–150 m) the bee uses a sickle dance, a half-way step between the round dance and the waggle dance. In both cases, duration and vigor of the dance indicate the quality of the source and the samples provide its scent. At these ranges, that's enough information for other bees to find it.

For more distant sources the bee employs the famous waggle dance shown in Figure 103 to signal both range and direction of the place of interest.[20] The bee runs straight ahead for a short distance, follows a semi-circle path back to the starting point, repeats the straight line course, and then makes a semi-circle in the opposite direction to return to the start. During the straight-line ("waggle-run") portions of the dance, the bee's body vibrates, producing a tail-wagging motion, and it beats its wings rapidly. All these actions encode parts of the message, but experiments prove that the length of the waggle run is directly related to the source's distance. The bee's orientation on the comb during the straight part of the dance indicates the direction to the source in relation to the position of the sun. That is, the bee adopts an angle, relative to the vertical, that matches the angle between the sun's current position and the source. Obviously, as the sun moves across the sky, the orientation of the dance will change during the day.[21] Clever stuff, and remember, it's all done in total darkness.

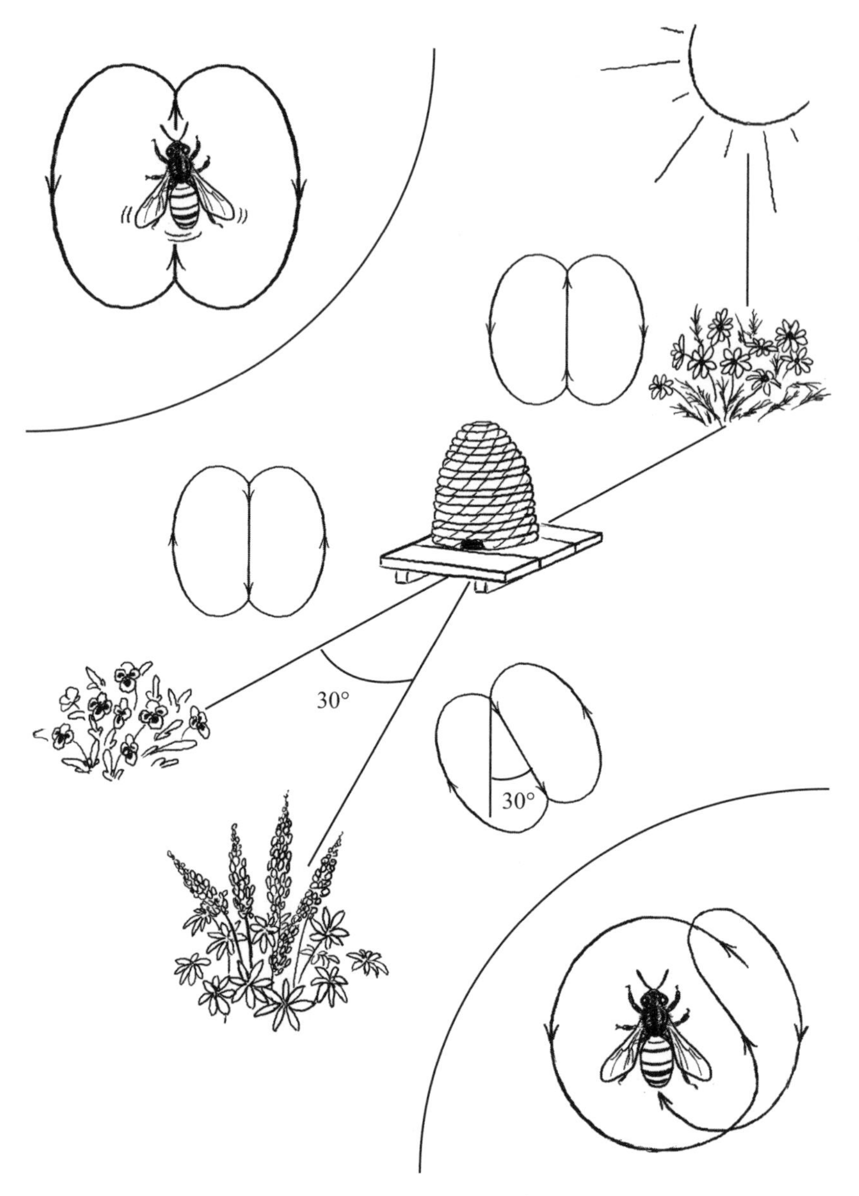

FIGURE 103 The waggle dance isn't the only way honeybees communicate information about food sources, but it is one we can interpret.

Language or Just Loud Noises?

If animals communicate, do they have language? It's a tricky question. And by "language," do we mean vocalizations only? Perhaps that's an unnatural distinction. Spoken human language is easier to follow if you can see the body language as well. Can we assert our superiority by stating that our spoken languages are not instinctive, but cultural—they have to be learned? But this is true for some animals: many songbirds have to learn, or be taught, their repertoire. Many also show clear regional dialects and accents.

Spoken human languages display *double articulation.*[22] We use words that are composed of elements of meaning, or morphemes (e.g., "tall" and "er"), which are themselves composed of elements of sound, or phonemes (e.g., "TIH," "AH," "LL," "UR"). A tiny set of phonemes (English has around 50) can produce thousands of unique words (English has at least half a million different words). Morphemes can be combined to make new words, and words can be combined to create an infinity of meaningful sentences. Can animals do this?

FIGURE 104 Barbary Macaques: talking or just communicating? We thank Anita McFadzean for this image.

Perhaps. There is emerging evidence that some animals, like those in Figure 104, can take individual elements of communication and assemble them into a sequence that means something totally different from any of its individual components.[23] They can create new "words" and even string them into "sentences" with apparent rules of syntax. Is this language or just an adaptation to allow the communications on which their survival depends? If physical limitations have restricted your range of basic communication elements, then just combine them to create new signals.

Perhaps the real difference is that human language has developed to let us exchange all manner of abstract, conceptual information. Few animals seem to share this need. Even those capable of conceiving new ideas, like those chimps stacking boxes to reach overhead bananas, don't then tell one another how they did it. They just do what they do, and others watch and copy.

Most animals don't have much to say; humans do. But we can be sure that whatever information animals need to communicate, they will have acquired the ability to communicate. If we want to understand what they are saying, we're going to have to watch them more closely.

Epilogue

We have explored many examples of animal adaptation in this book. We have seen how natural selection has exquisitely honed the form and functionality of animal designs, so that they are well suited to live and reproduce in a particular habitat and ecological niche. Thus, for example, binocular vision has evolved to aid predators, be they chimpanzees or jumping spiders.

The solutions provided by evolution do not have to be perfect to convey a survival advantage. Thus, we can live with the fact that our optical wiring is the wrong way around. Presumably, any evolutionary path that we would have to follow to correct this "defect" would initially take us down a road that leads away from exquisite adaptation to our environment. During this journey we would be less fit to survive, and so, even though the result would be optically superior eyes, it will not happen, because natural selection would weed out the travelers on this path.

Natural selection does not see into the future. It cannot plan. It simply chooses from the variability of life that exists at a given moment, so that the best-adapted life forms survive to reproduce at the expense of less-well-adapted forms. Our book is not concerned with the biochemical mechanisms of adaptation but only with the physical constraints that must always apply. Thus, we have seen that the acuity of simple eyes is limited by lens design and photoreceptor density. To take another example: hollow bones with a circular cross-section are mathematically optimum for resistance to bending. So, if selection works as we have described—to make an animal fit for survival—then skeletal forms must bow to this engineering diktat (if resistance to bending is a helpful characteristic). Real animal bones are, indeed, roughly circular in cross-section and hollow. Of course, constraints beyond engineering will modify our results, perhaps significantly. Sexual preference may matter more than resistance to bending during the selection of antlers in deer populations, for example. However—and here we have the central theme of our book—in many cases we can see that engineering considerations are sufficiently important in nature to merit the view that animals, even the Great Apes who wrote and who read this book, are works of engineering.

Notes

Further Reading

References

Acknowledgments

Index

Notes

1. Go with the Flow

1. Generally we use metric units in this book, hence the meters (1 m = 39.37 inches). One exception to this rule is units of heat; we use calories, because these are more familiar to most readers than the metric unit of Joules. Note that "calories" means "small calories," 1,000 of which makes a kilocalorie (kcal), also known as a "large calorie" or "food calorie." That chocolate bar you just ate may have cost you 300 calories on your diet sheet, but to a physicist or biologist, it contained 300 kcal of energy.

2. Wilson (1992) discusses trophic pyramids. "Trophic" comes from *trophe,* the ancient Greek word for food. Imagine a vertical food chain, such as grass-caterpiller-bird-cat (most chains are in fact more complicated webs, as we will see). Thousands of such vertical chains can be sliced horizontally to divide our animal diners into different types: primary producers, herbivores, small carnivores, or large carnivores. This horizontal organization is what we call the "trophic pyramid." In this section we refer to terrestrial examples, but remember that half our atmospheric oxygen comes from (usually microscopic) phytoplankton—a diverse group of photosynthesizing algae that occupy the surface layer of the oceans. In fact, the littoral regions of the world—the shallow offshore waters—contain the highest density of biomass, due to a happy combination of sunlight and well-oxygenated water. An example of an oceanic food chain is phytoplankton-krill-Right Whale.

3. Nothing is perfectly efficient. Your refrigerator puts out more heat than it extracts from its interior; your CD player converts into sound only a fraction of the electrical power that it consumes; your car converts only about a quarter of the energy of gasoline into useful energy of motion. Incidentally, we checked the human version of photosynthesis for energy efficiency: solar panels. These vary in efficiency from about 6% to 20% (depending upon price, and upon the technology used) so that solar panels are nominally 6–20 times as efficient as photosynthetic plants. On the other hand, solar power plants are less efficient (below 1%) because only part of the power plant area consists of panels.

4. For large carnivore caloric requirements, see Carbone et al. (2007). For population densities, see Creel and Creel (1997). See Hairston et al. (1960) for the classic paper on trophic levels. We will soon encounter the more complex *food webs;* these blur the distinction of trophic levels and somewhat weaken the case for energy flow as the major organizing principle of animal life.

5. The American-Italian Nobel-winning physicist Enrico Fermi was well known among physicists for this type of rough calculation. He once famously estimated the number of piano tuners in New York City in this way. Of course his prediction did not hit the bull's-eye, but it was (and this is the point) the right order of magnitude. The figures for Polish fox population densities are from Goldyn et al. (2003).

6. More accurately, we have demonstrated only that body size (daily energy consumption) limits the population density, and we have demonstrated this for only a couple of species. The idea applies much more widely than shown here. We lack the space for more detailed studies, and anyway (here is a recurrent theme), our aim is to explain the core concepts rather than prove them by repeated application.

7. The Kit Fox data are from Girard (2001). The gastrointestinal tract energy consumption figure is that of Cant et al. (1996). Yodzis (1984) examines the length of food chains.

8. Dinosaur predators greatly exceed the 1 ton upper limit, but these creatures were not mammals. In addition they had a lower metabolic rate, which drives up the maximum size of large predators, according to the energy-optimization theory of Carbone et al. (2007), summarized here. Note that herbivore maximum size is not limited by these energy considerations.

9. For these studies, see Stenseth et al. (1997).

10. The quadruped and hexapod gaits found in nature are discussed by Collins and Stewart (1993a,b). The walk-trot transition in horses is reported by Griffin et al. (2004b), and the energy efficiency of the horse's transverse gallop is demonstrated by Minetti et al. (1999).

11. Pfau et al. (2006) reports the study of horse tendons.

12. The units are arbitrary. If, say, the youngster's cube is 1 cm on a side, then the surface area is 6 cm^2 and the volume is 1 cm^3. If he generates two units of heat per minute for every cubic centimeter of his volume and loses 1 unit of heat per minute for every square centimeter of his surface area, then you can see that he loses four units of

heat per minute (2/3 per square centimeter of surface area). You may check that Mom loses eight units of heat per minute (only 1/3 unit per square centimeter), and Pop loses none.

13. Water is more effective than air at conducting heat, so Allen's Rule is particularly evident among sea mammals. Walruses, whales, sea lions, and the like tend to have a rounder shape than do *poikilothermic* (cold-blooded) fish in the same waters.

14. Stated differently, the area cubed is proportional to volume squared. The surface area A and the body weight W of our cubic *C. ridiculus* animals are related as $A^3 = 216V^2$. For different animal shapes, the constant factor (here, 216) changes but the square-cube relationship holds. (The math won't get any harder than this, we promise.)

15. Nature is more complex than this simple example of scaling suggests. Thus, for example, elephants adjust their gaits so that the heavy loads that arise during locomotion are spread more evenly than is the case for the equivalent gait of a lighter animal. Scale also changes the effects of gravity on an animal. Here the English geneticist J. B. S. Haldane, from his famous 1920s essay *On Being the Right Size,* explains in layman's terms: "You can drop a mouse down a thousand-yard mine shaft; and, upon arriving at the bottom, it gets a slight shock and walks away. A rat is killed, a man is broken, a horse splashes" (Haldane, 1985).

16. Kleiber's Law is now widely reported in both the research literature and popular accounts. See, for example, West et al. (1997, 1999), Dawkins (2004), and West and Brown (2005).

17. More than 600 species of placental mammals have been included, for example. The constant of proportionality that connects M_b to $W^{3/4}$ is different for different animal types (insects, fish, reptiles, or mammals), but the 3/4 scaling exponent (or something fairly close to 3/4, such as 2/3) appears to be common across different types.

18. One of myriad reasons for data variability is food availability within an ecosystem (Mueller and Diamond, 2001).

19. This telling comparison is due to Hill et al. (2004).

20. See, for example, Owen-Smith (1988), who presents data for herbivores in an African nature reserve, showing how biomass per square kilometer increases with individual body size.

21. For a summary of the experimental data pertaining to scaling, and of the main theories, see Schmidt-Nielsen (2005). West and Brown (2005) review the fractal theory of West et al. (1997). Bejan (2005) reviews his constructal law theory, which maximizes access to network fluid. A derivation of Murray's law, which assumes minimum cost for pumping fluid, is provided by Sherman (1981). The scaling data are picked over in detail by Alexander (2005), Dawson (2005), Speakman (2005), and Weibel and Hoppeler (2005).

22. The quote about fractal geometry is from Ho-Kim et al (1991, p. 262). True fractal networks have an infinite number of layers (smaller and smaller Hs in the network of Figure 6b), but networks in nature have only a finite number of layers, stopping at the length scale appropriate to a cell. So, biological distribution networks are not truly fractal, but they approximate fractals over a limited range of length scales.

Vascular networks that occupy 6–7% of body volume must be particularly efficient, according to Dawkins (2004), given that this is the volume found in many animal species. The three circulation networks are the pulmonary circulatory system (between lungs and heart), coronary circulatory system (feeds the heart itself), and systemic circulatory system (rest of the body).

23. The quote is from West and Brown (2005, p. 1585).

24. For blood, the situation is different, as shown in Chapter 2. Liquids are incompressible, and the requirement is therefore that volume flow rate, instead of pressure, is constant throughout the network.

25. The graph of Figure 7 shows the normalized radius, volume, and area of the pipeline network. Here "normalized" means that we are plotting the *n*th radius r_n divided by the first radius r_1; similarly we plot A_n/A_1 and V_n/V_1.

26. For a discussion of these claims, see West et al. (1997, 1999) and West and Brown (2005).

27. It is straightforward to construct log-log plots, such as that of Figure 5, supporting the prediction of heartbeat rate dependence on weight. Data for such plots can be found in, for example, Seymour and Blaylock (2000) for mammals, and Terres (1980) for birds.

28. See Klarsfeld and Revah (2004).

29. For a discussion, see Attwell and Laughlin (2001).

30. See, for example, Aiello and Wheeler (1995).

31. Tucker (1966) acquired the mouse data, whereas the mouse-opossum response to torpor is reported by Bozinovic (2007).

32. Hummingbirds must sleep, therefore they cannot eat, therefore they cannot maintain normal metabolic rate, therefore they become torpid or die. The hummingbird energy requirement data are from McKechnie (2002). The early American ornithologist Alexander Wilson wrote of hummingbird torpor in his book *American Ornithology* (1832, p. 182): "No motion of the lungs could be perceived. . . . The eyes were shut, and, when touched by the finger, [the bird] gave no signs of life or motion." Question: if hummers become torpid every night, then how do they brood their eggs? Part of the answer, for some species, involves high ambient temperatures. For other possible answers, see Vleck (1981).

33. See Carpenter et al. (1993).

34. Many cold-climate ectotherms also hibernate; during hibernation as much as 65% of their body water turns to ice. See Storey and Storey (1992) for more on this subject.

2. Structural Engineering

1. Skeletons alone are complicated enough. You were born with about 300 bones in your small body. Actually, at birth your skeleton is cartilaginous and not bony. Cartilage is less strong but more flexible. As you age, the cartilage is mostly replaced by

bone; some of the bones fuse, so that, by the time you became an adult, the number of independent bones in your body is reduced to about 200, and they make up about a seventh of your weight.

2. Most figures for biodiversity in this chapter are from E. O. Wilson (1992). Wilson also provides the scaling argument, which can be summarized as follows. Species are supremely adapted to their ecological niches, and so more niches means more species. Ecological niches can be physical niches—cracks and crevices. A species that exploits such niches must be physically smaller than the niche. If the number of niches per unit area of Earth's surface is inversely proportional to length scale (this is not obvious), then the number of species varies as $W^{-1/3}$.

3. The evolutionary cause and effect may just as easily be presented the other way around: cephalization led to bilateral symmetry. Whichever is the case, the two concepts evolved together.

4. We hope it is unnecessary here to point out that evolutionary "design" is not the same as airplane design, because airplanes require a designer. Evolution does not, repeat does not, require a designer. The mechanism of evolution is natural selection of random changes. When writing about evolution it is usual, though incorrect, to use such words as "design" ("bilateral symmetry is a good design") or "purpose" ("the purpose of eyes is to sense light"). This usage is a question of literary style only, which we follow for ease of explanation; it is not intended to imply that life on Earth has been designed or that it serves a purpose. Two very different ways of explaining this sensitive subject can be found in the popular writings of Stephen J. Gould and of Richard Dawkins. For the Cambrian lineage of bilateral symmetry, see, for example, Davidson and Erwin (2006). For the development of body plans, see Raff (1996).

5. Chitin is the invertebrate's equivalent of keratin, though chemically it more closely resembles cellulose. Like keratin, it is a good idea that has found widespread application across many animal groups. Humans apply chitin to numerous medical and industrial processes.

6. Larger animals may have external shells but not true exoskeletons. Turtles have no problem with heat dissipation, because they are aquatic reptiles. The armadillo gets rid of heat the same way as a dog: vasodilation and panting. These creatures (a reptile and a mammal) are vertebrates, and their external armor is made of bone, not chitin. Bone grows, so turtles and armadillos do not have to shed their armor. Mollusks have the same type of growth problem, which they overcome by adding material to the edge of their spiral shells, so that the spirals become larger.

7. The hinge joints of your elbows restrict movement less than those of your knees —and support less load. The two bones in your lower arm permit greater freedom of movement for your hands than would be the case for a single bone in the lower arm. Note how you can rotate your hand 180° without moving your elbow—now try the same trick with your feet. Here is an example of trade-off in skeleton design between functionality and strength.

8. Much of this bone optimization research has been conducted by R. McNeill Alexander and coworkers. It has been summarized in many textbooks and semi-technical accounts that are more accessible to nonspecialists than the original literature, for example, Wainwright et al. (1982), Alexander (1996), Currey (2002), Whiting and Zernicke (2008), and Vogel (2009). This type of optimization calculation—harnessing engineering lore to aid the understanding of animal structure—has been applied to other areas, such as the shape of teeth (see Freeman and Lemen, 2007). Impact loading may be a more significant limitation than muscle weight.

9. See, for example, Whiting and Zernicke (2008).

10. Bats belong to the order Chiroptera, which translates—very appropriately—as "hand wing." Note how thin the bones are in the bat wing; this is possible because of the scaling laws discussed in Chapter 1 (strength-to-weight ratio falls with body weight as $W^{-1/3}$).

11. For example, wing loading (weight W per unit area of wing) has to be low for flying birds. Area increases as $W^{2/3}$, you may recall from Chapter 1, and so wing loading increases as $W^{1/3}$. So, small is beautiful if you're a bird—usually.

12. See Ehrlich et al. (1988) and references therein for a more detailed account of skeletal and other adaptations for flight. The skeleton of the Magnificent Frigatebird (*Fregata magnificens*) supports a wing span of 2.1–2.4 m (more than 7 ft), yet is lighter than the bird's feathers.

13. There are five main types of feathers plus a cushion-load of specialized feathers. All are made from keratin, as are horns and scales (from which feathers evolved). Terres (1980) provides a clear nontechnical account.

14. Capillary radius, capillary blood flow rate, and red blood cell size are also pretty much constant across animal classes. See Dreyer and Ray (1910), Schmidt-Nielsen (1984), Vogel (2003), and references therein. The mathematical theory of West, Enquist, and Brown, discussed in Chapter 1, predicts that circulatory networks are most efficient when they occupy a fixed fraction of the body volume. See Dawkins (2004) and references therein.

15. The continuity equation is the reason pulmonary blood pressure is so much less than systemic blood pressure (about one-seventh): the pulmonary system is much shorter—less volume—and yet the flow rate must be the same. Hence, flow speed and thus blood pressure must be lower than the systemic values.

16. Murray's Law (discussed in Chapter 1) states that blood flowing through a tube requires minimum pumping power if the flow velocity is proportional to the cube of vessel radius (see Vogel, 1996). Such a law means, for our binary network model, that $k = 0.79$, which is very close to the value we derived from actual data. This observation tells us that the human circulatory system is close to optimum, in the sense that the power requirement is a minimum. Actually, the same result holds for many animals, not just humans. Evolution has optimized circulation system parameters.

17. The Magnus effect is responsible for the sideways curve of a baseball and the

lift of a golf ball with backspin, and it is usually cited as the reason for axial streaming. This may not be the true reason, however (S. Vogel, personal communication, October 2009). The physical law that expresses resistance to flow in terms of pipe radius is the *Hagen-Poiseuille Law;* it states that resistance is inversely proportional to the fourth power of pipe radius. If you suppose that all these named laws and effects reflect a large amount of effort invested in fluid flow research, then you suppose correctly.

18. All life on Earth is related, of course, through a common genetic code. What we mean here is that you are more closely related to a giraffe than to a geranium or a jellyfish.

19. Human hearts are smaller than average for mammals, as a percentage of body weight (the average mammalian heart is 0.6% of body weight, ours is about half as much), which, combined with our height, may explain why we feel dizzy when we suddenly get up off the sofa. We push the envelope for brain-above-heart build without specialist adaptations. Giraffe's hearts are about 2.3% of body weight—up there but not truly exceptional: whale and hummingbird hearts are relatively bigger. Where giraffes stand out, so to say, is in the blood pressure requirements they place on their hearts. See Vornanen (1989), Pedley et al. (1996), and Vogel (2003, 2009).

20. The *rete mirabile* is a vascular network that interrupts the continuity of flow in an artery or vein. It exists in several vertebrate animals (e.g., diving mammals) and serves different functions (e.g., temperature regulation).

21. See Pedley et al. (1996), Brook and Pedley (2002), Vogel (2003), and Mitchell et al. (2006). Incidentally, giraffe circulatory system research has influenced the debate about dinosaur physiology. The average blood pressure of modern reptiles is about half that of ours, but the great size of some dinosaurs suggests that they may have been as hypertensive as their modern descendents, the birds.

3. A Moving Experience

1. See Holmes et al. (2006). The mathematics of this article describes the dynamics of $2n$-legged animal locomotion. The authors also discuss possible neuromechanical control architectures.

2. Readers who crave mathematical details are referred to, for example, Collins and Stewart (1993a,b) and Golubitsky et al. (1998, 1999). These discuss the mathematics of legged motion, and how such motion may be controlled by a network of neurons called a "central pattern generator." This network generates muscle movement without sensory feedback.

3. When we are drunk, our feedback mechanism does not work properly. Note what is required for successful feedback: a sensor that detects pendulum angle (in humans, the balance sensor is in our ears—an evolutionary leftover from our fishy ancestors), and a mechanism for altering the angle. In drunken humans (so we have been informed . . .), both these facets are impaired, and thus we stagger.

4. The quote is from Hassan (2005, p. 484), who provides a description of how we enlist our muscles to respond to feedback information. There are many examples of negative feedback in animal movement, and a few examples of *positive feedback,* in which the perturbation is reinforced instead of reduced. Thus, the urge to push during defecation or during the birthing of young is amplified, once begun.

5. This step time may change a little, from one step to the next, if the step length is automatically adjusted as a result of the feedback mechanism discussed in the previous section. For standing, our feedback mechanism seeks to return the pendulum to an upright position ($a = 0$ in Figure 16a), whereas for walking it seeks to return the angle to a constant nonzero value, because we are leaning in the walk direction. The simple feedback model of the previous section can easily cope with this extension.

6. For a nonmathematical discussion of the Froude number in the context of animal gaits, see Vogel (2009). The Froude number is relevant when gravitational forces act (it is the ratio of inertial force to gravitational force). Other dimensionless numbers take over when other forces act. Thus, if viscous forces are important (swimming or flying fast), then the Reynolds number is what matters: animals of different sizes are dynamically similar at the same Reynolds number. Other dimensionless numbers (Strouhal number, advance ratio, etc.) define the conditions for dynamic similarity when other types of force dominate. See Alexander (2003, 2005) and Bejan and Marden (2006) for detailed technical accounts of scaling in the context of animal gaits. See Geyer et al. (2006) for the biomechanics of human walking and running.

7. Alexander (1977) plots the ratio of stride length to leg length against the square root of the Froude number for many mammal species (and ostriches), obtaining a straight line. Vogel (2009) estimates the Froude number at gait transitions in a different way than we do.

8. For detailed physical models of quadruped gaits and experimental data, see, for example, Minetti et al. (1999), Griffin et al. (2004a), and Walter and Carrier (2007). Stability, as well as energy expenditure, may be a key factor influencing gait transition. Consider your natural transition from a stiff-legged gait on a gentle incline to a bent-legged one as the slope increases. See Srinivasan and Ruina (2006) for a demonstration of walking as the energetically optimum gait at low speed and of running at high speed.

9. See Minetti (1998) for the relevance of skipping. It is a natural extension (to higher speeds) of walking, perhaps especially in a low-gravity environment. Astronauts on the moon found that hopping is better than walking: we can understand this from the Froude number, which increases in a low-*g* environment—see Vogel (2009) for a discussion of this and of pacing. Walter and Carrier (2007) describe the different quadruped gaits.

10. One of us has written a book about the physics of sailing. Sailboats involve, arguably, the most complicated of man-made flying machines, in that the airfoil—the sail—is of variable shape. Readers who wish to learn more will find a nonmathemati-

cal explanation of airplane and sailboat aerodynamic lift in the appendix to Denny (2008b). Due attention has been paid to the pitfalls inherent in the simplified aerodynamic theories. It is clear that such experts as National Aeronautics and Space Administration (NASA) scientists share our scepticism: their website on aerodynamics contains many pages debunking misleading but very persistent "theories."

11. Vogel (2009) notes that airplane development in the early days was hampered by engineers' study of bird flight: we now know that bird flight is inherently unstable, whereas airplane flight is designed to be as stable as possible. Birds have sophisticated control mechanisms that permit them to fly in an aerodynamically unstable manner. There is a rough analogy here with the next generation of fighter aircraft, which will have swept-forward wings instead of the familiar swept-back wings seen today. Swept-back wings are aerodynamically stable—slight variations in flight parameters will not result in sudden and violent changes to the flight trajectory. Swept-forward wings permit much greater maneuverability, but require next-generation computer control capabilities (as well as strong composite material construction). Alexander (2005) discusses airplane models (fixed-wing and rotary) of bird flight.

12. From $W \propto Av^2$ and $A^3 \propto V^2$ (from Chapter 1, where V is volume), and the observation that weight is proportional to volume (assuming constant density), we obtain $v^6 \propto W$.

13. Indeed, insects flap their wings more rapidly than do birds. Bumblebees, for example, flap at a frequency of $f = 150$ Hz (Hertz, or beats per second). Computational fluid dynamics studies of bumblebee flight show that it, too, is unstable. However, the timescale associated with instability (e.g., the time it would take to roll over sideways) is much longer than the bee's reaction time, so it can easily adjust its flying motion to maintain a stable trajectory. See Sun and Xiong (2005).

14. The frequency scaling rule applies much more widely than we have suggested here. It applies, for example, to legged locomotion (where frequency refers to step frequency) and to fish swimming (fin oscillations), as well as to fast-flying birds. The auks are one group of such birds that stands out from the crowd, in that their wing beats are much faster than other birds of the same weight. (Auks, like many diving birds, have dense bones—not hollowed out to reduce weight—so that they can dive without fighting a large buoyancy force. So, auks are heavier than other birds with similar wing spans. Their flight in air, though not under water, is laborious by comparison.) However, within this group it seems that different species of auks obey the scaling rule pretty closely, with $f \sim W^{-0.19}$. So, if we say that bird wing-beat frequency is related to weight as $f = aW^{-b}$, then a is larger for diving birds than for other birds, but b is about the same ($b \approx -1/6$). See Elliott et al. (2004) and Pennycuick (2001).

15. See Rayner et al. (2001) for a summary of intermittent flight mode characteristics and benefits, and for a theoretical analysis.

16. See Hambly et al. (2004) for details of these wing-clipping experiments.

17. The pelican experiment is reported in Weimerskirch et al. (2001). The appropri-

ately named Fish (1999) extends the discussion of formation flying to include shoals of fish. Migration flight dynamics for different bird species is discussed by, for example, Rayner et al. (2001), Bolshakov et al. (2003), and Wlodarczyk et al. (2007). There is a helpful and detailed summary of migration routes and strategies for North American bird species on the U.S. Fish and Wildlife Service (n.d.) website.

18. We said earlier that flying birds had to be small. Soaring birds are an exception—they have to be large. The reason is this: small size is driven by a requirement for low wing loading, but this is outweighed, so to speak, in soaring birds by the need for a very large lift force. (The price paid is a high stall speed. Albatrosses are obliged to fly fast, and in fact they look ridiculous at low speeds when taking off and landing.) Large lift is provided by long, narrow wings. It is no coincidence that albatross wings are very similar to those of man-made gliders.

19. Physiological energy expenditure of albatross flight has been estimated via heart beat rate. It has been found, for example, that during soaring, the average heart rate is little more than when resting on land. See Weimerskirch et al. (2000). For albatross dynamic soaring, see Strut (1883), Wood (1973), Pennycuick (1982), Weimerskirch et al. (2000), and Sachs (2005). A simple technical explanation can be found in Denny (2009). Airplanes have been designed to exploit the ground effect, and the ground effect is exploited by such birds as skimmers—see Withers and Timko (1977). The benefits of wave lift for albatross flight have been calculated by Sheng et al. (2005). The effectiveness of dynamic soaring for a radio-controlled model albatross is apparent in the Youtube video "Dynamic Soaring" (n.d.). Here the model is hand-launched into the wind; apart from this initial boost, the power required for flight is provided by the wind.

20. An example of increased maneuverability at the cost of reduced stability is provided by the hammerhead shark. The strange-shaped head acts like the canard wing in some fighter planes, which are notoriously unstable, but such wings enable rapid change of direction.

21. The gas bladders of many fish are below their center of mass, creating instability and requiring active control (this is why dead fish float upside down). However, having a gas bladder enables the fish to exert very little effort when not moving. Consequently, swimming probably requires less power per unit speed or per unit distance traveled than either flying or legged locomotion. See Vogel (2009). For discussions on the detailed function of different fin groups, see Drucker and Lauder (2000) and Wilga and Lauder (2004). The latter paper emphasizes the instability of fish swimming by comparison with aircraft flight.

22. Scaling in swimming fish is discussed in Drucker (1996) and Drucker and Jensen (1996). The theory of scaling that claims to underpin all forms of locomotion—running, flying, and swimming—can be found in Bejan and Marden (2006).

23. This analogy is perhaps bad: a snake cannot contract and extend its body length (as a caterpillar does) because of its vertebral column.

24. See Walton et al. (1990).

25. The mucus shear explanation of gastropod locomotion is now widely accepted. See Lauga and Hosoi (2006). The first mathematical model exploiting this idea was proposed by Mark W. Denny (1981)—no relation to your author. The use of adhesive locomotion by robots has led to the construction of "robosnails." See Chan et al. (2005).

4. A Mind of Its Own

1. Robotic arms are inspired by human arms but do not easily replicate their structure: most industrial robots need more moving parts than an animal limb to achieve the same results.

2. See Nagrath (2006).

3. See Ross (2004) for its use in engineering.

4. See Nathan (1997) for a highly readable classic on nervous systems, still in print after 40 years.

5. Even so, these transmissions are usually still not as fast as in vertebrates. See data in, for example, Bullock and Horridge (1965). For more about neurons, see Nathan (1997), Slaughter et al. (2001), and Morris and Fillenz (2003).

6. This steady state is often, somewhat misleadingly, referred to as "homeostasis."

7. For a comprehensive account of this complex subject that is accessible to nonspecialists, see Damasio (1999).

8. See, for example, Durie (2005) for an accessible account.

9. Constrained by eye separation and resolution, in humans this method only works out to about 7 m. Beyond that you're judging range based on experience and visual clues, like object size and shadows.

10. See Bruce et al. (2003).

11. Kolb (2003) discusses retinal function.

12. See Denny (2007).

13. See Hall and Llinas (2001).

14. The exception is animals using echolocation, analyzed in Chapter 9. Echolocating bats, whales, and dolphins measure range and angle with a high degree of accuracy, but we doubt that their brains represent these parameters as numbers to be plugged into mathematical algorithms.

15. After a few rounds, substitute a similar-looking ball with a much higher degree of drag. Note how the brain wrongly anticipates its trajectory. Then marvel at how quickly it adapts.

16. Pursuit curves followed by echolocating bats are discussed in Ghose et al. (2006).

17. See Justh and Krishnaprasad (2006).

18. For example, see Shaffer et al. (2004) on how to catch Frisbees with camera-equipped dogs.

5. Built for Life

1. Gould and Gould (2007) discuss the link between building behavior and the development of intelligence.

2. See, for example, Hansell (2000).

3. See Vogel (2003).

4. Experiments have shown prairie dog burrows to be grossly overventilated, refreshing their entire air supply every few minutes. This overkill suggests there's some additional function, possibly that of sucking external smells down into the home so the prairie dogs know what's happening outside. See Hansell (2000) and Vogel (2003).

5. See Hansell (2005).

6. See Vollrath and Selden (2007) and Selden et al. (2008).

7. We doubt you would get a research grant for such an experiment. Human infants do spontaneously stack things, hinting at some sort of instinct to build.

8. See Bruinsma (1979).

9. Hansell (2007) describes a young male Village Weaver Bird (*Ploceus cucullatus*) failing to even get its elaborate hanging nest started.

10. Vogel (2003) covers all you need to know about materials science, but if you insist on more, go to Gordon (1978.)

11. Humans also use animal urine—urea acts as an excellent binding agent.

12. See Hansell (2007).

13. For spider silk production and properties, see Gosline et al. (1999). The mixture of rigid and flexible components also discourages the propagation of cracks. For more on the mechanical properties of silk, see Matsumoto et al. (2007).

14. Such a spiral is known to mathematicians as an *Archimedes spiral.* If the spider's leg is damaged, then the shape of its web changes, because the measuring apparatus has changed. See Vollrath (1987).

15. See Griffin (1994).

16. For environment-altering construction, see Gutiérrez and Jones (2006). Beavers living in natural ponds and lakes are less prone to dam building than river beavers: if conditions don't favor a lodge, they will dig a burrow in the pond bank. European beavers are less likely to build dams and lodges than are their North American cousins.

17. In fact, any trees in the area flooded by a beaver dam die within a couple of years. In some areas drowning surpasses beaver browsing as a cause of tree death. See Hyvönen and Nummi (2008).

18. See Wright et al. (2002).

19. Vermicomposting takes this process further: "farmed" worms turn assorted organic waste into a rich compost.

20. Darwin's (n.d.) article on worms and archaeology is available online. Other tunnelers can be more destructive—rabbit infestations are threatening many ancient sites in Europe.

21. Commercial hives have a plug in the top that the beekeeper opens to allow wind-powered pressure-driven ventilation, freeing up more bees for making honey.

22. See Turner (2000).

23. For ant nest ventilation, see Kleineidam et al. (2001).

24. For termite architecture, see Turner and Soar (2008).

25. See Turner (2005).

26. For more on tools, see Hansell (2007, chapter 12) or Griffin (1994, chapter 5).

27. Intriguingly, when faced with a completely new challenge, chimps do seem capable of inventing a solution. The classic experiment involves a banana hung just out of reach and a series of boxes which, if stacked, will get the chimp within range of it. Most chimps can figure out the solution with no prior experience. What they don't seem to do is explain this fix to one another. That would require communication, the subject of Chapter 12.

6. Simple Complexity

1. See Jost et al. (2004).

2. For cockroach behavior, see Jeanson et al. (2005). A computational approach is provided by Levy (1992).

3. One motivation behind studies of emergent behavior in living species is our desire to build machines that can emulate it.

4. Any kind of animal aggregation has to balance an assortment of needs—more individuals in the group means less food for each but a better chance of finding a mate. More risk of attracting predators may be offset by increased likelihood that, today at least, it's someone else who gets eaten.

5. Robotics engineers also use the cockroach's physical characteristics as a model for small robot locomotion. One result is a cockroach-like robot which uses pheromone trails to "herd" cockroaches to a desired location, where they can, for example, be readily exterminated. See Caprari et al. (2005).

6. Other, simpler creatures, such as corals, live in aggregations often referred to as "colonies," but these animals don't cooperate.

7. Possibly the ugliest mammals in existence, although no doubt someone loves them, mole rats have a single fertile female, the queen, who breeds exclusively with a small number of male consorts. The other mole rats do all the work. It is worth noting that, physiologically, breeding mole rats are no different from the nonbreeding ones—breeding would appear to be suppressed by the queen's bullying behavior triggering hormonal changes in the others. So, are humans eusocial? As a species, no, but we do occasionally develop transient cultures whose populations are segregated into different castes that perform different tasks, often with different breeding rights.

8. That's why a male bee is called a "drone" and has a very limited sort of life, whereas a breeding male termite is termed a "king" and is the constant companion of the queen.

9. See Wehner and Muller (2006).

10. See Steel (1991).

11. See, for example, Deneubourg et al. (1991) for robotic ant behavior and Kugler et al. (1990) and Ladley and Bullock (2005) for termite mound building.

12. Some species are less adept than others: the bottom ants drown and the bridge is built over their bodies. Ants don't worry about construction casualties.

13. The smallest known bird is the aptly named Bee Hummingbird of Cuba, less than 5 cm long and weighing on average 1.8 g, slightly lighter than a U.S. dime (Piper 2007). The heaviest insect ever recorded was an exceptional pregnant New Zealand Giant Weta, tipping the scales at 71g. The species with the heaviest typical weight is probably the South American beetle *Titanus giganteus* (Walker 2001). All these figures are disputed by other contenders.

14. Reynolds (1987) started this model; Heppner (1990) expands it. See Bajec et al. (2005) and Möskon et al. (2007) for the addition of fuzzy-logic control.

15. In some cases it also improves their navigational abilities, as discussed in Chapter 11.

16. But not impossible. There are many well-documented cases from both Europe and North America of solitary wolves regularly killing elk or moose. However, as a general rule, a single wolf will not risk taking on a large herbivore unless the victim is already half dead from old age or disease.

17. See Manela and Campbell (1995).

18. See Korf (1992) for the original of this algorithm.

19. So why not hang back and let the rest of the pack take the risk before moving in to grab your share of the spoils? The problem would fill a whole book. As a rule of thumb, among social predators, the dominant eat first and best. Hanging back might work sometimes, but if it worked most of the time, evolution would favor it. More likely, wolf packs would long ago have starved to death as each wolf waited for someone else to kill something.

20. There is some dispute over the issue of whether wolves ever kill people. Some authorities claim there has never been an authenticated case of a healthy wolf killing a healthy human adult. Others claim they have always eaten people, with the paucity of recent cases being more to do with the shortage of wolves in the modern world. We restrict ourselves to noting that the wolf is a highly successful pack hunting predator noted for its opportunism and that one of the beauties of nature is that life adapts to circumstances. Don't take stupid chances.

21. Perhaps 95% of wolf hunts end in failure as the prey escapes. We did say that emergent behavior produced suboptimal solutions: not perfect but good enough. Wolves don't need to succeed every time.

22. See Turing (1952).

23. For tail spots, see Murray (1988, 2001). For zebra patterns, see Graván and Lahoz-Beltra (2004).

24. See Urdy et al (2010). Hutchinson (1990) suggests that the "new" shell simply follows the curve of the old shell whilst expanding to accommodate the organism's incremental growth.

7. A Chemical Universe

1. A sense of touch is equally widespread, but bumping into things doesn't tell you as much about the world as chemical sensing.

2. See Schopf (1999): the molten Earth barely had time to cool before something moved on the face of the waters.

3. Simultaneous comparison requires a minimum of two cells. Individual neurons in the human brain can apparently retain short-term memories, so bacterial memory seems plausible. This mechanism won't store fond recollections of its childhood, but probably involves a decaying electrical signal preserving short-term information. See Sidiropoulou et al. (2009).

4. Thus food loses its flavor when a bad cold fills your nasal passages with mucus.

5. What we perceive as flavor is a combination of taste, smell, temperature, texture, and probably several other things as well. Hold your nose, close your eyes, and try cold mashed potato, then cold mashed apple: most people can't tell the difference. Unsurprisingly, experiments (and experience) suggest that we are more sensitive to food odorants when we are hungry—see Pearson and Olla (1977).

6. Or, rather, it can detect chemical cues associated with the presence of certain tumors. This ability does not make the dog's diagnosis reliable—see Gordon et al. (2008).

7. The downstream signal processing differs substantially between vertebrates and invertebrates, but the receptor neurons are very similar.

8. See Keller et al. (2007).

9. People who lack a sense of smell are said to be "anosmic." They seem to get by, although dining out can't be that much fun. See Firestein (2001) for a general discussion of olfaction.

10. Proving the whole lock–key mechanism won Richard Axel and Linda B. Buck a Nobel prize in 2004. That's how big a deal this discovery was. See their Nobel Lectures, Axel (2005) and Buck (2005).

11. See Malnic et al. (1999) and Buck (2005).

12. See Wilson and Mainen (2006), Wilson (2007) and also Di Natale et al. (2008).

13. See Uchida and Mainen (2008) for a rather technical discussion.

14. In humans it seems that parts of the brain associated with visual processing (especially shape recognition) are also involved in olfactory processing. Or vice versa.

15. See, for example, Engen (1991) and Engen et al. (1991).

16. For the similarity of different animal olfactory senses, see Eisthen (2002). For insect olfactory sensing, see Benton (2006) and Sato et al. (2008).

17. A useful, though imprecise, rule of thumb is that, among closely related ani-

mals, the longer muzzle has the better sense of smell, simply because there's room for more olfactory epithelium. The best-smelling birds include the nocturnal, ground-dwelling kiwi; scavengers like vultures, who are attuned to the stench of lunch; and possibly, some of the marine tubenoses (i.e., albatrosses and petrels, of which the former, at least, use scent as a primary food locating sense).

18. For seal smell sensitivity, see Kowalewsky et al. (2006); Manton et al. (1972) discuss turtle underwater olfaction. See Stensmyr et al. (2005) for crab olfactory adaptations.

19. See Reznik (1990).

20. See Wyatt (2003).

21. The VNO is lacking altogether in larval amphibians; most turtles, crocodilians, and aquatic mammals; birds; most bats; and many primates, including humans, although there are vestigial traces of it during embryonic development.

22. See Baxi et al. (2006) for a discussion of all things vomeronasal.

23. Molecular diffusion through the fluid can be important over very short distances, but the time taken for the odorants to diffuse increases as the square of the distance—twice as far takes four times as long. See Vogel (1996).

24. If you're swimming or flying through the medium, working out which way *is* upstream may require visual cues from the environment to separate your motion from that of the current.

25. For an accessible overview, see Vickers (2000). For insects and robots, see Willis (2005).

26. Hagfish—ugly, slimy creatures that seem half eel and half leech—may not be vertebrates. The jury's still out on that one. They have a spinal cord and a skull, but no actual vertebrae or, indeed, any other bones. For an overview of taste physiology, see Dulac (2000). For taste evolution, see Northcutt (2004).

27. Cows are grazing herbivores, and many plant species discourage being eaten by dosing their leaves with low-volatility toxic chemicals. Few plants have leaves that smell bad, but many are quite poisonous. Identifying the edible ones is the taste buds' primary function.

28. See Finger (2006) for a general discussion of taste buds.

29. For the latest ideas on taste buds, see Northcutt (2004).

30. *Umami* is a Japanese term roughly translating as "savory"; see Nelson et al. (2002) for more about the umami taste. For taste perception in general, see Bradbury (2004).

31. See Smith and Margolskee (2006).

32. And they can occur in strange places. Thus flies taste with their feet, among other things, which means they really *do* know what they are standing in.

33. See Prabhakar and Peng (2004).

34. Insects can expel carbon-dioxide through one set of tubes while sucking in oxygen through another, but a similarly-sized lung-equipped vertebrate can shift a much

bigger volume of gas in a given time, which is one of the things that limits the size of bugs. For that, we should be grateful.

8. Sound Ideas

1. Humans can feel vibrations of very small amplitude (about 100 nm, or 10^{-7} m). Sensitivity depends on vibration frequency; we are most sensitive to vibrations at about 200 Hz (cycles per second). Birds feel vibrations through their feet—it's good to know who else may be in the tree they are perching on—and can sense 20 nm amplitude vibrations. Their optimum frequency is 800 Hz. Toads and fish can sense water waves propagating across the surface of a pond—perhaps indicating potential prey that has dropped onto the water and is struggling to get out. For a summary of human and bird mechanoreception, see *Britannica* (1998); insect adaptations are discussed by Huber et al. (1989); Speck and Barth (1982) investigate spider sensitivity to web vibrations.

2. How biologists are able to gather animal audiogram data, in a manner that is reliable and portable across species, will not concern us, though it is not hard to imagine the difficulties. As usual, we are interested in animals as examples of engineering, so we are more interested in the capabilities of animal acoustical structures than in their biological construction or use. Also as usual, we make one or two exceptions to this rule, by presenting some truly amazing examples of particular animal feats of hearing. The quoted upper frequency limit for human hearing decreases with age. The units of frequency are Hz (Hertz), meaning cycles per second, and kHz (kilohertz) meaning thousands of cycles per second. An octave is a factor of two in frequency, so that a voice singing at 4 kHz is one octave above a voice singing at 2 kHz and three octaves above a voice singing at 500 Hz.

3. In fact, as with many insects, bee hearing is specialized. It is likely that the honey bee auditory system has evolved to help them decode the waggle-dance of bees who have just returned to the hive and who wish to communicate the location of a new food source. See Tsujiuchi et al. (2007) and Chapter 12.

4. For general reviews of audition in animals, see Heffner (1998, 2007) and Heffner and Heffner (2003). For human directional hearing capabilities, see Hartmann (1999). Some of the great mass of data on animal hearing is a little overwhelming. Thus we learn that pigeons are able to discern the difference between Bach and Stravinsky compositions. Unlike humans, they consider Vivaldi compositions to more closely resemble those of Stravinsky than of Bach. This research is widely reported in the literature—see Bright (1993). Quite where it leads us, we cannot say. Should you wish to read more technical research on this leitmotif, see, for example, Steele (2003) on rats and Mozart.

5. The reindeer experiment is reported in Flydal et al. (2001). The data that establish the connection between visual field and sound localization is convincingly demonstrated in Heffner and Heffner (2003). The field of best vision and sound-localization

ability of a number of mammal species are plotted on a graph; these fall around a straight line, indicating that mammals localize sound direction only as accurately as is necessary to direct their eyes to the source.

6. You might think that song learning and matching has more to do with heredity or physiological similarity than with hearing capabilities, but research has demonstrated that this is not the case. For example, Mockingbirds (*Mimus polyglottos*) can very accurately mimic the Whip-Poor-Will (*Caprimulgus vociferous*). They hear the Whip-Poor-Will song much better than we do, and reproduce it well, even though they are a different species. An example of bird song that extends beyond their own hearing range is that of hummingbird vocalizations. See Stap (2005) for a readable nonspecialist review of bird song and the difficulties that must sometimes be overcome to record it. Hear a wren song played at full speed and half speed at www.birds.cornell.edu/brp/the-science-of-sound-1/what-is-a-spectrogram.

7. Here is how a statistician would explain this feat. If gannets can discern, say, 5 different notes each of duration 5 ms, then in 1/10 s there are 5^{20} different note combinations, or about a hundred million million. Enough for even the largest gannet colony. People can discern notes distinctly only when they are at least 10 times longer (50 ms), and so in 1/10 s we could spot only 5^2, or 25 combinations, at most.

8. Cats and humans can do much better than mice (predators tend to have directional hearing): feline sound localization accuracy is about 5°, whereas humans can manage an impressive 1.3°. Data are from Heffner and Heffner (2003).

9. Several species of mice, hedgehogs, and bats use the intensity technique for sound source location. There is another technique (phase difference), which we will get to soon enough; humans, elephants, cats, certain bats, weasels, squirrels, and others use *both* techniques. One complication that arises when using intensity differences between the sound arriving at the two ears is this: how do we account for the shadowing effect of the head? One ear may be exposed to the sound source directly, but the other may be in the acoustical shadow. The influence of head shape and composition on intensity, and how intensity varies with angle, is very complicated, but we (and others) seem to sort it out quite well. See Hartmann (1999) for details. The list of animals that use the intensity and/or phase techniques comes from Heffner and Heffner (2003).

10. The visible external ear tufts that some owls display are just that—display. They are not real pinnae. The facial disk acts as a more effective antenna. Removing the ruff feathers from this disk causes barn owls to lose their ability to estimate elevation, but it does not affect their azimuth estimation. The data for owl hearing comes from the extensive experimental work of Knudson and Konishi in the 1970s, following the discovery of acoustic prey location in owls by Roger Payne (S. Vogel, personal communication, 2009), which taught us much of what we know about owl hearing. See Knudsen (1980) and Konishi (1983). Ehrlich et al. (1988) includes a section on owl hunting.

11. A wave from directly in front of our owl arrives at both ears simultaneously. A wave from the left arrives at the left ear first; the delay before arriving at the right ear

increases as the angle increases, reaching a maximum when the sound source is to the side. Owls hunt at night, fixing the direction of their prey (typically a mouse) from the noise it emits when rustling through grass. Owls point their heads toward the sound source and know when it is directly in front because of the simultaneous arrival time at both ears. Owls hearing is (unusually for birds) more sensitive than ours (ten times more sensitive, for frequencies below 2 kHz, as their audiogram shows), and so they can hear a mouse rustling in the grass from some distance. They have special "whisper mode" flight feathers to reduce competing sounds. At the last moment of their swoop, just before they pounce, they fold their wings back and bring their talons forward. As if to tell us that they do all this blind, owls actually close their eyes at this crucial time (perhaps to avoid spearing their eyes by unseen blades of grass).

12. Pigs use phase difference to fix a sound within 5°; cats use both phase and intensity to achieve the same accuracy. Dogs manage about 8°. A typical diurnal raptor (hawk, falcon, or eagle) achieves 8–12°. Horses and cattle are less accurate, at 25° and 30°, respectively. Humans use timing differences to locate sounds with frequencies below 1.5 kHz and intensity differences for frequencies above 3 kHz. We are not so good at locating sounds with frequencies between these two values. Data are from Ehrlich et al. (1988), Hartmann (1999), and Heffner and Heffner (2003).

13. Data are from Heffner and Heffner (2003). Note that owls use timing (not phase difference), and so they must be able to discriminate time difference of about 5 μs to achieve the accuracy they do. This ability strongly suggests they are using a correlation processing algorithm. This level of sophistication occurs in a brain that weighs only 2.2 g (1/12 of an ounce).

14. See Miles et al. (1995), Robert et al. (1996), and Robert (2001). For a summary of the physics involved in the acoustic reception of this fly, see Denny (2008a).

15. See Martinez et al. (2004) and, for a popular account, Holmes and Bhattacharya (2004).

16. See Konishi (1983).

17. See Nelson (2003) for a study of acoustic attenuation through Florida scrub. He finds that intensity (in dB) reduces linearly with frequency and with distance from the source, over the frequency range of interest to animals and birds.

18. See Naguib and Wiley (2001) for a discussion on ranging (distance estimation) by animals using acoustical cues. Acoustic transmissivity also depends on humidity, which has been shown to be important to the vocalization of lemurs in a Madagascar forest (S. Vogel, personal communication, 2009).

9. Animal Sonar

1. A good lawyer could make a case for the radar capabilities of electric eels and certain other creatures that develop electric fields: they are transmitting energy that interacts with the environment, sending it back to the "receiver" in modified form.

The differences between transmitted and received signals are processed to yield information about the environment, which is the essence of active remote sensing. We will meet electric eels again.

2. Griffin and his colleagues announced their findings at a time in world history when radar was being developed and deployed in war for the first time (the year was 1940, and radar was a key factor in the Battle of Britain, fought during that summer). Some of Griffin's colleagues were incredulous that such small and not-too-intelligent creatures as bats might be capable of remote sensing accomplishments that compared with our own cutting-edge technology. In fact, we now know that bats are better than us at sonar processing, though humans are catching up fast. Bat sonar capabilities are discussed in Dawkins (1986), Fenton (1991, 2003), and Denny (2007). Biosonar in general is reviewed in Nactigall and Moore (1988) and Moss (1999).

3. See, for example, Fletcher and Tarnopolsky (1999).

4. See Cranford et al. (1997).

5. Technically, this region is still part of its nose. The Sperm Whale has the largest nose of any animal ever known to have existed, taking up roughly one-third of its overall length and one-tenth of its total mass. Quite why it is so big is still uncertain, but we are rooting for sonar as part of the explanation. Animals rarely evolve big structures without a good reason.

6. We met the technical term "target" earlier: a target is the object that the remote sensor would like to detect. If the target does not want to be detected (such as the bug in Figure 67), then it is *noncooperative;* if it wants to be seen (e.g., an airliner approaching air traffic control radar), then it is a *cooperative target.*

7. Echolocators have developed this "disconnect" to a high degree, requiring millisecond timing accuracy to synchronize the blanking with their rapid outgoing pulses.

8. These, of course, are not toothed whales, but they do seem to use a kind of low-frequency sonar to (1) find krill blooms on which they feed and (2) perform underwater terrain mapping. So, do other nontoothed whales (e.g., Humpbacks) use echolocation, or are they just talking? The jury is still out on this one—whale biologists are still debating the issue. See, for example, Au et al. (2001) and Mercado and Frazer (2001).

9. Inhomogeneities in the density of the medium (e.g., humidity in the atmosphere over an ocean or the salinity gradient in seawater) give rise to refraction effects. Surfaces give rise to reflections. Both these phenomena cause the wave direction to change, and they are included under the umbrella term "anomalous propagation." This term sounds like a bizarre new way of reproducing, but in fact it simply refers to waves that do not travel in straight lines.

10. The significance of attenuation for bat echolocation is discussed in Lawrence and Simmons (1982).

11. The technical terms are "illuminate" for radar and "ensonify" for sonar, but "paint" is a more descriptive and general term, and it is commonly used by radar systems engineers.

12. The manner in which echolocation clutter can confuse a bat is discussed in Wong (2001).

13. See Martin et al. (2004) for more on Oilbird vision. Oilbirds were named for their fat chicks, or squabs, which used to be harvested for their odorless oil. Chinese bird's nest soup is made from cave swiftlets' nests. For an overview of echolocation in these two birds (indeed, for an authoritative introduction on anything birdy), see Terres (1980).

14. The quote is from Fulton (2005, p. 2). Despite the claims on behalf of dolphins, we place bat echolocation in the lead, by a nose, on the basis of miniaturization: the sophisticated echolocation gadgetry of a dolphin occupies a cubic foot of space, whereas that of a microchiropteran bat occupies much less than a cubic inch.

15. Through their echolocation signals, Sperm Whales and Blue Whales compete for title of loudest animal on the planet. Sperm Whales' signals are high frequency; Blue Whales' are low frequency. See Møhl et al. (2003) for echolocation in Sperm Whales.

16. It is amazing, because dolphins, like us, don't register such small time intervals. Our nervous system operates in the millisecond range, and yet much shorter time intervals can be detected. We will explore later how this apparent contradiction is resolved by clever signal processing (in fact, by correlation processing, which we met in Chapter 8). Human radar and sonar, and echolocating bats, also estimate range via time delay. For an account of how the big brown bat does it, see Simmons et al. (1998).

17. In fact, the dolphin's hearing system extends beyond its ears. The lower jaw contains fat-filled cavities that are sensitive to high-frequency sounds, channeling these to the "detectors" in the inner ears.

18. We do not claim that whales and dolphins use this technique in the same way as radar and sonar engineers do, but they might. That they swap around their various echolocation clicks when investigating a target is very suggestive to any systems engineer. Also, their changing of frequencies and pulse types—stirring the pot of echolocation clicks—tells us that they are looking for patterns or for distinguishing telltale features of a target. These features are best picked out by correlation processing.

19. The bandwidth of Bottlenose Dolphin's echolocation pulses may be 100 kHz, and so these creatures can divide the spectral landscape into a thousand or so separate compartments, providing them with detailed information about the frequency spectrum of their echoes and enabling them to filter these echoes with precision. See Branstetter et al. (2003) and Dobbins (2007).

20. The evolution of bat echolocation capabilities is discussed in Novacek (1985) and Speakman (2001).

21. Here we include clutter as part of the noise, though you will appreciate from what we have said about clutter that it sometimes looks very different from true random noise. Sometimes, indeed, clutter can look very much like a target, which greatly complicates the processing of radar and sonar signals. One of us has spent a significant fraction of his career as a radar and sonar systems analyst just examining clutter and

thinking about how to process (i.e., get rid of) it. Despite these complications, which arise from the great variability of clutter, it is true that integration damps down the effects of noise and of many kinds of clutter.

22. Echolocation calls from some bats are as loud as a smoke alarm. Bat echolocation calls are way above our hearing, fortunately, so we are not woken up by nocturnal bats hunting for food. Some bats are described as "whisperers"—their echolocation calls are very quiet. Whisperers hunt in crowded environments, such as tree canopies, where loud calls would be bad, because the clutter power from echolocation pulses reflecting from leaves and branches would increase, smothering the target echo. For bat navigation, see Verboom et al. (1999).

23. We have already seen that big antennas have narrow beamwidths, and small antennas have wider beamwidths, for a given echolocation signal wavelength.

24. In general, the Arctiidae family of moths, to which the tiger moth belongs, and the Noctuid moths make use of these acoustical warfare techniques. For some of these creatures, the functional ears are located on the side of the thorax and not in their antennae (S. Vogel, personal communication, 2009).

25. For an account of bat hunting, see Fullard (1997).

26. Identification of a target means, in the context of bat echolocation, detecting the species of a target moth or bug. Classification means detecting that the target is some type of moth or some type of bug.

27. Hyperbolic frequency modulation (also known as *linear period modulation*) is discussed by Altes (1990) and Collins and Atkins (1999). The adaptations of bat brains for processing FM signals have been investigated by Sanderson and Simmons (2000).

28. Observations of Orcas hunting by echolocation show that they can detect salmon as far away as 100 m in good conditions. See Au et al. (2004). (One hundred meters is short range for a dolphin echolocator, and dolphins are searching for smaller fish.) If it is raining, then the noise on the sea surface drowns out the echolocation signals and reduces detection range to less than 40 m. Seals are larger targets than salmon, and so an Orca should be able to detect a seal well over 100 m away in good conditions.

29. Seals are amphibious, and so their hearing has to work in both air and water, which leaves less room for specialization than in cetaceans. Seals would need either a separate sonar system or have to accept degraded performance in air. In fact, there is evidence that, although their hearing is generally better in water, it is still pretty good in air. Interestingly, some seals have been observed apparently listening to their pups with one ear in the air and the other in the water. The inference (from their subsequent behavior) is that they can judge the pup's distance based on different arrival times in the different media. At present the evidence for the pinnipeds favors a combination of good eyesight and excellent hearing to explain how they sense the underwater world. We also note their well-developed whiskers. See Schusterman et al. (2000) and Ahlborn (2004).

30. For more on penguin hunting capabilities, see Poulter (1969) and Ropert-Coudert et al. (2000).

31. There may be at least two species. Rats navigating darkened laboratory mazes have been observed to squeak a lot, and their ability to detect openings in the walls appears to correlate with the squeaks, perhaps suggesting that they are using use their high-pitched squeaks to examine the walls, as do shrews. The interesting question with rats is not how it would work, but is it instinctive or learned behavior?

32. Kish (2009) provides an introduction.

10. Seeing the Light

1. The quotation and data concerning the evolution of eyes come from Land and Fernald (1992, p. 1). There is a third type of eye, based on the reflection of light by mirrors instead of refraction through lenses, which we do not discuss here. A scallop may have 50 such eyes. Telescopes are classified according to the same two optical principles: refractors or reflectors. There is an intriguing theory that suggests the Cambrian Explosion (a rapid proliferation of animal body-plan designs that occurred about half a billion years ago) happened because of the evolution of eyes. This "light switch" theory is explained in Parker (1998, 2003).

2. The seawater attenuation data can be found in many military sources, because of its consequences for communications with submarines and for undersea remote sensing. See, for example, Bindi et al. (2005).

3. Aquatic eyes must rely on the lens for refraction; the cornea causes no refraction, because the refractive index of seawater is almost the same as that of the cornea. For raptors shaping their corneas, see Raymond (1985) and Jones et al. (2007).

4. The largest eye is probably that of a deep-sea squid. It has a 40 cm (16 inch) diameter. These creatures live at several hundred meters depth, where even visible light is severely attenuated. See Land and Nilsson (2002).

5. Angular resolution of 0.017° means that we can see a 0.3 mm object at a distance of 1 m, or a golf ball at 100 yards. Vision researchers prefer a different unit for acuity—you will encounter this unit (cycles per degree, or cpd) if you pursue many of the references for this chapter. Ten cpd corresponds to an angular resolution of 0.1°; 30 cpd corresponds to 0.033°.

6. There is not a one-to-one correlation between each photodetecting cell and an optic axon: if there were, the optic nerve would have to be an inch in diameter. In fact it is much smaller than this, because there is a great deal of preprocessing that takes place in the retina, before the signal is transmitted along the optic nerve.

7. See Schneeweis and Schnapf (1995).

8. See Ewer (1997).

9. Unsurprisingly, the angular resolution determined by foveal density is about the same as the angular resolution determined by the lens optics for most animals. (There would be little point in having a very high density of cones, for example, if the lens was not up to the task of producing a well-focused image on the retina.) Owls

have no cones—their foveae are made up of rods, so their high-resolution vision is not in color. For eagle cone data and acuity, see Raymond (1985).

10. Data about cat color vision is from Ewer (1997). See Cheney and Marshall (2009) for an example of fish color vision. For color vision in birds, see Goldsmith (2006) and Jones et al. (2007).

11. For the artistic appreciation of pigeons, see, for example, Monen et al. (1998). For the utility of color vision, see Noë and Thompson (2002).

12. Human eyes are the most intensely studied of all eyes. A convenient and comprehensive online source is provided in University of Texas at Houston (n.d.). Binocular vision means that both eyes can see an object under scrutiny. Because the two eyes see the object from slightly different angles, they help provide depth perception and a three-dimenstional representation of the image.

13. Data rate and bandwidth are pretty much the same thing to a communications engineer. Imagine a signal consisting of many waves of different frequencies (a wideband signal). Each wave can be thought of as containing one piece of information (the wave phase), and so signal bandwidth is the same as signal information rate. The mouse data are from Nikonov et al. (2006).

14. This is the data flow from the eyes. The data flow to the mouse's brain is much less than the rate we have found, reduced significantly by all the hard-wired preprocessing than goes on in the retina.

15. TSAT data were obtained from GlobalSecurity.org (n.d.).

16. For kestrels seeing ultraviolet vole pee, see Viitala et al. (1995).

17. See Jones et al. (2007). S. Vogel (personal communication, 2009) has found that eyeball mass scales as body weight to the power 0.60 for mammals.

18. The energy expended in flight can be estimated by multiplying drag force by path length. So, if a spiral trajectory requires $\frac{1}{2}$ the drag force and $1\frac{1}{2}$ times the path length of a straight trajectory, then the energy expended is $\frac{3}{4}$ that of the straight trajectory. The intriguing phenomenon of spiralling raptor flight paths is reported by Tucker (2000) and Tucker et al. (2000).

19. We are grateful to Aaron Sloman (personal communication, 2009) for information on this subject. See also Sloman (2001). Bach (n.d.) contains many optical illusions, some of them animated, which impressively demonstrate that, if not how, our brains do a lot of optical processing.

20. For cephalopod vision and vision-related characteristics, see, for example, Sivak (1982), Shahar et al. (1996), and Hanlon (2007).

21. See Land (1969) and Harland and Jackson (2000). Thomas Shahan (n.d.) maintains a website with many excellent photographs of jumping spiders.

22. We can indeed roughly equate interommatidial angle with resolution angle. Thus, for example, the interommatidial angle for bees is 1° elevation and 2° azimuth, whereas the angular resolution is 2.5° in both directions. For the dragonfly *Sympetrum striolatus,* a calculation similar to that we carried out for the fly yields an interomma-

tidial angle of 0.8°, whereas the resolution angle these insects achieve is 0.4°—the best of any insect. See Harland and Jackson (2000), Horridge (2003), and Rutowski et al. (2009). For the size of a compound eye with the resolution of a human eye, see Land and Nilsson (2002).

23. See Chiao et al. (2000) for shrimp color vision.

24. See Grace et al. (2001).

25. See Bakken and Krochmal (2007).

26. For example, nerve cells generate an action potential (see Chapter 4), causing a change of about 130 mV over a millisecond. Platypus electroreception is discussed in detail by Pettigrew (1999). See Dawkins (2004) for a nontechnical summary. Speculation on the signal processing performed by Platypuses can be found in Proske and Gregory (2003).

11. There and Back Again

1. For a detailed account of many of the traditional methods of tracking and measuring bird migrations, see Emlen (1975). For an example using the latest technology, see Hünerbein et al. (2001). The oldest tracking method is bird banding. Although very inefficient (only 0.1% of banded songbirds are ever recovered—the proportion is higher for waterfowl), this technique has provided some spectacular results that first alerted us to the long-distance nature of bird migration. For example, a Lesser Yellowlegs (*Tringa flavipes*) banded in Massachusetts was recovered 5 days later and 3,000 kilometers away in the West Indies. In 1966 a Peregrine Falcon (*Falco peregrinus*) banded in the Northwest Territories of Canada was shot 14,400 km away in Argentina. See Terres (1980) for more such figures.

2. The Arctic Tern is easily confused with the Common Tern (*S. hirundo*) in parts of the world where their ranges overlap, such as Great Britain. Consequently a British birder—to avoid the embarrassment of misidentifying a species—refers to both species as the "comic tern."

3. The dragonfly claim is made by Anderson (2009).

4. It may be that the Polynesians who first settled in Hawaii, and who were no mean navigators, got there by following migrating Golden Plovers. Another group of Lesser Golden Plovers migrates from Argentina to Alaska. For a popular account of many more examples of bird migrations, see Bright (1993).

5. For an account of the Manx Shearwater displacement experiment, see James (1986).

6. For a more detailed account of this starling experiment, see, for example, Perdeck (1967) and Lund (2002).

7. Caged birds often exhibit "migratory restlessness" (or *Zugenruhe*), which makes them spend a disproportionate fraction of time at a compass point in their cages that corresponds to the migration direction. After some days this direction may change, as

the birds try to obey their genetic instructions. The Blackcap experiments are reported by Day (1991).

8. Directional orientation using the sun as a compass is utilized by many types of animal, including crustaceans (sand fleas), insects (ants, bees, beetles, and butterflies), mammals (bats), many bird species (e.g., starlings, swallows, warblers, pigeons, and ducks) and fish (salmon), even though refraction through the water surface distorts apparent sun direction. Amphibians and reptiles also use sun-and-clock techniques. See Dingle (1996).

9. In theory an animal could use the height of the sun at noon to indicate latitude. There is no evidence that the sun is utilized in this manner to determine position, though there is plenty of evidence to show that it is used to determine direction. One example of many: radio-tagged Blue-Winged Teal circled to gain height, upon release, and headed off in a definite direction only when above the clouds.

10. Also we find that the mean distance R increases proportionally as mean path length increases, unsurprisingly (not shown in Figure 92). The simulation results show that the positional error of the ant—where she thinks her nest is—grows with the mean path length and more slowly with the number of paths. Both these results are to be expected. The surprise is that the positional error does not grow with angle estimation error or with path length estimation error (for this special case where the goal—the intended finish location—is the same as the start location).

11. See Dingle and Drake (2007).

12. Birds prepare extensively for their migrations by building up large reserves of fat—they may lose 30–40% of body weight during the journey. Their hormone levels and metabolic rates change during migration, triggered by day length or temperature. Many raptors live at bottlenecks along the migrants' flyways and time their breeding season to coincide with prey species migration (e.g., Eleanora's Falcon on several Mediterranean islands). One advantage of the short breeding season in the far north is reproductive synchrony: many millions of young fledge at about the same time, so that there is some safety in numbers during the fall flight south.

13. Here are some specific illustrative examples from the bird world. The Whinchat (*Saxicola rubetra*) and the Siberian Stonechat (*S. maura*) are long-distance migrants, whereas the closely related *S. rubicola* and *S. torquata* are partial migrants or are sedentary. American Kestrel males remain on their breeding grounds longer than do the females, and they migrate shorter distances, so they can return sooner to claim territory. Crossbills are not regular migrants, but they irrupt in large numbers from time to time, moving wherever there is a local abundance of their food (pine seeds). Herring Gulls migrate shorter and shorter distances as they get older. Daytime migrants travel more slowly, because they must stop to feed, and generally they follow winding coastlines, where a reliable source of insect food is available. Wind speed and direction have a significant influence on the timing and direction of migration: birds are very unwilling to fly long distances into a headwind, for example.

14. The shearwaters knew the location of home but not their own location or, initially, the direction of home. To be sure that dead-reckoning is not utilized on the outward journey by birds that are the subject of displacement experiments, some pigeons have been displaced while anesthetized—yet they found their way to within a kilometer or so of home quickly, while blindfolded. So, these pigeons must have been able to determine their position at the release site without landmarks or knowledge of the outward journey, and they must have known when they were near home without being able to see it. See, for example, Walker et al. (2002).

15. The polarization sense can be thrown off by severe atmospheric disturbances, such as the cloud of particles spewed into the atmosphere by a volcanic eruption. See Bright (1993). When it works, sensing polarized light may assist those birds displaced to unfamiliar territory by investigating biologists. The birds may be able to estimate their longitude by comparing the local sunrise/sunset time with their internal clock, which is calibrated to home territory sunrise/sunset time. In other words, they can estimate their longitude by their jet lag.

16. The navigation capabilities of the Savannah Sparrow are discussed in Able and Able (1996).

17. It has been shown, for example, that the Garden Warbler (*Sylvia borin*) utilizes celestial cues, but that these alone are not sufficient for it to navigate. Many of the *Sylvia* spp. of Old World warblers and many New World songbirds, such as the Indigo Bunting, are capable of stellar navigation.

18. Stellar navigation of birds is discussed by Emlen (1975) and Dingle (1996).

19. See Døving and Stabell (2003). We will see in the next section that salmon make use of olfactory cues to return to their natal streams; it is hypothesized that sensing the velocity differences of different layers of water assists them in tracing odors to their source.

20. The information about pigeon sensitivity to infrasound is from Kreithen and Quine (1979). Details of the interference by the Concorde, and of the proposed acoustic avian map theory to follow, are provided by Hagstrum (2000).

21. Returning salmon use a combination of different odors that are carried in different layers of water traveling at different speeds and directions to pick out their home stream from other nearby streams. The layering of water provides a fine vertical structure to the odor patterns. See Dingle (1996) and Døving and Stabell (2003).

22. The anomalies can help or hinder navigation, depending on whether they form part of an animal's cognitive map. Young Loggerhead Turtles find their own way from beaches to the sea directed by ocean waves, but thereafter they navigate the open ocean in a 13,000 km round trip via magnetic map and compass. See Trivedi (2001). One unintentional displacement experiment transported sea turtles 1,000 km from Nicaragua to the Florida Keys (fisherman captured the turtles and marked their shells prior to selling, but the turtles escaped). The same turtles were later found back in their home waters. However, birds encountering a strong magnetic anomaly in Sweden

(presumably for the first time) became disoriented, as shown by radar tracking. See Alerstam (1987). Magnetic maps may be particularly important underwater, because visibility is too poor for visual landmarks.

23. See Walker et al. (2002).

24. See Dingle (1996). For the butterfly magnetic sense, see also Srygley et al. (2006). For Loggerhead Turtles, see Lohmann et al. (1999); for amphibians, see Sinsch (2006); for *Tenebrio* beetles, see Vácha et al. (2007); for bats, see Wang et al. (2007); for alligators, see Rodda (1984).

25. The quote is from Dingle (1996, p. 212).

26. Magnetite crystals have been found in the brains of humans, which gives rise to intriguing speculation about the possibility that we might have a magnetic sense. Currently there is no experimental evidence to support this hypothesis. See Finney (1995). Historically humans have made use of larger lumps of magnetite for orientation—these are the "lodestones" of ancient seafarers.

27. See Lohmann et al. (1999).

28. The experiment involving European Robins is reported in Wiltschko et al. (2002); the two distinct mechanisms of magnetoreception in birds are reported in, for example, Wiltschko et al. (2005, 2006). Yes, the Wiltschkos have (for four decades) been significant contributors to our understanding of birds' magnetic senses.

29. See Wiltschko and Wiltschko (1999) and Wiltschko et al. (2002).

30. Details of the Monarch Butterfly's time-compensated sun compass are to be found in Sauman et al. (2005). For more on the coupling of magnetic and visual senses in birds, see Wiltschko et al. (2003) and Muheim (2006). Songbirds recalibrating their magnetic sense via twilight cues is discussed in Cochran et al. (2004). See Lohmann and Lohmann (1998) for the role of geomagnetism in sea turtle navigation.

12. Talk to the Animals

1. Dawkins (2006) gives one discussion of what constitutes animal communication.

2. In a surprisingly large number of animal species, both vertebrate and invertebrate, females can produce young via parthenogenesis, a form of asexual reproduction in which the young develop without any input from a male.

3. In crocodiles embryonic young synchronize their hatching by calling to one another from the egg. The parents also hear these calls and dig up the eggs. The combined effect is that they all hatch together, giving the young the benefit of safety in numbers and focused parental care as well as minimizing the amount of predator-attracting new-baby calls. Similar behavior is seen in many bird species, as mentioned in Chapter 8.

4. Bradbury and Vehrencamp (1998) cover most aspects of animal communication.

5. For all things pheromonal, see Wyatt (2003).

6. Biron (2003) discusses various forms of bioluminescence, including those used for communication. There are more than 170 species of firefly in North America alone.

7. This analysis is more fully described in Fletcher (2004).

8. The whole business of animal bioacoustics is well covered by Fletcher (2007).

9. Rattlesnakes are an interesting example of a vertebrate using stridulation. They compensate for the lack of a decent resonator by having highly efficient muscles. We are reliably informed that some can rattle away for hours.

10. The more massive the vocal chords are, the lower the frequency at which they vibrate. For humans, the mass is greater in males. It also increases when you have a cold, so your voice gets deeper.

11. It's only in the past few decades that science has recognized how widespread and important vibration through the substrate—the thing you're standing on—is as a communication mechanism for many disparate species across the whole animal kingdom. What did we *think* the critters were doing all this time? Now, fortunately, there is an excellent book on the subject: Hill (2008).

12. See Turner (2000) for a discussion.

13. In this case human experience differs from the natural world. We developed broadcast electrical communication systems (radio) and only then adapted them as sensor systems (radar).

14. See Hopkins (1999) and Stoddard (2006) for more on this topic.

15. One reason can be communication. For some animals a conspicuous waste of energy can be a good way of showing a potential mate just how fit and virile they are. Conspicuous consumption in humans can fulfil the same role.

16. Searcy and Nowicki (2005) discuss all manner of deception and unreliability among animals.

17. We prefer not to consider camouflage as a form of communication, because its purpose is to deny information to another, not convey it.

18. See Tautz et al. (2001) for more on honeybee dances.

19. This process is described by Thom et al. (2007).

20. Karl von Frisch won a Nobel Prize for his work on honeybee communications. See Frisch (1976) for the details.

21. North Carolina State University's apiculture program has a superb interactive simulation of the waggle dance (see North Carolina State University n.d.), where you can experiment with the parameters that drive the performance.

22. See Laver (1994).

23. See Arnold and Zuberbühler (2008).

Further Reading

The References provide a much-pruned selection of primary sources for those readers who may wish to pursue in more detail the topics discussed in this book. Any attempt to be comprehensive is doomed to failure, given the broad sweep of our subject matter and the large number of research papers written on animal biology (there have been more than 6,000 technical papers written about shrews alone, so you see the scale of our problem). In addition to these mostly primary sources, in this section we supply some secondary references for further reading for each chapter. These sources have been included because they are informative and well-written popular science—a good read.

1. Go with the Flow

A classic ecology book by a world authority and intended for the lay reader is Edward O. Wilson's *The Diversity of Life* (London: Penguin, 1992). The trophic pyramid is one of many topics discussed. Richard Dawkins has written widely and well about evolution; here we pick out his *The Ancestor's Tale* (Boston: Houghton Mifflin, 2004), in which Kleiber's Law is elucidated.

2. Structural Engineering

A splendid general reference for anything birdy is *The Audubon Society Encyclopedia of North American Birds* by John K. Terres (New York: Alfred A. Knopf, 1980). Of relevance to this chapter, see the sections on skeletons and feathers.

3. A Moving Experience

Much of the subject matter of this chapter (and more) is discussed, with clear diagrams replacing most equations, in Steve Vogel's *Comparative Biomechanics: Life's Physical World* (Princeton, N.J.: Princeton University Press, 2003).

4. A Mind of Its Own

For an accessible account of the whole human nervous system, with diversions into other creatures, a good place to start is Peter W. Nathan's *The Nervous System* (4th edition, Hoboken, N.J.: Wiley Blackwell, 1997).

5. Built for Life

There are surprisingly few books about animal architects, and most seem to be written by the same people. A good introduction is Mike H. Hansell's *Built by Animals: The Natural History of Animal Architecture* (Oxford: Oxford University Press, 2007).

6. Simple Complexity

Books on emergent behavior tend to start with examples related to eusocial insects and then plunge straight into what this means for human society and engineering without any further discussion of animals. A useful treatment of the basic concepts is Steven Johnson's *Emergence: The Connected Lives of Ants, Brains, Cities, and Software* (London: Penguin Books, 2002).

7. A Chemical Universe

An accessible introduction to smell (and other, nonchemical, senses) can be found in *Seeing, Hearing and Smelling the World* (Chevy Chase, Md.: Howard Hughes Medical Institute, 1995, pp. 46, 54). Any good general book that covers the nervous system should have chapters on the chemical senses. If you want to go deeper, try *Fundamental Neuroscience* by Larry R. Squire and Floyd Bloom (Oxford: Academic Press, 2008, chapter 4).

8. Sound Ideas

General books on animal acoustic sensing are thin on the ground: most books either cover all the senses or are technical tomes on specific aspects or creatures. *The Acoustic Sense of Animals* by William C. Stebbins (Cambridge, Mass.: Harvard University Press, 1987) gives a good overview.

9. Animal Sonar

There is a huge amount of information on the web about echolocation. An informative and accessible account of bat echolocation is maintained by the University of Maryland: http://www.bsos.umd.edu/psyc/batlab. A detailed and technical website on all aspects of dolphin biosonar is maintained by James Fulton: http://www.hearingresearch.net/files/dolphinbiosonar.htm.

10. Seeing the Light

Many books cover vision in animals. Try *Animal Eyes* by Michael F. Land and Dan-Eric Nilsson (Oxford: Oxford University Press, 2002) or, for the most abundant visual systems on the planet, *Invertebrate Vision* by Eric Warrant and Dan-Eric Nilsson (Cambridge: Cambridge University Press, 2006).

11. There and Back Again

The popular literature of migration has been extended magnificently in recent years by a number of films. *Winged Migration* (2001) and *March of the Penguins* (2005) both contain stunning photography and convey, if little else, the effort involved in long-distance bird migrations.

12. Talk to the Animals

There are many good books on animal communication, some of them cited in Chapter 12. For an overview of what it is and why it happens, try Marc D. Hauser's *The Evolution of Communication* (Cambridge, Mass.: MIT Press, 1997).

References

Able, K. P., and M. A. Able. 1996. "The flexible migratory orientation system of the Savannah Sparrow (*Passerculus sandwichensis*)." *Journal of Experimental Biology* 199: 3–8.

Ahlborn, B. K. 2004. *Zoological Physics: Quantitative Models of Body Design, Actions, and Physical Limitations of Animals.* Berlin: Springer-Verlag.

Aiello, L. C., and P. Wheeler. 1995. "The expensive tissue hypothesis: The brain and the digestive system in human and primate evolution." *Current Anthropology* 36: 199–221.

Alerstam, T. 1987. "Bird migration across a strong magnetic anomaly." *Journal of Experimental Biology* 130: 63–86.

Alexander, R. M. 1977. "Mechanics and scaling of terrestrial locomotion," in *Scale Effects in Animal Locomotion,* ed. T. J. Pedley. London: Academic Press, pp. 93–110.

———. 1996. *Optima for Animals.* Princeton, N.J.: Princeton University Press.

———. 2003. *Principles of Animal Locomotion.* Princeton, N.J.: Princeton University Press.

———. 2005. "Models and the scaling of energy costs for locomotion." *Journal of Experimental Biology* 208: 1645–1652.

Altes, R. A. 1990. "Radar/sonar acceleration estimation with linear-period modulated waveforms." *IEEE Transactions AES* 26: 914–924.

Anderson, R. C. 2009. "Do dragonflies migrate across the western Indian Ocean?" *Journal of Tropical Ecology* 25: 347–358.

Arnold, K., and K. Zuberbühler. 2008. "Meaningful call combinations in a non-human primate." *Current Biology* 18: R202–R203.

Attwell, D., and S. B. Laughlin. 2001. "An energy budget for signaling in the grey matter of the brain." *Journal of Cerebral Blood Flow & Metabolism* 21: 1133–1145.

Au, W. W. L., A. Frankel, D. A. Helweg, and D. H. Cato. 2001. "Against the humpback whale sonar hypothesis." *IEEE Journal of Oceanic Engineering* 26: 295–300.

Au, W. W. L., J. K. B. Ford, J. K. Horne, and K. A. N. Allman. 2004. "Echolocation signals of free-ranging killer whales (*Orcinus orca*) and modeling of foraging for chinook salmon (*Oncorhynchus tshawytscha*)." *Journal of the Acoustical Society of America* 115: 901–909.

Axel, R. 2005. "Scents and sensibility: A molecular logic of olfactory perception," in *The Nobel Prizes 2004,* ed. T. Frängsmyr. Stockholm: Nobel Foundation, pp. 234–256.

Bach, M. n.d. "81 optical illusions and visual phenomena." http://www.michaelbach.de/ot/.

Bajec, L. I., N. Zimic, and M. Mraz. 2005. "Simulating flocks on the wing: The fuzzy approach." *Journal of Theoretical Biology* 233: 199–220.

Bakken, G. S., and A. R. Krochmal. 2007. "The imaging properties and sensitivity of the facial pits of pit vipers as determined by optical and heat transfer analysis." *Journal of Experimental Biology* 210: 2801–2810.

Baxi, K. N., K. M. Dorries, and H. L. Eisthen. 2006. "Is the vomeronasal system really specialized for detecting pheromones?" *Trends in Neurosciences* 29: 1–7.

Bejan, A. 2005. "The constructal law of organization in nature: Tree-shaped flows and body size." *Journal of Experimental Biology* 208: 1677–1686.

Bejan, A., and J. H. Marden. 2006. "Unifying constructal theory for scale effects in running, swimming and flying." *Journal of Experimental Biology* 209: 238–248.

Benton, R. 2006. "Visions & reflections: On the origin of smell: Odorant receptors in insects." *Cellular and Molecular Life Sciences* 63: 1579–1585.

Bindi, V., J. Strunk, J. Baker, R. Bacon, M.G. Boensel, F. E. Shoup, III, and R. Vaidyanathan. 2005. "Littoral undersea warfare in 2025." Naval Postgraduate School Report NPS-97-06-001. Also available at http://www.nps.edu/Academics/Programs/SEA/docs/2005/SEAreport.pdf.

Biron, K. 2003. "Fireflies, dead fish and a glowing bunny: A primer on bioluminescence." *BioTeach Journal* 1: 19–26.

Bolshakov, C. V., V. N. Bulyuk, A. Mukhin, and N. Chernetsov. 2003. "Body mass and fat reserves of Sedge Warblers during vernal nocturnal migration: Departure vs. arrival." *Journal of Field Ornithology* 74: 81–89.

Bozinovic, F., J. L. Muñoz, D. E. Naya, and A. P. Cruz-Neto. 2007. "Adjusting energy expenditures to energy supply: Food availability regulates torpor use and organ size in the Chilean mouse-opossum *Thylamys elegans.*" *Journal of Comparative Physiology B* 177: 393–400.

Bradbury, J. 2004. "Taste perception: Cracking the code." *PLoS Biology* 2: e297.

Bradbury, J., and S. L. Vehrencamp. 1998. *Principles of Animal Communication.* Sunderland, Mass.: Sinauer Associates.

Branstetter, B. K., S. J. Mevissen, L. M. Herman, A. A. Pack, and S. P. Roberts. 2003. "Horizontal angular discrimination by an echolocating bottlenose dolphin (*Tursiops truncatus*)." *Bioacoustics* 14: 15–34.

Bright, M. 1993. *The Private Life of Birds.* London: Transworld, chapter 4.

Britannica. 1998. "Mechanoreception." Standard edition CD-ROM.

Brook, B. S., and T. J. Pedley. 2002. "A model for time-dependent flow in (giraffe jugular) veins: Uniform tube properties." *Journal of Biomechanics* 35: 95–107.

Bruce, V., P. R. Green, and M. A. Georgeson. 2003. *Visual Perception: Physiology, Psychology and Ecology,* 4th edition. Hove, Sussex: Psychology Press.

Bruinsma , O. H. 1979. Ph.D. thesis, Wageningen University, Wageningen, The Netherlands. Described in *Self-Organization in Biological Systems,* S. Camazine, J.-L. Deneubourg, N. R. Franks, J. Sneyd, G. Theraulaz, and E. Bonabeau. Princeton, N.J.: Princeton University Press, 2001.

Buck, L. B. 2005. "Unraveling the sense of smell," in *The Nobel Prizes 2004,* ed. T. Frängsmyr. Stockholm: Nobel Foundation, pp. 267–283.

Bullock, T. H., and G. A. Horridge. 1965. *Structure and Function in the Nervous Systems of Invertebrates.* London: W. H. Freeman.

Cant, J. P., B. W. McBride, and W. J. Croom, Jr. 1996. "The regulation of intestinal metabolism and its impact on whole animal energetics." *Journal of Animal Science* 74: 2541–2553.

Caprari, G., A. Colot, R. Siegwart, J. Halloy, and J.-L. Deneubourg. 2005. "Animal and robot mixed societies: Building cooperation between microrobots and cockroaches." *IEEE Robotics & Automation Magazine* 12: 58–65.

Carbone C., A. Teacher, and J. M. Rowcliffe. 2007. "The costs of carnivory." *PLoS Biology* 5: e22. doi: 10.1371/journal.pbio.0050022.

Carpenter, F. L., M. A. Hixon, C. A. Beuchat, R. W. Russell, and D. C. Paton. 1993. "Biphasic mass gain in migrant hummingbirds: Body composition changes, torpor, and ecological significance." *Ecology* 74: 1173–1182.

Chan, B., N. J. Balmforth, and A. E. Hosoi. 2005. "Building a better snail: Lubrication and adhesive locomotion." *Physics of Fluids* 17: 113101. doi: 10.1063/1.2102927.

Cheney, K. L., and N. J. Marshall. 2009. "Mimicry in coral reef fish: How accurate is this deception in terms of color and luminance?" *Behavioral Ecology* 20: 459–468. doi: 10.1093/beheco/arp017.

Chiao, C.-C., T. W. Cronin, and N. J. Marshall. 2000. "Eye design and color signaling in a stomatopod crustacean *Gonodactylus smithii.*" *Brain, Behavior and Evolution* 56: 107–122.

Cochran, W. W., H. Mouritsen, and M. Wileski. 2004. "Migrating songbirds recalibrate their magnetic compass daily from twilight cues." *Science* 304: 405–408.

Collins, J. J., and I. Stewart. 1993a. "Coupled nonlinear oscillators and the symmetries of animal gaits." *Journal of Nonlinear Science* 3: 349–392.

———. 1993b. "Hexapodal gaits and coupled nonlinear oscillator models." *Biological Cybernetics* 68: 287–298.

Collins, T., and P. Atkins. 1999. "Non-linear frequency modulation chirps for active sonar." *IEEE Proceedings F, Radar Signal Processing* 146: 312–316.

Cranford, T. W., W. G. van Bonn, M. S. Chaplin, J. A. Carr, T. A. Kamolnick, D. A. Carder, and S. H. Ridgway. 1997. "Visualizing dolphin sonar signal generation using high-speed video endoscopy." *Journal of the Acoustical Society of America* 102: 3213.

Creel, S., and N. M. Creel. 1997. "Lion density and population structure in the Selous Game Reserve: Evaluation of hunting quotas and offtake." *African Journal of Ecology* 35: 83–93.

Currey, J. D. 2002. *Bones: Structure and Mechanics.* Princeton, N.J.: Princeton University Press.

Damasio, A. R. (ed.). 1999. *The Scientific American Book of the Brain.* New York: Lyons Press.

Darwin, C. n.d. *The Formation of Vegetable Mould,* chapter 4. Available at http://www.darwin-literature.com.

Davidson, E. H., and D. H. Erwin. 2006. "Gene regulatory networks and the evolution of animal body plans." *Science* 311: 796–800.

Dawkins, R. 1986. *The Blind Watchmaker.* Bath, UK: Longman, chapter 2.

———. 2004. *The Ancestor's Tale.* Boston: Houghton Mifflin.

———. 2006. *The Selfish Gene: Edition 30.* Oxford: Oxford University Press.

Dawson, T. H. 2005. "Modeling of vascular networks." *Journal of Experimental Biology* 208: 1687–1694.

Day, S. 1991. "Migrating birds use genetic maps to navigate." *New Scientist* 1765: 21.

Deneubourg, J.-L., S. Goss, N. Franks, A. Sendova-Franks, C. Detrain, and L. Chretien. 1991. "The dynamics of collective sorting: Robot-like ants and ant-like robots," in *From Animals to Animats: Proceedings of the First International Conference on Simulation of Adaptive Behavior,* eds. J. A. Meyer and S. W. Wilson. Cambridge, Mass.: MIT Press, pp. 356–365.

Denny, M. 2007. *Blip, Ping and Buzz: Making Sense of Radar and Sonar.* Baltimore: Johns Hopkins University Press.

———. 2008a. "Physics between a fly's ears." *European Journal of Physics* 29: 1051–1057.

———. 2008b. *Float Your Boat: The Science of Sailing.* Baltimore: Johns Hopkins University Press.

———. 2009. "Dynamic soaring: Aerodynamics for albatrosses." *European Journal of Physics* 30: 75–84.

Denny, M. W. 1981. "A quantitative model for the adhesive locomotion of the terrestrial slug, *Ariolimax columbianus*." *Journal of Experimental Biology* 91: 195–217.

Di Natale, C., E. Martinelli, R. Paolesse, A. D'Amico, D. Filippini, and I. Lundstrom. 2008. "An experimental biomimetic platform for artificial olfaction." *PLoS One* 3: e3139.

Dingle, H. 1996. *Migration: The Biology of Life on the Move.* Oxford: Oxford University Press.

Dingle, H., and V. A. Drake. 2007. "What is migration?" *Bioscience* 57: 113–121.

Dobbins, P. 2007. "Dolphin sonar—Modeling a new receiver concept." *Bioinspiration and Biomimetics* 2:19–29.

Døving, K. B., and O. B. Stabell. 2003. "Trails in open water: Sensory cues in salmon migration," in *Sensory Processing in Aquatic Environments,* eds. S. P. Collin and N. J. Marshall. New York: Springer-Verlag, pp. 39–52.

Dreyer, G., and W. Ray. 1910. "The blood volume of mammals as determined by experiments upon rabbits, guinea-pigs, and mice; And its relationship to the body weight and to the surface area expressed in a formula." *Proceedings of the Royal Society of London* 201: 133–160.

Drucker, E. G. 1996. "The use of gait transition speed in comparative studies of fish locomotion." *American Zoologist* 36: 555–566.

Drucker, E. G., and J. S. Jensen. 1996. "Pectoral fin locomotion in the Striped Surfperch. II. Scaling swimming kinematics and performance at a gait transition." *Journal of Experimental Biology* 199: 2243–2252.

Drucker, E. G., and G. V. Lauder. 2000. "A hydrodynamic analysis of fish swimming speed: Wake structure and locomotor force in slow and fast labriform swimmers." *Journal of Experimental Biology* 203: 2379–2393.

Dulac, C. 2000. "The physiology of taste, vintage 2000." *Cell* 100: 607–610.

Durie, B. 2005. "Doors of perception." *New Scientist* 2484: 34.

Dynamic Soaring. n.d. Youtube video of model albatross flight. http://www.youtube.com/watch?v=orekaYTsY3I.

Ehrlich, P. R., D. S. Dobkin, and D. Wheye. 1988. *The Birder's Handbook.* New York: Simon and Schuster.

Eisthen, H. L. 2002. "Why are olfactory systems of different animals so similar?" *Brain, Behavior, and Evolution* 59: 273–293.

Elliott, K. H., M. Hewett, G. W. Kaiser, and R. W. Blake. 2004. "Flight energetics of the Marbled Murrelet, *Brachyramphus marmoratus*." *Canadian Journal of Zoology* 82: 644–652.

Emlen, S. T. 1975. "The stellar orientation system of a migratory bird." *Scientific American* (August): 102–111.

Engen, T. 1991. *Odor Sensation and Memory.* Santa Barbara, Calif.: Greenwood Publishing Group.

Engen, T., M. M. Gilmore, and R. G. Mair. 1991. "Odor memory," in *Smell and Taste in Health and Disease,* eds. T. V. Getchell, R. L. Doty, L. M. Bartoshuck, and J. B. Snow. New York: Raven Press, pp. 315–328.

Ewer, R. 1997. *Carnivores.* Ithaca, N.Y.: Cornell University Press.

Fenton, M. B. 1991. "Seeing in the dark." *Bats* 9: 9–13.

———. 2003. "Eavesdropping on the echolocation and social calls of bats." *Mammal Review* 33: 193–204.

Finger, T. E. 2006. "Evolution of taste: From single cells to taste buds." *Chemosense* 9: 2–6.

Finney, B. 1995. "A role for magnetoreception in human navigation?" *Current Anthropology* 36: 500–506.

Firestein, S. 2001. "How the olfactory system makes sense of scents." *Nature* 413: 211–218.

Fish, F. E. 1999. "Energetics of swimming and flying in formation." *Comments in Theoretical Biology* 5: 283–304.

Fletcher, N. H. 2004. "A simple frequency scaling rule for animal communication." *Journal of the Acoustical Society of America* 115: 2334–2338.

———. 2007. "Animal bioacoustics," in *Springer Handbook of Acoustics,* ed. T. D. Rossing. New York: Springer, pp. 473–490.

Fletcher, N. H., and A. Tarnopolsky. 1999. "Acoustics of the avian vocal tract." *Journal of the Acoustical Society of America* 105: 35–49.

Flydal, K., A. Hermansen, P. S. Engers, and E. Reimers. 2001. "Hearing in reindeer (*Rangifer tarandus*)." *Journal of Comparative Physiology A* 187: 265–269.

Freeman, P. W., and C. A. Lemen. 2007. "The trade-off between tooth strength and tooth penetration: Predicting optimal shape of canine teeth." *Journal of Zoology* 273: 273–280.

Frisch, Karl von. 1976. *Bees: Their Vision, Chemical Senses, and Language.* Ithaca, N.Y.: Cornell University Press.

Fullard, J. H. 1997. "Predator and prey: Life and death struggles." *Bats* 9: 5–7.

Fulton, J. T. 2005. *Processes in Biological Hearing.* Corona del Mar, Calif.: Jearing Concepts, Appendix L, p. 2. Available at http://www.hearingresearch.net.

Geyer, H., A. Seyfarth, and R. Blickhan. 2006. "Compliant leg behaviour explains basic dynamics of walking and running." *Proceedings of the Royal Society B* 273: 2861–2867.

Ghose K., T. K. Horiuchi, P. S. Krishnaprasad, and C. F. Moss. 2006. "Echolocating bats use a nearly time-optimal strategy to intercept prey." *PLoS Biology* 4: e108.

Girard, I. 2001. "Field cost of activity in the kit fox *Vulpes macrotis.*" *Physiological and Biochemical Zoology* 74: 191–202.

GlobalSecurity.org. n.d. Transformational Communications Satellite data. http://www.globalsecurity.org/space/systems/tsat.htm.

Goldsmith, T. H. 2006. "What birds see." *Scientific American* (July): 69–75.

Goldyn, B., M. Hromada, A. Sarmacki, and P. Tryjanowski. 2003. "Habitat use and diet of the red fox *Vulpes vulpes* in an agricultural landscape in Poland." *European Journal of Wildlife Research* 49: 191–200.

Golubitsky, M., I. Stewart, P.-L. Buono, and J. J. Collins. 1998. "A modular network for legged locomotion." *Physica D* 115: 56–72.

———. 1999. "Symmetry in locomotor central pattern generators and animal gaits." *Nature* 401: 693–695.

Gordon, J. E. 1978. *Structures: Or Why Things Don't Fall Down.* New York: Plenum Press.

Gordon, R. T., C. B. Schatz, L. J. Myers, M. Kosty, C. Gonczy, J. Kroener, M. Tran, P. Kurtzhals, S. Heath, J. A. Koziol, N. Arthur, M. Gabriel, J. Hemping, G. Hemping, S. Nesbitt, L. Tucker-Clark, and J. Zaayer. 2008. "The use of canines in the detection of human cancers." *Journal of Alternative and Complementary Medicine* 14: 61–67.

Gosline, J. M., P. A. Guerette, C. S. Ortlepp, and K. N. Savage. 1999. "The mechanical design of spider silks: From fibroin sequence to mechanical function." *Journal of Experimental Biology* 202: 3295–3303.

Gould, J. L., and C. G. Gould. 2007. *Animal Architects: Building and the Evolution of Intelligence.* New York: Basic Books.

Grace, M. S., O. M. Woodward, D. R. Church, and G. Calisch. 2001. "Prey targeting by the infra-red imaging snake *Python:* Effects of congenital and experimental visual deprivation." *Behavioural Brain Research* 119: 23–31.

Graván, C. P., and R. Lahoz-Beltra. 2004. "Evolving morphogenetic fields in the zebra skin pattern based on Turing's morphogen hypothesis." *International Journal of Applied Mathematics and Computer Science* 14: 351–361.

Griffin, D. R. 1994. *Animal Minds.* Chicago: University of Chicago Press, chapter 4.

Griffin, T. M., R. P. Main, and C. T. Farley. 2004a. "Biomechanics of quadrupedal walking: How do four-legged animals achieve inverted pendulum-like movements?" *Journal of Experimental Biology* 207: 3545–3558.

Griffin, T. M., R. Kram, S. J. Wickler, and D. F. Hoyt. 2004b. "Biomechanical and energetic determinants of the walk–trot transition in horses." *Journal of Experimental Biology* 207: 4215–4223.

Gutiérrez, J. G., and C. G. Jones. 2006. "Physical ecosystem engineers as agents of biogeochemical heterogeneity." *BioScience* 56: 227–236.

Hagstrum, J. T. 2000. "Infrasound and the avian navigational map." *Journal of Experimental Biology* 203: 1103–1111.

Hairston, N. G., F. E. Smith, and L. B. Slobodkin. 1960. "Community structure, population control and competition." *American Naturalist* 94: 421–425.

Haldane, J. B. S. 1985. *On Being the Right Size.* Oxford: Oxford University Press.

Hall, D. L., and J. Llinas (eds.). 2001. *Handbook of Multisensor Data Fusion.* Boca Raton, Fla.: CRC Press.

Hambly, C., E. J. Harper, and J. R. Speakman. 2004. "The energetic cost of variation in wing span and wing asymmetry in the zebra finch *Taeniopygia guttata.*" *Journal of Experimental Biology* 207: 3977–3984.

Hanlon, R. 2007. "Cephalopod dynamic camouflage." *Current Biology* 17: R400–R404.

Hansell, M. H. 2000. *Bird Nests and Construction Behavior.* Cambridge: Cambridge University Press.

———. 2005. *Animal Architecture.* Oxford: Oxford University Press.

———. 2007. *Built by Animals: The Natural History of Animal Architecture* Oxford: Oxford University Press.

Harland, D. P., and R. R. Jackson. 2000. "'Eight-legged cats' and how they see—A review of recent research on jumping spiders (Araneae: Salticidae)." *Cimbebasia* 16: 231–240.

Hartmann, W. M. 1999. "How we localize sound." *Physics Today* (November): 24–29.

Hassan, Z. 2005. "The human motor control system's response to mechanical perturbation: Should it, can it and does it ensure stability?" *Journal of Motor Behavior* 37: 484–493.

Heffner, H. E. 1998. "Auditory awareness." *Applied Animal Behaviour Science* 57: 259–268.

———. 2007. "Hearing ranges of laboratory animals." *Journal of the American Association for Laboratory Animal Science* 46: 11–13.

Heffner, H. E., and R. S. Heffner. 2003. "Audition," in *Handbook of Research Methods in Experimental Psychology,* ed. S. Davis. Oxford: Blackwell, pp. 413–440.

Heppner, F. H. 1990. "Of flocks and chaos." *Bioscience* 40: 429–430.

Hill, P. S. M. 2008. *Vibrational Communication in Animals.* Harvard: Harvard University Press.

Hill, R. W., G. A. Wyse, and M. Anderson. 2004. *Animal Physiology.* Sunderland, Mass.: Sinauer Associates.

Ho-Kim, Q., N. Kumar, and C. S. Lam. 1991. *Invitation to Contemporary Physics.* Singapore: World Scientific, p. 207.

Holmes, P., D. Koditschek, and J. Guckenheimer. 2006. "The dynamics of legged locomotion: Models, analyses and challenges." *SIAM Review* 48: 207–304.

Holmes, R., and S. Bhattacharya. 2004. "Early hominid ears primed for speech." *New Scientist* (June 22): 12–19. Available at http://www.newscientist.com/article/dn6053-early-hominid-ears-primed-for-speech.html.

Hopkins, C. D. 1999. "Signal evolution in electric communication," in *The Design of Animal Communication,* eds. M. D. Hauser and M. Konishi. Cambridge, Mass.: MIT Press, pp. 461–491.

Horridge, G. A. 2003. "Visual resolution of gratings by the compound eye of the bee *Apis mellifera*." *Journal of Experimental Biology* 206: 2105–2110.

Huber, F., T. E. Moore, and W. Loher (eds.). 1989. *Cricket Behavior and Neurobiology.* Ithaca, N.Y.: Cornell University Press.

Hünerbein K., W. Wiltschko, and E. Rüter. 2001. "Flight tracks of homing pigeons measured with GPS." *Journal of Navigation* 54: 167–175.

Hutchinson, J. M. C. 1990. "Gastropod shell form via apertural growth rates." *Journal of Morphology* 206: 259–264.

Hyvönen, T., and P. Nummi. 2008. "Habitat dynamics of beaver *Castor canadensis* at two spatial scales." *Wildlife Biology* 14: 302–308.

James, P. C. 1986. "How do Manx Shearwaters *Puffinus puffinus* find their burrows?" *Ethology* 71: 287–294.

Jeanson, R., C. Rivault, J.-L. Deneubourg, S. Blanco, R. Fournier, C. Jost, and G. Theraulaz. 2005. "Self-organized aggregation in cockroaches." *Animal Behaviour* 69: 169–180.

Jones, M. P., K. E. Pierce, and D. Ward. 2007. "Avian vision: A review of form and function with special consideration to birds of prey." *Journal of Exotic Pet Medicine* 16: 69–87.

Jost, C., S. Garnier, R. Jeanson, M. Asadpour, J. Gautrais, and G. Theraulaz. 2004. "The embodiment of cockroach behaviour in a micro-robot," in *Proceedings of the 35th International Symposium on Robotics,* March 23–26, Paris, France.

Justh, E. W., and P. S. Krishnaprasad. 2006. "Steering laws for motion camouflage." *Proceedings of the Royal Society A* 462:3629–3643.

Keller, A., H. Zhuang, Q. Chi, L. B. Vosshall, and H. Matsunami. 2007. "Genetic variation in a human odorant receptor alters odour perception." *Nature* 449: 468–472.

Kish, D. 2009. "Echo vision: The man who sees with sound." *New Scientist* 2703: 31–33.

Klarsfeld, A., and F. Revah. 2004. *The Biology of Death: Origins of Mortality.* Ithaca, N.Y.: Cornell University Press.

Kleineidam, C., R. Ernst, and F. Roces. 2001. "Wind-induced ventilation of the giant nests of the leaf-cutting ant *Atta vollenweideri.*" *Naturwissenschaften* 88: 301–305.

Knudsen, E. I. 1980. "Sound localization in birds," in *Comparative Studies of Hearing in Vertebrates,* eds. A. N. Popper and R. R. Fay. New York: Springer-Verlag, pp. 289–322.

Kolb, H. 2003. "How the retina works." *American Scientist* 91: 28–35.

Konishi, M. 1983. "Night owls are good listeners." *Natural History* (September): 56–59.

Korf, R. E. 1992. "A simple solution to pursuit games," in *Proceedings of the Eleventh International Workshop on Distributed Artificial Intelligence, Glen Arbor, Mich.* Ann Arbor: University of Michigan, pp. 183–194.

Kowalewsky, S., M. Dambach, B. Mauck, and G. Dehnhardt. 2006. "High olfactory sensitivity for dimethyl sulphide in harbour seals." *Biology Letters* 2: 106–109.

Kreithen, M. L., and D. B. Quine. 1979. "Infrasound detection by the homing pigeon: A behavioral audiogram." *Journal of Comparative Physiology A* 129: 1–4.

Kugler, P. N., R. E. Shaw, K. J. Vincente, and J. Kinsella-Shaw. 1990. "Inquiry into intentional systems I: Issues in ecological physics." *Psychological Research* 52: 98–121.

Ladley, D., and S. Bullock. 2005. "The role of logistic constraints in termite construction of chambers and tunnels." *Journal of Theoretical Biology* 234: 551–564.

Land, M. F. 1969. "The structure of the retinae of the principal eyes of jumping spiders (Salticidae: Dendryphantinae) in relation to visual optics." *Journal of Experimental Biology* 51: 443–470.

Land, M. F., and R. D. Fernald. 1992. "The evolution of eyes." *Annual Review of Neuroscience* 15: 1–29.

Land, M. F., and D.-E. Nilsson. 2002. *Animal Eyes.* Oxford: Oxford University Press.

Lauga, E., and A. E. Hosoi. 2006. "Tuning gastropod locomotion: Modeling the influence of mucus rheology on the cost of crawling." *Physics of Fluids* 18: 113102. doi: 10.1063/1.2382591.

Laver, J. 1994. *Principles of Phonetics.* Cambridge: Cambridge University Press.

Lawrence, B. D., and J. A. Simmons. 1982. "Measurements of atmospheric attenuation at ultrasonic frequencies and the significance for echolocation by bats." *Journal of the Acoustical Society of America* 71: 585–590.

Levy, S. 1992. *Artificial Life.* London: Vintage Books.

Lohmann, K. J., and C. M. F. Lohmann. 1998. "Sea turtle navigation and the determination of geomagnetic field features." *Journal of Navigation* 51: 10–22.

Lohmann, K. J., J. T. Hester, and C. M. F. Lohmann. 1999. "Long distance navigation in sea turtles." *Ethology, Ecology and Evolution* 11: 1–23.

Lund, N. 2002. *Animal Cognition.* Hove, UK: Routledge, 2002, chapters 2, 3.

Malnic, B., J. J. Hirano, T. Sato, and L. B. Buck. 1999. "Combinatorial receptor codes for odors." *Cell* 96: 713–723.

Manela, M., and J. A. Campbell. 1995. "Designing good pursuit problems as testbeds for distributed AI: A novel application of genetic algorithms," in *From Reaction to Cognition: Selected Papers from the 5th European Workshop on Modeling Autonomous Agents in a Multi-Agent World, MAAMAW '93,* eds. C. Castelfranchi and P. Müller. Lecture Notes in Artificial Intelligence 957. Berlin: Springer-Verlag, pp. 231–252.

Manton, M., A. Karr, and D. W. Ehrenfeld. 1972. "Chemoreception in the migratory sea turtle, *Chelonia Mydas.*" *Biological Bulletin* 143: 184–195.

Martin, G. R., L. M. Rojas, Y. Ramírez, and R. McNeil. 2004. "The eyes of oilbirds (*Steatornis caripensis*): Pushing at the limits of sensitivity." *Naturwissenschaften* 91: 26–29.

Martinez, I., M. Rosa, J.-L. Arsuaga, P. Jarabo, R. Quam, C. Lorenzo, A. Gracia, J.-M. Carretero, J.-M. Bermúdez de Castro, and E. Carbonell. 2004. "Auditory capacities in Middle Pleistocene humans in the Sierra de Atapuerca in Spain." *Proceedings of the National Academy of Sciences USA* 101: 9976–9981. doi: 10.1073/pnas.0403595101.

Matsumoto, A., H. J. Kim, I. Y. Tsai, X. Wang, P. Cebe, and D. L. Kaplan. 2007. "Silk," in *Handbook of Fiber Chemistry*, ed. M. Lewin. Boca Raton, Fla.: CRC Press, pp. 383–404.

McKechnie, A. 2002. "Avian facultative hypothermic responses: A review." *Condor* 104: 705–724.

Mercado, E., and L. N. Frazer. 2001. "Humpback whale song or humpback whale sonar? A reply to Au et al." *IEEE Journal of Oceanic Engineering* 26: 406–415.

Miles, R. N., D. Robert, and R. R. Hoy. 1995. "Mechanically coupled ears for directional hearing in the parasitoid fly *O. ochracea*." *Journal of the Acoustical Society of America* 98: 3059–3070.

Minetti, A. E. 1998. "The biomechanics of skipping gaits: A third locomotion paradigm?" *Proceedings of the Royal Society B* 265: 1227–1235.

Minetti, A. E., L. P. Ardigo, E. Reinach, and F. Saibene. 1999. "The relationship between mechanical work and energy expenditure of locomotion in horses." *Journal of Experimental Biology* 202: 2329–2338.

Mitchell, G., S. K. Maloney, D. Mitchell, and D. J. Keegan. 2006. "The origin of mean arterial and jugular venous blood pressures in giraffes." *Journal of Experimental Biology* 209: 2515–2524.

Møhl, B., M. Wahlberg, P. T. Madsen, A. Heerfordt, and A. Lund. 2003. "The monopulsed nature of sperm whale clicks." *Journal of the Acoustical Society of America* 114: 1143–1154.

Monen, J., E. Brenner, and J. Reynaerts. 1998. "What does a pigeon see in a Picasso?" *Journal of the Experimental Analysis of Behavior* 69: 223–226.

Morris, R., and M. Fillenz. 2003. *Neuroscience—Science of the Brain: An Introduction for Young Students.* Liverpool: British Neuroscience Association.

Möskon, M., F. H. Heppner, M. Mraz, N. Zimic, and L. I. Bajec. 2007. "Fuzzy model of bird flock foraging behavior," paper presented at the 6th EUROSIM Congress on Modelling and Simulation, Sep. 9–13 2007, Ljubljana, Slovenia.

Moss, C. F. 1999. "Echolocation" in *Encyclopedia of the Cognitive Sciences*, eds. R. Wilson and F. Keil. Cambridge, Mass.: MIT Press, pp. 253–255.

Mueller, P., and J. Diamond. 2001. "Metabolic rate and environmental productivity: Well-provisioned animals evolved to run and idle fast." *Proceedings of the National Academy of Sciences USA* 98: 12550–12554.

Muheim, R. 2006. "Polarized light cues underlie compass calibration in migratory songbirds." *Science* 313: 837–839.

Murray, J. D. 1988. "How the leopard gets its spots." *Scientific American* (March): 80–87.

———. 2001. *Mathematical Biology.* New York: Springer-Verlag.

Nactigall, P. E., and P. W. B. Moore (eds.). 1988. *Animal Sonar: Processes and Performance.* New York: Plenum.

Nagrath, I. J. 2006. *Control Systems Engineering,* 4th edition. New Delhi: New Age International.

Naguib, M., and R. H. Wiley. 2001. "Estimating the distance to a source of sound: Mechanisms and adaptations for long-range communication." *Animal Behaviour* 62: 825–837.

Nathan, P. W. 1997. *The Nervous System,* 4th edition. Hoboken, N.J.: Wiley Blackwell.

Nelson, B. S. 2003. "Reliability of sound attenuation in Florida scrub habitat and behavioral implications." *Journal of the Acoustical Society of America* 113: 2901–2911.

Nelson, G., J. Chandrashekar, M. A. Hoon, L. Feng, G. Zhao, N. J. Ryba, and C. S. Zuker. 2002. "An amino-acid taste receptor." *Nature* 416: 199–202.

Nikonov, S. S., R. Kholodenko, J. Lem, and E. N. Pugh, Jr. 2006. "Physiological features of the S- and M-cone photoreceptors of wild-type mice from single cell recordings." *Journal of General Physiology* 127: 359–374.

Noë, A., and E. T. Thompson (eds.). 2002. *Vision and Mind: Selected Readings in the Philosophy of Perception.* Cambridge, Mass: MIT Press.

North Carolina State University. n.d. Bee dance tutorial. http://entomology.ncsu.edu/apiculture/Dance_tutorial.html.

Northcutt, R. G. 2004. "Taste buds: Development and evolution." *Brain, Behavior, and Evolution* 64: 198–206.

Novacek, M. J. 1985. "Evidence for echolocation in the oldest known bats." *Nature* 315: 140–141.

Owen-Smith, R. N. 1988. *Megaherbivores: The Influence of Very Large Body Size on Ecology.* New York: Cambridge University Press.

Parker, A. R. 1998. "Colour in Burgess Shale animals and the effect of light on evolution in the Cambrian." *Proceedings of the Royal Society of London: Biological Sciences* 265: 967–972.

———. 2003. *In the Blink of an Eye.* London: Simon and Schuster.

Pearson, W. H., and B. L. Olla. 1977. "Chemoreception in the blue crab, *Callinectes sapidus.*" *Biological Bulletin* 153: 346–354.

Pedley, T. J., B. S. Brook, and R. S. Seymour. 1996. "Blood pressure and flow rate in the giraffe jugular vein." *Philosophical Transactions of the Royal Society of London* 351: 855–866.

Pennycuick, C. J. 1982. "The flight of petrels and albatrosses (Procellariiformes) observed in south Georgia and its vicinity." *Philosophical Transactions of the Royal Society of London B* 300: 76–106.

———. 2001. "Speeds and wingbeat frequencies of migrating birds compared with calculated benchmarks." *Journal of Experimental Biology* 204: 3283–3294.

Perdeck, A. C. 1967. "Orientation of Starlings after displacement to Spain." *Ardea* 55: 194–204.

Pettigrew, J. D. 1999. "Electroreception in monotremes." *Journal of Experimental Biology* 202: 1447–1454.

Pfau, T., T. H. Witte, and A. M. Wilson. 2006. "Centre of mass movement and mechanical energy fluctuation during gallop locomotion in the Thoroughbred racehorse." *Journal of Experimental Biology* 209: 3742–3757.

Piper, R. 2007. *Extraordinary Animals.* Westport, Conn.: Greenwood Press, pp. 113–115.

Poulter, T. C. 1969. "Sonar of penguins and fur seals." *Proceedings of the California Academy of Sciences* 36: 363–380.

Prabhakar, N. R., and Y.-J. Peng. 2004 "Peripheral chemoreceptors in health and disease." *Applied Physiology Journal* 96: 359–366.

Proske, U., and E. Gregory. 2003. "Electrolocation in the platypus—Some speculations." *Comparative Biochemistry and Physiology A* 136: 821–825.

Raff, R. A. 1996. *The Shape of Life.* Chicago: University of Chicago Press.

Raymond, L. 1985. "Spatial visual acuity of the eagle *Aquila audax;* A behavioral, optical and anatomical investigation." *Vision Research* 25: 1477–1491.

Rayner, J. M. V., P. W. Viscardi, S. Ward, and J. R. Speakman. 2001. "Aerodynamics and energetics of intermittent flight in birds." *American Zoologist* 41: 188–204.

Reynolds, C. W. 1987. "Flocks, herds, and schools: A distributed behavioral model." *Computer Graphics* 21: 25–34.

Reznik, G. K. 1990. "Comparative anatomy, physiology and function of the upper respiratory tract." *Environmental Health Perspectives* 85: 171–176.

Robert, D. 2001. "Innovative biomechanics for directional hearing in small flies." *Biological Bulletin* 200: 190–194.

Robert, D., R. N. Miles, and R. R. Hoy. 1996. "Directional hearing by mechanical coupling in the parasitoid fly *Ormia ochracea.*" *Journal of Comparative Physiology A* 179: 29–44.

Rodda, G. H. 1984. "The orientation and navigation of juvenile alligators: Evidence of magnetic sensitivity." *Journal of Comparative Physiology A* 154: 649–658.

Ropert-Coudert, Y., K. Sato, A. Kato, J.-B. Charrassin, C.-A. Bost, Y. le Maho, and Y. Naito. 2000. "Preliminary investigations of prey pursuit and capture by king penguins at sea." *Polar Bioscience* 13: 101–112.

Ross, T. J. 2004. *Fuzzy Logic with Engineering Applications.* Hoboken, N.J.: John Wiley and Sons.

Rutowski, R. L., L. Gislén, and E. J. Warrant. 2009. "Visual acuity and sensitivity increase allometrically with body size in butterflies." *Arthropod Structure and Development* 38: 91–100.

Sachs, G. 2005. "Minimum shear wind strength required for dynamic soaring of albatrosses." *Ibis* 247: 1–10.

Shahan, T. n.d. Photographs of jumping spiders. http://www.flickr.com/photos/7539598@N04/.

Sanderson, M. I., and J. A. Simmons. 2000. "Neural responses to overlapping FM sounds in the inferior colliculus of echolocating bats." *Journal of Neurophysiology* 83: 1840–1855.

Sato, K., M. Pellegrino, T. Nakagawa, T. Nakagawa, L. B. Vosshall, and K. Touhara. 2008. "Insect olfactory receptors are heteromeric ligand-gated ion channels." *Nature Letters* 452: 1002–1006.

Sauman, I., A. D. Briscoe, H. Zhu, D. Shi, O. Froy, J. Stalleicken, Q.Yuan, A. Casselman, and S. M. Reppert. 2005. "Connecting the navigational clock to sun compass input in the monarch butterfly brain." *Neuron* 46: 457–467.

Schmidt-Nielsen, K. 1984. *Scaling—Why Is Animal Size so Important?* Cambridge: Cambridge University Press.

———. 2005. "Scaling functions to body size: Theories and facts." *Journal of Experimental Biology* 208: 1573–1574.

Schneeweis, D. M., and J. L. Schnapf. 1995. "Photovoltage of rods and cones in the macaque retina." *Science* 268: 1053–1055.

Schopf, J. W. 1999. "Deep divisions in the Tree of Life—What does the fossil record reveal?" *Biological Bulletin* 196: 351–355.

Schusterman, R. J., D. Kastak, D. H. Levenson, C. J. Reichmuth, and B. L. Southall. 2000. "Why pinnipeds don't echolocate." *Journal of the Acoustical Society of America* 107: 2256–2264.

Searcy, W. A., and S. Nowicki. 2005. *The Evolution of Animal Communication: Reliability and Deception in Signaling Systems.* Princeton, N.J.: Princeton University Press.

Selden, P. A., W. A. Shear, and M. D. Sutton. 2008. "Fossil evidence for the origin of spider spinnerets, and a proposed arachnid order." *Proceedings of the National Academy of Sciences USA* 105: 20781–20785.

Seymour, R. S., and A. J. Blaylock. 2000. "The principle of Laplace and scaling of ventricular wall stress and blood pressure in mammals and birds." *Physiological and Biochemical Zoology* 73: 389–405.

Shaffer, D. M., M. Scott, M. E. Krauchunas, and M. K. McBeath. 2004. "How dogs navigate to catch Frisbees." *Psychological Science* 15: 437–441.

Shahar, N., P. Rutledge, and T. Cronin. 1996. "Polarization vision in cuttlefish: A concealed communication channel?" *Journal of Experimental Biology* 199: 2077–2084.

Sheng, Q., D. Wu, and L. Zhang. 2005. "Aerodynamic forces acting on an albatross flying above sea-waves." *Applied Mathematics and Mechanics* 26: 1222–1229.
Sherman, T. F. 1981. "On connecting large vessels to small—The meaning of Murray's Law." *Journal of General Physiology* 78: 431–453.
Sidiropoulou, K., F.-M. Lu, M. A. Fowler, R. Xiao, C. Phillips, E. D. Ozkan, M. X. Zhu, F. J. White, and D. C. Cooper. 2009. "Dopamine modulates an mGluR5-mediated depolarization underlying prefrontal persistent activity." *Nature Neuroscience* 12: 190–199.
Simmons, J. A., M. J. Ferragamo, and C. F. Moss. 1998. "Echo-delay resolution in sonar images of the big brown bat, *Eptesicus fuscus*." *Neurobiology* 95:12647–12652.
Sinsch, U. 2006. "Orientation and navigation in Amphibia." *Marine and Freshwater Behaviour and Physiology* 39: 65–71.
Sivak, J. G. 1982. "Optical properties of a cephalopod eye (the short-finned squid, *Illex illecebrosus*)." *Journal of Comparative Physiology A* 147: 323–327.
Slaughter, M., J. Nyquist, and B. E. Evans. 2001. *Basic Concepts in Neuroscience: A Student's Survival Guide.* New York: McGraw-Hill Professional.
Sloman, A. 2001. "Evolvable biologically plausible visual architectures," in *Proceedings of British Machine Vision Conference,* eds. T. Cootes and C. Taylor. Manchester, UK: British Machine Vision Association, pp. 313–322.
Smith, D. V., and R. F. Margolskee. 2006. "Making sense of taste." *Scientific American* (special edition): 84–92.
Speakman, J. R. 2001. "The evolution of flight and echolocation in bats: Another leap in the dark." *Mammal Review* 31: 111–130.
———. 2005. "Body size, energy metabolism and lifespan." *Journal of Experimental Biology* 208: 1717–1730.
Speck, J., and F. G. Barth. 1982. "Vibration sensitivity of pretarsal slit sensilla in the spider leg." *Journal of Comparative Physiology A* 148: 187–194.
Srinivasan, M., and A. Ruina. 2006. "Computer optimization of a minimal biped model discovers walking and running." *Nature* 439: 72–75.
Srygley, R. B., R. Dudley, E. G. Oliveira, and A. J. Riveros. 2006. "Experimental evidence for a magnetic sense in Neotropical migrating butterflies (Lepidoptera: Pieridae)." *Animal Behaviour* 71: 183–191.
Stap, D. 2005. *Birdsong.* New York: Scribner.
Steele, K. M. 2003. "Do rats show a Mozart effect?" *Music Perception* 21: 251–265.
Steels, L. 1991. "Toward a theory of emergent functionality," in *From Animals to Animats: Proceedings of the First International Conference on Simulation of Adaptive Behavior.,* eds. J. A. Meyer and S. W. Wilson. Cambridge, Mass.: MIT Press, pp. 451–461.
Stenseth, N. C., W. Falck, O. N. Bjørnstad, and C. J. Krebs. 1997. "Population regulation in snowshoe hares and Canadian lynx: Asymmetric food web

configurations between hare and lynx." *Proceedings of the National Academy of Sciences USA* 94: 5147–5152.

Stensmyr, M. C., S. Erland, E. Hallberg, R. Wallen, P. Greenaway, and W. S. Hansson. 2005. "Insect-like olfactory adaptations in the terrestrial giant robber crab." *Current Biology* 15: 116–121.

Stoddard, P. K. 2006. "Plasticity of the electric organ discharge waveform: Contexts, mechanisms, and implications for electrocommunication," in *Communication in Fishes,* eds. F. Ladich, S. P. Collin, P. Moller, B. G. Kapoor, and N. H. Enfield. Enfield, N.H.: Science Publishers, pp. 623–645.

Storey, K. B., and J. M. Storey. 1992. "Natural freeze tolerance in ectothermic vertebrates." *Annual Review of Physiology* 54: 619–637.

Strut, J. (Lord Rayleigh) 1883. "The soaring of birds." *Nature* 27: 534–535.

Sun, M., and Y. Xiong. 2005. "Dynamic flight stability of a hovering bumblebee." *Journal of Experimental Biology* 208: 447–459.

Tautz, J., J. Casas, and D. Sandeman. 2001. "Phase reversal of vibratory signals in honeycomb may assist dancing honeybees to attract their audience." *Journal of Experimental Biology* 204: 3737–3746.

Terres, J. K. 1980. *The Audubon Society Encyclopedia of North American Birds.* New York: Alfred A. Knopf.

Thom, C., D. C. Gilley, J. Hooper, and H. E. Esch. 2007. "The scent of the waggle dance." *PLoS Biology* 5: e228.

Trivedi, B. P. 2001. "Magnetic map found to guide animal migration." *National Geographic Today* (October 12). http://news.nationalgeographic.com/news/2001/10/1012_TVanimalnavigation.html.

Tsujiuchi, S., E. Sivan-Loukianova, D. F. Eberl, Y. Kitagawa, and T. Kadowaki. 2007. "Dynamic range compression in the honey bee auditory system toward waggle dance sound." *PLoS ONE* 2: e234.

Tucker, V. A. 1966. "Diurnal torpor and its relation to food consumption and weight changes in the California pocket mouse *Perognathus californicus.*" *Ecology* 47: 245–252.

———. 2000. "The deep fovea sideways vision and spiral flight paths in raptors." *Journal of Experimental Biology* 203: 3745–3754.

Tucker, V. A., A. E. Tucker, K. Akers, and J. H. Enderson. 2000. "Curved flight paths and sideways vision in peregrine falcons (*Falco peregrinus*)." *Journal of Experimental Biology* 203: 3755–3763.

Turing, A. M. 1952. "The chemical basis of morphogenesis." *Philosophical Transactions of the Royal Society of London B* 237: 37–72.

Turner, J. S. 2000. *The Extended Organism.* Cambridge, Mass.: Harvard University Press.

———. 2005. "Extended physiology of an insect-built structure." *American Entomologist* 51: 36–38.

Turner, J. S., and R. M. Soar. 2008. "Beyond biomimicry. What termites can tell us about realizing the living building," in *Proceedings of the First International Conference on Industrialized, Intelligent Construction (I3CON)*, Loughborough, UK: Loughborough University Media Services.

Uchida, C. N., and Z. F. Mainen. 2008. "Odor concentration invariance by chemical ratio coding." *Frontiers in Systems Neuroscience* 1: 1–6.

University of Texas at Houston. n.d. Neuroscience Online website. Includes a comprehensive survey of human vision. http://neuroscience.uth.tmc.edu/s2/ii14-1.html.

Urdy, S., N. Goudemand, H. Bucher, and R. Chirat. 2010. "Allometries and the morphogenesis of the molluscan shell: A quantitative and theoretical model." *Journal of Experimental Zoology Part B: Molecular and Developmental Evolution* 314B: 280–302.

U.S. Fish and Wildlife Service. n.d. Wildlife website. Includes "Migration of birds." http://www.npwrc.usgs.gov/resource/birds/migratio/.

Vácha, M., D. Drstkova, and T. Puzová. 2007. "*Tenebrio* beetles use magnetic inclination compass." *Naturwissenschaften* 95: 761–765. doi: 10.1007/s00114-008-0377-9.

Verboom, B., A. M. Boonman, and H. J. G. A. Limpens. 1999. "Acoustic perception of landscape elements by the pond bat (*Myotis dasycneme*)." *Journal of Zoology* 248: 59–66.

Vickers, N. J. 2000. "Mechanisms of animal navigation in odor plumes." *Biological Bulletin* 198: 203–212.

Viitala, J., E. Korplmäki, P. Palokangas, and M. Koivula. 1995. "Attraction of kestrels to vole scent marks visible in ultraviolet light." *Nature* 373: 425–427.

Vleck, C. M. 1981. "Hummingbird incubation: Female attentiveness and egg temperature." *Oecologia* 51: 199–205.

Vogel, S. 1996. *Life in Moving Fluids—The Physical Biology of Flow*, 2nd edition. Princeton, N.J.: Princeton University Press.

———. 2003. *Comparative Biomechanics: Life's Physical World*. Princeton, N.J.: Princeton University Press.

———. 2009. *Glimpses of Creatures in Their Physical Worlds*. Princeton, N.J.: Princeton University Press.

Vollrath, F. 1987. "Altered geometry of webs in spiders with regenerated legs." *Nature* 328: 247–248.

Vollrath, F., and P. A. Selden. 2007. "The role of behavior in the evolution of spiders, silks, and webs." *Annual Review of Ecology, Evolution, and Systematics* 38: 819–846.

Vornanen, M. 1989. "Basic functional properties of the cardiac muscle of the common shrew (*Sorex araneus*) and some other small mammals." *Journal of Experimental Biology* 145: 339–351.

Wainwright, S. A., W. D. Biggs, J. D. Currey, and J. M. Gosline. 1982. *Mechanical Design in Organisms.* Princeton, N.J.: Princeton University Press.

Walker, M. M., T. E. Dennis, and J. L. Kirschvink. 2002. "The magnetic sense and its use in long distance navigation by animals." *Current Opinion in Neurobiology* 12: 735–744.

Walker, T. J. 2001. University of Florida book of insect records. http://entomology.ifas.ufl.edu/walker/ufbir/.

Walter, R. M., and D. R. Carrier. 2007. "Ground forces applied by galloping dogs." *Journal of Experimental Biology* 210: 208–216.

Walton, M., B. C. Jayne, and A. F. Bennett. 1990. "The energetic cost of limbless locomotion." *Science* 249: 524–527.

Wang, Y., Y. Pan, S. Parsons, M. Walker, and S. Zhang. 2007. "Bats respond to polarity of a magnetic field." *Proceedings of the Royal Society B* 274: 2901–2905.

Wehner, R., and M. Muller. 2006. "The significance of direct sunlight and polarized skylight in the ant's celestial system of navigation." *Proceedings of the National Academy of Sciences USA* 103: 12575–12579.

Weibel, E. R., and H. Hoppeler. 2005. "Exercise-induced maximal metabolic rate scales with muscle aerobic capacity." *Journal of Experimental Biology* 208: 1635–1644.

Weimerskirch, H., T. Guionnet, J. Martin, S. A. Shaffer, and D. P. Costa. 2000. "Fast and fuel efficient? Optimal use of wind by flying albatrosses." *Proceedings of the Royal Society of London* 267: 1869–1874.

Weimerskirch, H., J. Martin, Y. Clerquin, P. Alexander, and S. Jiraskova. 2001. "Energy savings in flight formation." *Nature* 413: 697–698.

West, G. B., and J. H. Brown. 2005. "The origin of allometric scaling laws in biology from genomes to ecosystems: Towards a quantitative unifying theory of biological structure and organization." *Journal of Experimental Biology* 208: 1575–1592.

West, G. B., J. H. Brown, and B. J. Enquist. 1997. "A general model for the origin of allometric scaling laws in biology." *Science* 276: 122–126.

———. 1999. "The fourth dimension of life: Fractal geometry and allometric scaling of organisms." *Science* 28: 1677–1679.

West, G. B., W. H. Woodruff, and J. H. Brown. 2002. "Allometric scaling of metabolic rate from molecules and mitochondria to cells and mammals." *Proceedings of the National Academy of Sciences USA* 99: 2473–2478.

Whiting, W. C., and R. F. Zernicke. 2008. *Biomechanics of Musculoskeletal Injury.* Champaign, Ill.: Human Kinetics.

Wilga, C. A. D., and G. V. Lauder. 2004. "Biomechanics of locomotion in sharks, rays and chimeras," in *Biology of Sharks and Their Relatives,* eds. J. C. Carrier, J. A. Musick, and M. R. Heithaus. Boca Raton, Fla.: CRC Press, pp. 139–164.

Willis, M. A. 2005. "Odor-modulated navigation in insects and artificial systems." *Chemical Senses* 30: 287–288.

Wilson, A. 1832. *American Ornithology,* vol. 1. London: Whittaker, Treacher and Arnot.

Wilson, E. O. 1992. *The Diversity of Life.* London: Penguin.

Wilson, R. I. 2007. "Neural circuits underlying chemical perception." *Science* 318: 584–585.

Wilson, R. I., and Z. F. Mainen. 2006. "Early events in olfactory processing." *Annual Review of Neuroscience* 29: 163–201.

Wiltschko, R., and W. Wiltschko. 1999. "Celestial and magnetic cues in experimental conflict," in *Proceedings of the 22nd International Ornithology Congress,* eds. N. J. Adams and R. H. Slotow. Johannesburg: Birdlife South Africa, pp. 988–1004.

Wiltschko, R., T. Ritz, K. Stapput, P. Thalau, and W. Wiltschko. 2005. "Two different types of light-dependent responses to magnetic fields in birds." *Current Biology* 15: 1518–1523.

Wiltschko, W., J. Traudt, O. Güntürkün, H. Prior, and R. Wiltschko. 2002. "Lateralization of magnetic compass orientation in a migratory bird." *Nature* 419: 467–470.

Wiltschko, W., U. Munro, H. Ford, and R. Wiltschko. 2003. "Magnetic orientation in birds: Non-compass responses under monochromatic light of increased intensity." *Proceedings of the Royal Society of London B* 270: 2133–2140.

Wiltschko, W., U. Munro, H. Ford, and R.Wiltschko. 2006. "Bird navigation: What type of information does the magnetite based receptor provide?" *Proceedings of the Royal Society of London B* 273: 2815–2820.

Withers, P. C., and P. L. Timko. 1977. "The significance of ground effect to the aerodynamic cost of flight and energetics of the black skimmer (*Rhyncops nigra*)." *Journal of Experimental Biology* 70: 13–26.

Wlodarczyk, R., P. Minias, K. Kaczmarek, T. Janiszewski, and A. Kleszcz. 2007. "Different migration strategies used by two inland wader species during autumn migration, case of Wood Sandpiper *Tringa glareola* and Common Snipe *Gallinago gallinago.*" *Ornis Fennica* 84: 119–130.

Wong, K. 2001. "Cluttered surfaces baffle echolocating bats." *Scientific American* (December). http://www.innovations-report.com/html/reports/life_sciences/report-6627.html.

Wood, C. J. 1973. "The flight of albatrosses (a computer simulation)." *Ibis* 15: 244–256.

Wright, J. P., C. G. Jones, and A. S. Flecker. 2002. "An ecosystem engineer, the beaver, increases species richness at the landscape scale." *Oecologia* 132: 96–101.

Wyatt, T. D. 2003.*Pheromones and Animal Behaviour: Communication by Smell and Taste.* Cambridge: Cambridge University Press.

Yodzis, P. 1984. "Energy flow and the vertical structure of real ecosystems." *Oecologia* 65: 86–88.

Acknowledgments

This book has been a large project for both authors. It has taken us two years to research and write and has benefited from the contributions, in many different ways, of many people around the world.

First and foremost we are very grateful to our artist, Anne Gutmann, and to Prof. Steve Vogel. Anne is a trained biologist as well as a gifted artist; her illustrations add greatly to our book. Steve has significantly improved our text by reading the draft version and making many, many comments that corrected errors, suggested additions, or made the text more clear. The enthusiasm and support of Anne and Steve throughout this project is much appreciated.

At Harvard University Press we are grateful for the assistance of Ann Downer-Hazell, Michael Fisher, Vanessa Hayes, and Anne Zarrella. For permission to reproduce images, we thank Wim Hoek, Brigitte and Norbert Holzl, Tom McDonald, Thomas Shahan, Aaron Sloman, Rich Swanner, and David Webster. We thank the *Proceedings of the National Academy of Sciences* for permission to reproduce the data in Figure 5, and Prof. Tim Pedley for the poem about giraffe circulatory systems in Chapter 2.

For providing research papers and answering our technical questions, we thank Jo Falls, Andrew Parker, Matjaz Perc, and Ian Stewart.

Index